D1427947

TREE–CROP INTERACTIONS

TREE–CROP INTERACTIONS
A Physiological Approach

Edited by

Chin K. Ong

International Centre for Research in Agroforestry
P.O. Box 30677
Nairobi
Kenya

and

Peter Huxley

Oxford, UK

CAB INTERNATIONAL

in association with the

International Centre for Research in Agroforestry

CAB INTERNATIONAL   Tel: +44(0)1491 832111
Wallingford   Fax: +44(0)1491 833508
Oxon OX10 8DE   E-mail: cabi@cabi.org
UK   Telex: 847964 (COMAGG G)

A catalogue record for this book is available from the British
Library.

ISBN 0 85198 987 X

Published in association with:

International Centre for Research in Agroforestry
P.O. Box 30677
Nairobi
Kenya

Typeset in Plantin by Techset Composition Ltd, Salisbury
Printed and bound in the UK at the University Press, Cambridge

Contents

Contributors

C.R. Black
Department of Physiology and
Environmental Science,
University of Nottingham,
Sutton Bonington Campus,
Loughborough,
Leicestershire LE12 5RD, UK

A.J. Brenner
School of Natural Resources and
Environment,
University of Michigan,
Dana Building,
430 E. University,
Ann Arbor,
Michigan 48109-1115,
USA

J.E. Corlett
Horticulture Research International,
Wellesbourne,
Warwick CV35 9EF, UK

C.T. de Wit
Department of Theoretical Production
Ecology,

Wageningen Agricultural University,
PO Box 430,
6700 AK Wageningen,
The Netherlands

D.P. Garrity
ICRAF, Southeast Asia Regional
Program,
PO Box 161, Bogor 16001,
Indonesia

J.J.R. Groot
AB-DLO, Haren,
The Netherlands

K. Hairiah
Brawijaya University Malang,
Indonesia

P. Huxley
Flat 4, 9 Linton Road,
Oxford OX2 6UH, UK

G. Lawson
Institute of Terrestrial Ecology (ITE),
Edinburgh, UK

F.M. Marshall
*Department of Physiology and
Environmental Science,
University of Nottingham,
Sutton Bonington Campus,
Loughborough,
Leicestershire LE12 5RD, UK*

C.K. Ong
*International Centre for Research in
Agroforestry,
ICRAF House,
PO Box 30677,
Nairobi,
Kenya*

R. Ranganathan
*Department of Theoretical Production
Ecology,
Wageningen Agricultural University,
PO Box 430,
6700 AK Wageningen,
The Netherlands
(Present address: ICRISAT Asia
Center,*

*Patancheru 502 324,
Andhra Pradesh,
India)*

A. Soumaré
*Institut d'Economie Rurale (IER),
BP 258,
Bamako,
Mali*

M. van Noordwijk
*International Centre for Research in
Agroforestry (ICRAF)
SE Asia,
PO Box 161,
Bogor 16001,
Indonesia*

J.S. Wallace
*Institute of Hydrology,
Wallingford,
Oxon OX10 8BB, UK*

Foreword

Agroforestry is rapidly being transformed from an empirical, largely anec-dotal collection of beliefs and practices into an emerging science in the field of natural resource management. This book is a major contribution towards this goal. The authors have applied principles of plant ecology and crop physiology to develop more precise approaches that quantify biological interactions in agroforestry systems. The various models developed, par-ticularly the tree–crop interactions equation, provide practical but rigorous approaches for both above- and below-ground processes.

The book focuses on two basic resources: water and light. Tree–crop interactions for nutrients is not treated in depth for a very simple reason: the dearth of data on this subject. Advances on competition for nutrients, 'safety nets' and other aspects are currently being made. This subject may justify a similar book on its own right.

ICRAF, as a global strategic research centre dealing with agroforestry, is pleased with the production of this book. Our thanks to the editors and authors for their hard work in producing this advance in the science of agroforestry.

Pedro A. Sanchez
Director General of ICRAF
Nairobi, Kenya

23 February, 1996

Preface

Is a separate scientific field of 'agroforestry' really necessary? Do we need more basic research to support agroforestry, or is enough known from other disciplines? Is agroforestry sufficiently intellectual and/or vocational a subject to teach in Colleges and Universities?... It is worth reflecting on these questions, because 20 years ago the answers were debatable. How do we feel now?

The last question is the easiest to answer first. Readers can, for example, judge the scope and content of this volume for themselves, but when ICRAF started there were doubts expressed by some as to whether sufficient material existed to teach agroforestry as a separate subject. Now there is a plethora of general teaching materials, but still relatively few on focused topics. The last 10 years, especially, have seen agroforestry emerge worldwide as a reputable subject for courses of all kinds (Lassoie *et al.*, 1994), and there is no question as to its vocational value.

The concept of agroforestry as a defined landuse, to be promoted, adapted and researched, emerged in the mid 1970s, although various aspects of it had attracted attention from field operators and individual scientists well before this. Among the international community the report by John Bene *et al.*, *Trees, Food and People* (1977), following the activities of his team which was commissioned by IDRC to investigate the forestry–agriculture interface in tropical low income countries, was a landmark document that resulted, among other things, in the formation of ICRAF in 1977. It had a perceptive message that alerted foresters and agriculturists alike, and it clearly envisaged a multidisciplinary approach, although it somewhat naively just assumed that 'Information collected (about trees used in combination with crops) will have to be tested on different soils in different ecological zones and disseminated...'. This rather optimistic message was to colour the approach to research for some time.

As was correctly pointed out, 'agroforestry', i.e. the intercropping of woody and non-woody plant components, was widely practised worldwide in one form or another (and probably had been since the Stone Age?). In temperate countries, though, the progress of both agriculture and forestry into intensively managed monocultures, developing largely as separate practices each supported by its own applied scientific discipline, did much to deflect and obscure this. No one doubted that a great deal of agroforestry was 'out there', but what exactly was being done? Where and by whom, and in what way? And as scientists how could we begin to answer such questions as 'Are we using the right trees and shrubs?', 'What proportion of trees to crops do we need?', and 'How best do we arrange and manage them?'.

Surprisingly, the answers to such seemingly simple practical questions were *not* at all clear. Nor did they immediately emerge from a perusal of ecological literature, or by asking agricultural scientists or foresters. Soil scientists were aware of, but had no specific solutions to, some of the long term implications. There were, and are, two basic reasons for this. First, scientists in these and related disciplines had not been asking exactly the right questions. For example, ecologists studying plant associations are often concerned with how species turn out to be 'winners' or 'losers'. In agroforestry we do not want *any* losers! Foresters and agriculturists have each, largely separately, investigated ways of improving output from their own plant types. Horticulturists and those concerned with temperate tree fruits[1] and tropical plantation crops were often dealing with analogous situations (trees with grass or cover crops), but usually with the goal of maximizing the outputs of the tree crops. So the scientific information available did not quite fit. Second, it soon became apparent that there *are* special problems in optimizing systems that combine plants of such disparate life cycles, stature and behaviour, as the chapters that follow will elaborate.

John Bene *et al.*'s report had, perhaps somewhat intuitively, arrived at the need for some kind of research thrust. Thus, ICRAF was established through a combination of frugality, prudence and incertitude as a 'Council' rather than a Research 'Centre'[2] to help mobilize foresters, agriculturists and others channel resources into a range of new and necessary initiatives. In those early days there was considerable doubt as to what extent ICRAF should itself become involved in 'hands on' research, although collaborative field experimentation through 'networking' was considered suitable. But to counsel or advise implies that you already know the answers. Did we, indeed?

---

[1] A much neglected source of useful information, not only with regard to tree responses but also systems management (e.g. Hogue and Neilson, 1987).
[2] It is now a Research Centre and part of the Consultative Group on International Agricultural Research (CGIAR).

My first assignment at ICRAF was to prepare a paper for a workshop. Having been an agroforester for one week, and with a background of horticulture, tropical agriculture and tree and crop physiology, my contribution was to urge everyone to utilize the vast body of existing scientific information available from all the cognate disciplines. Sixteen years later I believe we have still not adequately done this, and this volume is a timely attempt, we hope, to explore and, critically, *to extend* what is known mainly from one topic, crop physiology. To evaluate and develop relevant strategic research in this way is a big step forward, and one that needs to be followed by those who can contribute in other disciplines.[3] However, the reader will note that the extent to which crop physiology can help to provide clear answers to agroforestry questions is still dependent on the further extension of theory, and the acquisition of a great deal more data to validate and to provide detailed case studies.

Finding solutions to the problems that practical agroforesters raise is one of the most exciting and challenging research tasks to emerge in recent years, especially for biophysical scientists. It requires one to review and extend ideas and concepts from a whole range of specific topics, and then to integrate them. In these circumstances, assembling some 'principles' makes for cost-effectiveness, especially when dealing with such a variety of different practices and where one is faced with such 'disorderly' and site-specific systems. The only alternative is a never ending 'trial and error' approach. This site-specificity dependence in agroforestry arises not only from the inherent multiplicity of outputs and services that agroforestry practices are often required to provide, and the mix of species that are needed in order to supply them, but also from the fact that the outcome of any woody–non-woody plant mixture will depend very much on the *interactions* between these components, which will be very site and management dependent.

Comprehending the nature and extent of the interactions at the 'tree–crop interface' (TCI) in various circumstances was clearly a key issue in understanding more about the way agroforestry systems functioned (although not the only one as Chapter 1 points out), and it was possible to start some investigations on this in a small way in ICRAF's first TCI programme (Huxley *et al.*, 1989). Only a small part of this was experimental, and it largely concentrated on providing observational material for developing research methodologies. Other organizations with more resources available (IITA, ICRISAT, CATIE, CSIRO, etc.) were soon also contributing to our understanding of the basic issues involved.

---

[3] As has been done already with another of ICRAF's 'Science and Practice of Agroforestry' Series, i.e. *Agroforestry for Soil Conservation* by Anthony Young (1989, with a revised edition due out in 1997).

However, and perhaps because of the largely unsubstantiated claims that were being made for agroforestry and so-called 'multipurpose trees' around this time, many scientists working in cognate fields were reluctant to become engaged in what they no doubt saw as the latest 'fad', and so were slow to reinforce the applied field work that was springing up through the 1980s.[4] A fairly broad-based workshop held at ICRAF in 1983 on 'Plant Research and Agroforestry' (Huxley, 1983) mustered some perceptive initial insights into relevant functional aspects, and the ITE's meeting on 'Trees as Crop Plants' (Cannell and Jackson, 1985) still makes extremely useful and relevant reading. However, it is only comparatively recently that serious consideration to agroforestry research issues are beginning to be addressed by the scientific community as a whole.

So, 'Yes', there still is a need to develop a comprehensive scientific approach to agroforestry. And now we can see even more clearly the identity of the problems to be solved, and that their scope and dimensions are such that the answers are *not* just waiting for us from existing research. Nevertheless, existing knowledge must form the starting points, and solutions to greater productivity and sustainability in agroforestry landuse will undoubtedly be greatly facilitated by continuing to explore these avenues, no longer as a supplement to adaptive work but as the foundation for it.

The specific aim of this volume is to outline and discuss how the principles of crop physiology can usefully be applied, adapted and/or extended to agroforestry. Not all possible topics are covered. For example, plant nutritional aspects still need to be addressed specifically, as do the interactive effects of trees in pastures/grasslands. The themes included are discussed, to a large extent, starting from what has been learned from the last 50 years of work with agricultural crops. As several authors point out there is a crucial need to develop more understanding between form and function, and between above- and below-ground interactions. 'That's nothing new!', you might rightly say (e.g. see the account in Balfour (1890) of the state of Botany as described in von Sach's book written in 1875), but to resolve the outcome of mixing complex multispecies canopies and root systems, and of managing them in many different ways, as we do in agroforestry, just cannot be fully achieved without such integration. If nothing else, and despite the progress that has been made, this brings home to us how far short we still are in achieving this crucial level of understanding at a practical level, despite the progress made theoretically (e.g. see Cannell and Dewar, 1994).

---

[4] Professor de Wit was one such that I wrote to in 1981, and it is sad that his recent death has deprived us of the benefit of his wisdom and experience in helping develop this field, but gratifying that he had, indeed, begun to address some thought to the problems and been able to make a welcome contribution to this volume.

It has been said that it is the customary fate of 'new' scientific truths to begin as heresies and end as superstitions. We must make sure that this does not happen with agroforestry and build a solid foundation of well-researched facts that can be readily applied by field practitioners. Whether it is best to grow plants as monocultures or intercrops and, in the latter case, whether or not these should consist of woody and non-woody plants will depend on a host of different biophysical and socio-economic circumstances that we need to understand and weigh carefully. Agroforestry has its parts to play in the spectrum of sensible, sustainable landuse options, but it is up to us to provide credible information about the biological opportunities and, above all, the feasibility of the choices. In this process we must not forget the many good reasons why trees and crops are often grown with such success separately. Perhaps the 'new truth' is that, to the scientific community, agroforestry is an opening of the mind to possibilities that practical landusers have long since accepted, and which scientists from many disciplines must now seek to interpret and evaluate at a basic level. To do this we need to re-evaluate old ideas and re-shape them into new tools.

References

Balfour, I.B. (reviser) (1890) *History of Botany (1530–1860)*, by Julius von Sachs (transl. Garnsey, H.E.F.). Clarendon, Oxford, 561 pp.

Bene, J.G., Beall, H.W. and Coté, A. (1977) *Trees, Food and People: Land Management in the Tropics*. IDRC, Ottawa, 52 pp.

Cannell, M.G.R. and Dewar, R.C. (1994) Carbon allocation in trees: a review of concepts for modelling. *Advances in Ecological Research* 25, 59–104.

Cannell, M.G.R. and Jackson, J.E. (1985) *Attributes of Trees as Crop Plants*. Natural Environment Research Council/Institute of Terrestrial Ecology, Abbots Ripton, 592 pp.

Hogue, E.J. and Neilson, G.H. (1987) Orchard floor vegetation management. *Horticultural Reviews* 9, 377–430.

Huxley, P.A. (ed.) (1983) *Plant Research and Agroforestry*. ICRAF, Nairobi, 617 pp.

Huxley, P.A., Darnhofer, T., Pinney, A., Akunda, E. and Gatama, D. (1989) The tree–crop interface: a project designed to generate experimental methodology. *Agroforestry Abstracts* 2, 127–145.

Lassoie, J.P., Huxley, P.A. and Buck, L.E. (1994) Agroforestry education and training: a contemporary view. *Agroforestry Systems* 28, 5–19.

Young, A. (1989) *Agroforestry for Soil Conservation*. CAB International, Wallingford/ICRAF, Nairobi, 276 pp.

Peter Huxley
Oxford

Acknowledgements

The editors and publishers would like to acknowledge the assistance of the Overseas Development Administration (UK), in providing a financial contribution towards the costs of publishing this book.

1

◆

A Framework for Quantifying the Various Effects of Tree–Crop Interactions

C.K. ONG

International Centre for Research in Agroforestry, P.O. Box 30677, Nairobi, Kenya

Introduction

> Agroforestry systems will be able to mimic or replicate many of the nutrient-cycling and favourable environmental influences found with forest ecosystems, while generating the exportable outputs achieved with agricultural systems.
>
> Kidd and Pimentel (1992)

Early assessments of the potential benefits of agroforestry were based largely on the assumption that it is possible to evaluate existing information on forestry and agriculture and extrapolate from it (Huxley, 1983; Nair, 1984), and partly based on the observations of traditional agroforestry systems which show increased growth of understorey vegetation (Charreau and Vidal, 1965; Felker, 1978). A number of negative effects have also been recognized, such as competition for moisture, excessive shading, allelopathy, but these have attracted much less attention from scientists. Most of the evidence of benefits and drawbacks of agroforestry continues to be qualitative or indirect, i.e. extrapolated from a wide range of systems, creating often unrealistic expectation of the benefits of agroforestry technologies. Fortunately, the volume of agroforestry research has grown rapidly since 1983, as various international and national institutes have become involved in both tropical and temperate regions. With many field experiments in progress the growing volume of evidence necessary to establish a scientific basis for the *quantitative* analysis of the various interactions that occur when trees and crops are grown together in a range of climatic and geographic regions is fast becoming available.

A scientific framework for a quantitative analysis of tree–crop interactions is needed for several reasons. First, it should provide a reliable method to determine which benefits are likely to be realized for a given agroforestry technology in a defined situation. Second, it should enable researchers to evaluate the relative importance of each interaction in order to guide them more precisely in the choice of research priorities. This is no trivial matter because agroforestry research requires a long term commitment in research resources and it is not easy to separate the complex interacting factors involved (Anderson and Sinclair, 1993). Third, the advantage of agroforestry cannot be quantified simply in terms of productivity alone, because some of the benefits are due to environmental improvements, e.g. resulting from erosion control and increased organic matter content, and these cannot be measured in only a few seasons. Finally, a quantitative approach is an important step in the quest for a fuller understanding of the complex mechanisms of tree–crop interactions, which should then offer the scientific basis for designing yet more productive and sustainable agroforestry systems.

The aim of this chapter is to provide a brief description of the individual effects of tree–crop interactions and to suggest how these may be quantified together. Other chapters in this book will examine how tree–crop interaction can be explained in terms of competition principles (Chapter 2) and in terms of a simple model of shading and fertility improvement (Chapter 3). Subsequent chapters in the book will explore in detail the physiological mechanisms involved in each interaction.

Main Types of Tree–Crop Interaction

Before considering methods to quantify the overall effects of tree–crop interactions it is useful to list the biophysical benefits and consequences that are commonly attributed to agroforestry systems to determine whether the evidence for each interaction is based on indirect or direct observations. The relative importance of each effect will no doubt depend on both the type of agroforestry system and the location of the site. For example, the effects of any soil fertility enrichment by agroforestry will be less obvious if fertilizer input is high. For the purpose of this chapter it is premature to include effects that have not been substantiated by field observations. For example, there is, as yet, no quantitative experimental evidence available concerning the effectiveness of agroforestry to control weeds or maintain sustainability (Table 1.1) although there are good reasons to expect such benefits on theoretical grounds. Another unresolved issue is the potential

Table 1.1. Main effects of tree–crop interactions.

| | Evidence | | |
Effects	Direct	Indirect	Source
1. Increased productivity	+	+	Ong, 1991
2. Improved soil fertility	+	+	Kang *et al.*, 1990
3. Nutrient cycling	+	+	Szott *et al.*, 1991
4. Soil conservation	+	+	Lal, 1989; Wiersum, 1991
5. Microclimate improvements	+	+	Monteith *et al.*, 1991
6. Competition	−	−	Ong *et al.*, 1991
7. Allelopathy	0,?	−	Rizvi, 1991; Tian and Kang, 1994
8. Weed control	0	+	Rizvi, 1991
9. Sustainability and stability	0	+	Sanchez, 1987; Young, 1991
10. Pests and diseases	0	−,+	Zhao, 1991

Positive effects are indicated by (+) and negative by (−): where evidence is not available it is indicated by (0). Only key or recent sources are quoted.

importance of allelopathy, which has been demonstrated repeatedly for some trees species in laboratory conditions but is doubtful in field conditions for various reasons (see critical review by Horsley, 1991).

There is ample evidence to show that the overall (biomass) productivity of an agroforestry system is generally greater than that of an *annual* system although not necessarily greater than that of a forestry or grassland system. The basis for the potentially higher productivity could be due to the capture of more growth resources, e.g. light or water (see Chapter 4, this volume), or due to improved soil fertility. Competition, which is a negative effect in this context, is usually a significant factor in simultaneous agroforestry systems, even when there is evidence of increased combined productivity by both components. It is fair to conclude that, at present, only the top six effects in Table 1.1 have been substantiated by field observations. Certainly, there is still an urgent need for research to acquire more 'hard evidence' before it is possible to attempt the formidable task of translating the 'promise' of agroforestry into sustainable landuse.

Soil fertility improvements

Many of the often quoted examples of soil fertility improvement are based on traditional agroforestry systems that have been established for many years. The potential for microsite enrichment by some trees is an extremely important aspect of agroforestry which has received considerable attention (Nair, 1984; Young, 1989). Surprisingly, most of these examples are based on widely scattered, slow-growing trees in arid or semi-arid environments such as *Faidherbia albida* in West Africa (Felker, 1978), *Prosopis cineraria* in Rajasthan, India (Singh and Lal, 1969) and *Pinus caribea* in the savanna of Belize (Kellman, 1979). These authors concluded that the accumulation of mineral nutrients is the result of a *long term* process of capture of precipitation or nutrient-rich litter. This argument implies that the ability of the trees to contribute directly to fertility enrichment is likely to be small at first since they are very slow growing until they have reached a considerable size. Recent studies of newly planted (5–10-year-old) *Faidherbia albida* stands in India and Africa confirm that microsite enrichment is, indeed, a slow process in such situations.

In contrast to the slow enrichment of soil fertility in traditional agroforestry systems, the relatively new concept of alley cropping (also called hedgerow intercropping or avenue cropping) using fast-growing, nitrogen fixing trees, e.g. *Leucaena leucocephala* and *Gliricidia sepium*, in the humid tropics can substantially increase soil fertility in 2–3 years (Kang *et al.*, 1990). A major feature of the alley-cropping concept is the capacity of the trees to produce a large quantity of biomass for green manure and the need for regular pruning to prevent shading in order to reduce competition with associated crops. Alley cropping has been demonstrated to be successful in relatively fertile soils but attempts to extend this technology to the acid infertile soils of the humid tropics (Szott *et al.*, 1991; Matthews *et al.*, 1992) or to the semi-arid tropics (Singh *et al.*, 1989) have been disappointing. The main constraints are poor tree growth, aluminium toxicity, low nutrient reserves and excessive competition with crops (Kang, 1993). Considerable progress has been made in selecting acid-tolerant fast-growing tree species such as *Senna reticulate*, *Senna spectabilis*, *Inga edulis* and *Calliandra calothrysus* but economical techniques for reducing tree–crop competition are still lacking. Thus far, it appears that alley cropping is not sustainable on acid, infertile soils without addition of chemical fertilizers, chiefly because of infertility of the soil and the insufficient recycling of nutrients from prunings at Yurimaguas, Peru, Northern Zambia and Claveria, Philippines (Maclean *et al.*, 1992; Matthews *et al.*, 1992; Szott and Kass, 1993).

Soil conservation

Contour hedgerows have consistently been demonstrated to be highly effective in controlling soil erosion even in as little as 18 months (Lal, 1989; Maclean *et al.*, 1992). The woody hedgerows provide a semi-permeable barrier to surface movement of water, while mulch from the trees reduces the impact of raindrops on the soil and minimizes splash and sheet erosion (Young, 1989). Mulching also provides an effective means of reducing soil evaporation and other improvement in microclimate, but these effects are seldom measured in agroforestry systems. Current emphasis is on the selection of tree species that provide a more effective physical barrier to erosion and produce mulch material that has a longer lasting protective role (Kiepe and Rao, 1994). Relatively little information is available, so far, on the influence of trees on the physical properties of soils in terms of infiltration rate, or bulk density and soil water storage capacity.

Microclimate improvements

The use of trees as shelterbelts in areas that experienced high wind or sand movement is a well-established example of microclimate improvement that resulted in improved yields (Reifsnyder and Darnhofer, 1989). Of course where the environmental conditions are already favourable for crop growth there is little advantage in reducing windspeed or in moderating air temperature (Monteith *et al.*, 1991). Even when there is a clear advantage in reducing windspeed to protect young seedlings there might be a possibility of negative effects due to competition for moisture between the roots of trees and crops during dry periods (Malik and Sharma, 1990). Maximum benefit of shelterbelts is observed when soil water supply is not limiting, especially where irrigation is possible.

The evidence for the beneficial effects of shade trees depends on the nature of the understorey crops. The clearest effect is reported for crops that require shading for normal growth, e.g. black pepper, turmeric, cacao (Nair, 1984). Recent analysis of *Paulownia* and tea in sub-tropical China suggests that tea production is slightly improved when shading is about 37%, but overall economic benefit is largely due to the production of additional timber from *Paulownia* (Yu *et al.*, 1991). Details of shelterbelt effects on microclimate and crop responses are described by Brenner (Chapter 5, this volume).

Competition

Competition between crops and trees for the same limiting growth resources is most obvious when they are grown in close proximity. However, the extent of below-ground competition is often not apparent (Singh *et al.*, 1989). Assessment of the extent of competition is thus complicated by possible interference between plots due to the proliferation of tree roots into nearby plots or due to the effect of shading especially with tall trees (Huxley *et al.*, 1989; Rao *et al.*, 1991). Another complication is the choice of an appropriate control for both trees and crops to provide a reliable basis for the assessment of competition on crop yields (Ong, 1991). For example, many studies of alley cropping used the yield of the alley crop where mulch is removed as the sole crop 'control'. This is clearly erroneous because the crop is still influenced by the hedgerows in these circumstances (Gichuru and Kang, 1989). More rigorous experimental design and precautions are necessary to ensure that the assessment of competition is free from inter-ference by other treatments. These aspects will be described in detail later in this chapter.

A simple but effective method for determining competition is to measure crop and tree yields along a transect across the tree–crop interface as proposed by Huxley (1985). However, to understand the causes of competition it is necessary to have some measurements of the soil and aerial environment in the transect. For example, a recent study of two tree species (*Vitellaria paradoxa, Parkia biglobosa*) scattered among three annual crops (cotton, sorghum and pearl millet) in south Mali, West Africa, showed that sorghum and pearl millet showed a 50–60% yield reduction under the trees, despite a significant increase in soil fertility (Kater *et al.*, 1992). However, cotton yield was relatively unaffected by *V. paradoxa* (−8%) and *P. biglobosa* (−16%). The use of a transect is particularly useful in such on-farm situations since it is difficult to obtain a sole crop control in scattered tree systems. These authors concluded that yield reduction was due to plant mortality caused by fungal attack or by shading, which was in agreement with an earlier study by Kessler (1992).

Methods for Quantifying Interactions in Mixtures

Land equivalent ratio

The concept of Land Equivalent Ratio (LER) is the most widely accepted index for evaluating the effectiveness of all forms of mixed cropping (Willey, 1979) and has been extended to agroforestry by some workers (Rao *et al.*, 1990, 1991). LER is the ratio of the *area* under sole cropping to the area under intercropping, at the same level of management, that gives an equal amount of yield. The sum of the fractions of the yields of the intercrops relative to their sole crop yields provides a measure of the overall effective-ness of the mixed system. Thus LER can be expressed as:

$$\text{LER} = \frac{X_i}{X_s} + \frac{Y_i}{Y_s} \tag{1.1}$$

where X and Y are the component crops in either an intercrop (i) or a sole crop (s) system. When the LER = 1, there is no advantage to intercropping over sole cropping. When LER < 1, more land is needed to produce a given yield by each component as an intercrop. As with most comparisons, the choice of the denominator or sole treatment of each crop should be the optimal treatment for the site. Another implicit assumption is that the optimum treatment is such that the relationship between yield and plant density is linear, otherwise the proportionality assumption is not valid. Because LER is used extensively in mixed cropping this point is worthy of further elaboration here.

For example, if the yield–population relationship of a species is given by a straight line then a calculation of LER based on any values between A and B is perfectly valid (Fig. 1.1). However if this relationship for a particular species is curvilinear (dotted line), then the calculation of LER is not straightforward. If B is chosen as the optimum population for, say, a sorghum cultivar and the intercropped sorghum is grown as in a population D then the expected partial LER is c. 0.5 for the ABC relationship but c. 0.75 for the parabolic curve. In such an analysis of the intercropping literature, Ong and Black (1994) have shown that failure to account for the 'yield–population effect' can result in misleading conclusions concerning the yield advantage of intercropping or agroforestry systems. For example, a sorghum–pigeonpea intercropping system is not more effective than sole pigeonpea in capturing radiation and it is more useful to express LER in terms of a yield–population response curve.

Notwithstanding, the literature on agroforestry and intercropping systems contains various examples of LER values exceeding 1, confirming that mixtures are usually more productive than sole stands (Ofori and

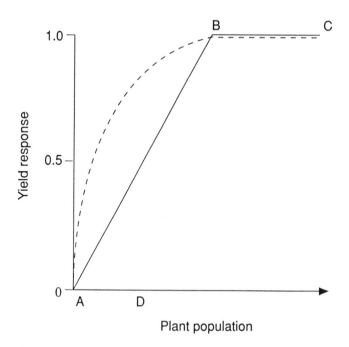

Fig. 1.1. Curvilinear (dashed line) and linear (AB) responses to plant population. Calculation of Land Equivalent Ratio at population D tends to overestimate the advantage of intercropping when the actual response is curvilinear and pure crop stand is between B and C.

Stern, 1987, give a list on intercropping systems). Table 1.2 provides a list of LER values of agroforestry work reported in India, which has concentrated mainly on *Leucaena leucocephala* systems. In virtually all the examples cited in India the tree component is the most dominant and accounts for the high total LER values. It is not clear whether the competitiveness of the tree is confined to *Leucaena* or is a common feature as the tree becomes taller and bigger than the associated crops.

Outside the Indian examples it is hard to find examples of agroforestry data that provide a sole tree comparison to allow a meaningful calculation of LER. Generally, it is assumed that it is more advantageous to combine trees and crops, due to the shortage of land.

Competition indices

There is a long history of ecological research on competition in mixed species, where the emphasis has been on the role of competition in determining the natural abundance of specific species (recently summarized in a volume edited by Grace and Tilman, 1990). Such studies analyse the effects of interspecific and intraspecific competition by varying the proportion of the species and/or the population of the mixture. This approach was extended by de Wit and his co-workers (de Wit and van den Bergh, 1965) to crop mixtures and crop–weed competition and more recently to agroforestry (see Chapter 2 in this volume for application to agroforestry).

An early competition index is the Relative Crowding Coefficient (RCC) (de Wit, 1960) which is a forerunner to the LER.

$$\text{RCC} = \frac{X_i}{X_s} \times \frac{Y_s}{X_s} \qquad (1.2)$$

using the same nomenclature as in LER, previously.

Table 1.2. Land Equivalent Ratios of agroforestry systems in India.

Trees/crops	LER			Source
	Crop	Tree	Total	
1. *Leucaena*–wheat–rice	0.81	0.52	1.33	Singh *et al.*, 1990
2. *Leucaena*–coriander	0.83	1.22	2.05	Singh *et al.*, 1990
3. *Leucaena*–rice–green gram	0.81	0.68	1.49	Singh *et al.*, 1990
	1.29	1.27	2.36	Singh *et al.*, 1990
4. *Leucaena*–groundnut	0.47	0.70	1.17	Rao *et al.*, 1990
5. *Leucaena*–millet	0.41	0.96	1.37	Corlett *et al.*, 1992
6. Perennial pigeonpea–groundnut	0.64	0.97	1.61	Odongo *et al.*, unpublished
	0.12	1.39	1.51	

When RCC values exceed 1 then species X is more competitive than species Y. This RCC concept is further extended to include the performance of the other species in the mixture by calculating the Relative Yield Total (RYT) by de Wit and van den Bergh (1965).

$$\text{RYT} = \frac{X_i}{X_s} + \frac{Y_i}{Y_s} \tag{1.3}$$

This is the same as LER except that it is only valid for that particular total planting density. Thus RYT is used to analyse results in replacement series trials and is less popular than the LER concept which has fewer restrictive conditions for its use. However, as pointed out earlier the LER concept is not completely free from the influence of the 'yield–population effect'. The RYT concept is thus not appropriate for the analysis of additive experiments that represent the bulk of intercropping and crop–weed experiments. Further development of the competition index by Spitters (1983) and Ranganathan (1993) to overcome the major limitations of RYT and LER is described in Chapter 2 of this volume.

Tree–crop interaction (TCI) equation

The limited value of existing intercropping methods of quantifying interactions with species mixtures has led to a more comprehensive and practical approach which includes the numerous interactions (Table 1.1) generally attributed to agroforestry systems and provides an initial overall appraisal of the total effect (Ong, 1995). For example, the effect of mulching is a major feature of most agroforestry systems which will have direct effect on soil fertility, soil evaporation, soil temperature and soil conservation. Changes in soil physical properties due to the extensive root system of the tree and associated changes in soil fauna cannot simply be quantified in terms of productivity alone and will be evident only after a few years. Some of those effects will have an immediate influence on crop yield but others will depend on the weather and on soil characteristics. However, at this stage, a general framework for the quantitative analysis of tree–crop interactions should include only the major interactions, which can be measured relatively easily, and should avoid minor or secondary interactions which would make the equation unwieldy and cumbersome.

A simple approach is to quantify the effects of each factor in relation to the performance of the control or sole crop (Co) in the absence of any interference by trees, assuming that the main factors are not confounded experimentally (Ong, 1995). For example the overall tree–crop interaction (I) on crop yield can be expressed as

$$I = F - C \pm M + P + L \tag{1.4}$$

where F is the benefit of tree prunings on soil fertility and microclimate of the soil surface, e.g. temperature and soil moisture, C is the crop yield reduction due to interspecific competition between the tree and crop, M the consequences of above-ground changes in temperature, light humidity and windspeed, P the consequence of changes in soil properties, and L is the reduction or avoidance of losses of nutrient or water, especially on sloping land. The effect of each factor is expressed as a percentage of the control crop (Co).

F can then be measured as the difference between the crop yield with added tree prunings (Cm) or without any (Co). Prunings can be placed on the soil surface (mulching) or incorporated into the soil depending on the expected role of the prunings. Nitrogen utilization by crops is usually lower with mulching (5–16% less) than with incorporation (Kang *et al.*, 1981). However, mulching is more desirable for soil and water conservation. Competition (C) can be estimated from the yield difference between Cm and Hm (the yield of crop grown in the alleys with prunings applied); or as the difference between Co and Ho (where the crop is grown in the alleys but with prunings removed). The estimate of C resulting from the two methods need not be the same because the presence of the prunings might reduce loss by soil evaporation, runoff and nutrient to benefit crop yield. Estimates of M, L and P are more difficult to obtain because they are rarely measured and are not easily separated from F and C. How to approach these issues will be dealt with later. However, with the four main treatments it should be possible to examine F and C, assuming that M, L and P are small.

Interference between plots

As in the case of LER calculations, an accurate measurement of Co is critical since the effect of each factor is expressed as a percentage of Co. A major source of error is introduced when the value of Co is influenced by adjacent plots, which is common in randomized block designs involving contiguous small plots of less than 6 × 6 m. Plot-to-plot interference is a major concern in agroforestry experiments since roots of trees can influence crop yield well beyond the effects of shading. For example at Machakos, Kenya, the roots of *Leucaena leucocephala* can reduce the yield of maize to 5 m away after 2 years of growth. A simple test of the reliability of Co is to calculate competition, C, using the difference between Co and Ho or Cm and Hm, if available. A further test is to install a root barrier or trench down to 0.5 m around the Co plot. Experience in Hyderabad, India, indicates that yield of sole groundnut doubled with trenching compared with Co in undisturbed plots (Rao *et al.*, 1991). Root barriers may be circumvented by tree roots after only one season, however.

Application of the equation

To determine the usefulness of the TCI equation we start by examining data that contain actual values of Co, Cm, Ho and Hm. The first set of data (P. Kiepe, unpublished) is from Machakos, Kenya, which is at an elevation of 1600 m above sea level with a bimodal rainfall of 760 mm. The trial consisted of Co, Cm and Ho and Hm plots with *Senna siamea* (syn. *Cassia siamea*) contour hedgerows spaced at 4 m between rows with intra-row 0.25 m on a slope of 14%. The crops were maize (long rains) followed by cowpea in the next season (short rains). Another set of treatments (Ho and Hm) was planted similarly with *Leucaena leucocephala* but was managed in the same way as the *Senna* hedges. Plot size was 10 × 40 m, therefore there was virtually no interference from adjacent treatments. Measurements of microclimate and soil physical properties have only recently begun therefore *M* and *P* data are not examined here. Losses by runoff and soil loss averaged about 3% per annum and 4 t ha^{-1} per annum, respectively, and appeared to have had no significant effect on crop yield so far. Therefore, in the analysis that follows attention is confined to how *C* and *F* progressed during the last 5 years when rainfall declined from 528 mm to 222 mm (Fig. 1.2). Absolute grain yield of maize in Co declined from 3.02 t ha^{-1} in the first year to 1.51 t ha^{-1} in the last year, almost proportionally to the amount of rainfall received.

In the first 3 years *Senna* had no detectable competition effects (*C*) on the maize (calculated as Co − Ho or Cm − Ho), whereas *Leucaena* was highly competitive (Co − Ho) after the first year, with *C* reaching a maximum of 75% in the final year. The advantage of mulching (*F*) with *Cassia* ranged from 0 to 33% and *F* was only less than *C* in 1991, resulting in an overall positive benefit of 30 down to 4%. Unfortunately there was no Cm treatment with *Leucaena* to estimate *C* in the presence of mulch or to estimate *F* for *Leucaena*.

The estimate of *I* for *Leucaena* based on Hm and Co indicated that the outcome was negative during the 5 years, becoming progressively worse and averaging −37% for the whole period. This simple analysis highlights the importance of competition in determining the effects of the hedges on crop yield. Furthermore, it demonstrates that it is not enough to select 'fast-growing' trees for contour hedgerows for soil conservation because the beneficial effect can be outweighed by competitive effects.

Quantifying the effects of microclimate

Unlike the effects of *F* and *C* which can be obtained directly by using appropriate treatments, the effects of microclimate on crop yield are usually estimated *indirectly* by a combination of microclimate measurements and interpretation of the process involved (Singh *et al.*, 1989; Corlett *et al.*,

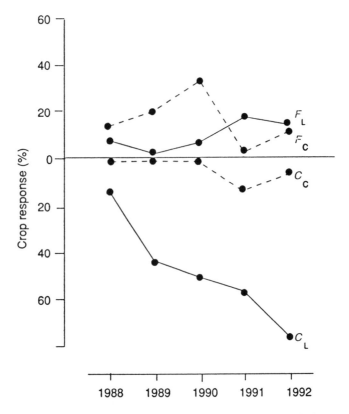

Fig. 1.2. Trend with age in competition (*C*) and fertility (*F*) of *Senna siamea* (C) and *Leucaena leucocephala* (L) contour hedgerows at Machakos, Kenya (P. Kiepe, unpublished).

1992). Such indirect estimates of these factors have certain drawbacks because measurement of these processes is expensive (mainly because of costly equipment) and requires a considerable expertise in the interpretation of the data. However, a combination of measurements and appropriate treatments can provide an effective method of quantifying each factor. An illustration of this approach is a recent study of the effects of the canopy of *Acacia* or *Faidherbia albida* on soil temperature at the ICRISAT Sahelian Centre, Sadore, Niger (Vandenbelt and Williams, 1992).

The radiation and soil temperature under a well-established *F. albida* were measured during the rainy season under the canopy (2 m from trunk) and in the open (5 m from the trunk) using a radiometer and soil thermocouples connected to a datalogger. The shade created by the tree resulted in substantially lower soil temperature. Maximum temperature during the afternoon (12.00 to 17.00 h) was generally 10°C lower in the shade than in the open.

In a companion experiment, a 1-m tall fence of millet stalks was erected in a north–south direction to provide a range of shading and duration. Pearl millet seeds were sown in rows perpendicular to the fence and soil temperature was measured at every 0.5 m from the fence. Soil surface temperatures followed the trend observed under the tree (Fig. 1.3). Millet survival and growth decreased linearly with distance from the fence and negatively with mean soil temperature (Fig. 1.4). These results suggest that the lower soil temperature induced by the shade of *F. albida* is a major benefit which can be quantified in terms of millet growth. Therefore *M* in the equation can be quantified in a similar approach as long as other factors that can be modified by shading, e.g. rainfall interception, soil evaporation, can be discounted.

The influence of temperature on plant development rate, unlike growth, is more simple to quantify because crop response to temperature is well documented and described (Squire, 1990). Development is defined as the progress from one discrete event to another, e.g. from sowing to flowering, and is usually measured in days. Development rate can be described by the equation:

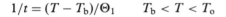

$$1/t = (T - T_b)/\Theta_1 \qquad T_b < T < T_o$$

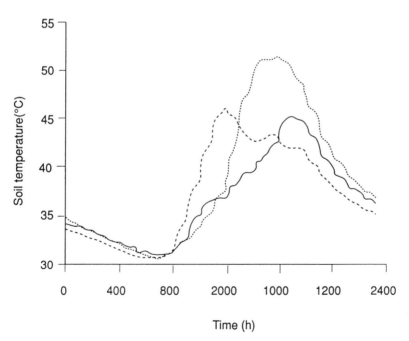

Fig. 1.3. Soil temperatures at 2 cm depth at two positions inside (solid and dashed lines), and at 5 m outside (dotted line) the canopy of a 56 cm diameter *Faidherbia albida* tree, Sadore, Niger (Vandenbelt and Williams, 1992).

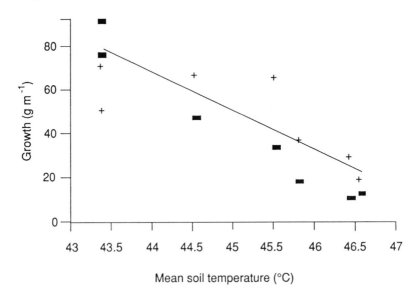

Fig. 1.4. Growth response of pearl millet cultivars (+ MBH 110, ▅ CIVT) to mean soil temperatures at Sadore, Niger (Vandenbelt and Williams, 1992).

where the rate of development is defined as the reciprocal of time (t) between the beginning and end of a specific development process. T is the temperature experienced by the plant which exceeds the base temperature (T_b), below which no development occurs, but is less than the optimum temperature (T_o), above which increasing temperatures reduce the rate of development. The thermal duration (Θ_1) is the number of degree days (°Cd) above T_b but below T_o necessary for the completion of the developmental phase and can be determined when temperature is the only variable controlling development. Development may cease when temperature exceeds a maximum (T_m) and, when temperatures exceed T_o, Θ_2 may be defined as the number of degree days below T_m but above T_o necessary for completion of the development phase.

Corlett *et al.* (1992) use the above equation to estimate the effect of temperature changes on the development of pearl millet at Hyderabad in India in a *Leucaena* alley-cropping system. They concluded that the small reduction in temperature delayed the flowering of millet by 2–3 days. Such a delay is beneficial for growth if moisture is not limiting but would be a disadvantage if water stress exists. The interactions between shelter effects and moisture status on crop yield are examined in more detail by Brenner (Chapter 5, this volume).

Quantifying the effects of P and L

The protective role of soil conservation structures and vegetation is usually measured in terms of the quantity of water, soil or nutrients lost in comparison with a 'control', i.e. a plot without any soil conservation or with a local practice. This approach has the advantage of providing valuable insights on how the processes of erosion works, and on how the protective role can be improved (Wiersum, 1991). However, the approach does not provide a *direct* way of estimating the consequence on crop yield when runoff or soil loss is prevented, assuming that it will not lead to undesirable effects, e.g. anaerobic conditions.

The conventional approach to quantifying the protective role of contour hedgerows is seen in a long term trial conducted at IITA, Ibadan, Nigeria, where alley-cropping research was pioneered (Lal, 1989). The annual total rainfall ranges from 1100 to 1300 mm over two growing seasons and the cropping system was a maize–cowpea rotation. The main treatments were contour hedgerows with *Leucaena leucocephala* and *Gliricidia sepium* at 2 m and 4 m spacings, and the 'control' was crop alone with no-till or plough-till. Runoff, soil and nutrient losses were measured. Over the 6-year period runoff in the plough-till crop control amounted to 17% of the rainfall during the maize season and 43% during the cowpea season. Soil loss averaged 4.3 t ha^{-1} year^{-1} in the maize season, but it was negligible in the cowpea season. All agroforestry systems and the no-till control reduced runoff to about 3% of total rainfall and soil loss to 0.5 t ha^{-1} during the maize season. Maize yield was only 10% lower compared with the two controls: that is, almost proportional to the area planted to trees. During the 6 years, maize yield in the control (no soil conservation) declined by 8.5% per year and cowpea yield by 10.7% per year but this period corresponded to a fall in rainfall. When expressed as yield loss per millimetre of rainfall each year the decline was substantially greater, about 14.5% each. A major difficulty is to determine whether the decline in crop yield is caused by nutrient loss by crop uptake or by erosion. This problem would be eliminated if a treatment is included that prevents erosion, assuming that there is no confounding with moisture supply to crops. A possible solution is the use of an inert mesh which prevents soil loss but allows water movement.

Because changes in P and L occur over different timescales it is theoretically possible to separate them over time. Initially, changes in P are relatively unimportant as trees have to produce a substantial root system or litter first. Furthermore, improvements in P, say, infiltration rate, are unlikely to produce an increased crop performance unless the roots have limited access to more below-ground resources or low infiltration rate has prevented recharge by rain. As with L it is not easy to quantify P simply by direct measurement of moisture content and infiltration rate in sole crop and agroforestry treatments. To estimate P by different treatments it is necessary to include treatments in which the tree shoot is removed and the potential nutrient enrichment of soil by the decay of below-ground biomass is overcome by fertilizer application. Such studies have not been undertaken yet, but significant changes in both P and L are potentially the most crucial contribution to the sustainability of agroforestry systems in marginal lands.

Consequences

The relation between C and F

The alley-cropping examples of how C and F vary with *Leucaena leucocephala* and *Senna siamea* at Machakos illustrate the relative importance of competition on the final outcome, I. Is there a connection between biomass production, which determines F and C? Preliminary analysis of multiseason and species variation can provide some clues. Regrettably, there are only very limited data to explore such relationships. A recent analysis of results from a few sites in Eastern Africa (Akyeampong *et al.*, 1996) shows that F and C are, indeed, closely related such that species with a high biomass production are usually the ones which are competitive (Fig. 1.5). Using such species may, therefore, lead to a small or negligible improvement in crop yield. This situation is not serious when the biomass of a tree can be removed for economic purposes, e.g. fodder, but it becomes a major deterrent to adoption by farmers if additional pruning is necessary to reduce competition with crops.

The trade-off between F and C is rarely explored in agroforestry studies because it is widely believed that high biomass production *per se* is desirable. This simplistic concept of tree ideotype is now being challenged because there is mounting evidence that a combination of fast-growing trees and crops usually leads to *lower* crop production. Indeed, the considerable literature on competition in plant ecology indicates that the combination of fast-growing species, whether annual or woody species, generally results in competition (Grace and Tilman, 1990). In sharp contrast is the encouraging experience with intercropping which reveals that a combination of a fast-growing and a slow-growing species is needed to achieve greater overall production (Willey *et al..*, 1986). Chapter 4 will describe the various mechanisms for producing complementarity in both intercropping and agroforestry.

Long term changes

So far only short term changes in various components of the equation have been discussed. Because of the recent origin of formal agroforestry research

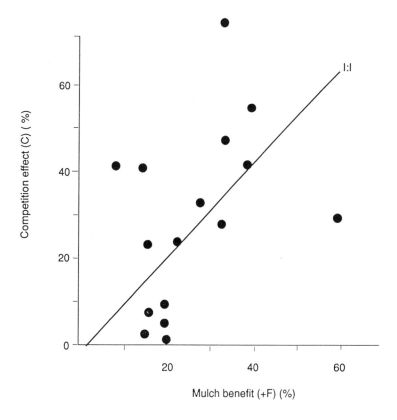

Fig. 1.5. The relation between competition (*C*) and fertility (*F*) for a range of species grown as alley cropping with maize (Ong, 1995).

it is not possible at this stage to determine the long term consequence on each component of the equation. The most urgent need is to determine the timing, rate and direction in which each component will vary over time. Such information is needed to answer practical questions like when would a certain agroforestry system be able to provide significant benefit in terms of crop yield or soil conservation, or whether it is necessary to modify tree or crop management in later years. The logical tool for integrating this information and for providing the answers to such questions is modelling. However, to be useful this will be dependent on the reliability of the critical inputs to the model.

Consider, for example, the contour hedgerow system with *L. leucocephala* at Machakos (Fig. 1.2) where *C* increased at about 12% each year during the study period. Extrapolation of the relationship indicates that competition began around 1987, 3 years after the trial was established.

There were no measurements during the first 3 years. In contrast, results from a nearby trial, established in 1989, with the same species and management, showed that competition ($I = 30\%$) began within 6 months and averaged 26% each year! The major difference between the sites was soil depth and, of course, the differences in tree age. In a more recent trial at Machakos, the effect of artificially manipulating soil depth (0.4 m and 2 m) was investigated using *Senna spectabilis* (syn. *Cassia spectabilis*) which showed that competition was less ($C = -20\%$) at 0.4 m and high (-80%) at 2 m soil depth. This preliminary analysis suggests that the competitiveness of the tree is enhanced when the roots have a greater access to below-ground resources. This example illustrates the importance of soil characteristics which may have a major effect on the timing, rate and direction of competition, and perhaps on the other components of the equation too.

Biophysical limits

There is now a growing realization that agroforestry, like most cropping systems, is not a panacea for all landuse problems and there might be environments in which the combination of trees and crops may be disadvantageous. This was amply demonstrated when alley cropping, which was developed for the humid tropics as an alternative to natural fallow, was extended to the semi-arid tropics, especially in the Indian sub-continent (Ong *et al.*, 1992). The Indian experience reveals that competition for moisture between trees and crops greatly outweighs the advantages of improved resource capture.

Several important lessons are relevant to *all* agroforestry development. First, alley cropping or improved fallows should not be considered as the solution for declining soil fertility in *all* environments. It is important to establish the biophysical limits (e.g. rainfall) within which the agroforestry technology will be advantageous as offering a potential, for example, for increased crop yield over local agricultural systems. Second, even within a site or station, we can expect contrasting results depending on, not only species and management schemes chosen, but also soil depth and initial fertility level.

Limitations and Opportunities

The tree–crop interaction equation is *one* approach to examine the relative importance of fertility and competition using the example of alley cropping. Alley cropping is relatively well known in the tropics and provides a useful starting point for analysis of spatial agroforestry systems. The link between nitrogen availability and competition in alley cropping has been approached by modelling. For instance, Haggar and Beer (1993) described a model of maize growth in alleys of *Erythrina poeppigiana* and *Gliricidia sepium* in Costa Rica, which explores the relationship between crop duration, relative crop productivity and regrowth of trees. The time dimension of crop and tree growth is an important determinant of tree–crop interaction which is explored in Chapter 7 (Huxley, this volume). The relationship between soil fertility and competition by shading in alley-cropping systems is examined in another model in Chapter 3 for a range of multipurpose tree species (van Noordwijk, this volume). The purpose of this model is to provide a general delineation of the superiority of alley cropping over other landuse systems. Both modelling approaches highlight the need to identify the processes and traits which influence both the positive and negative effects of intercropping. These processes and traits are explored in detail in subsequent chapters.

There are several limitations on the use of the tree–crop interaction equation for other agroforestry systems such as the widely spaced trees in the parkland systems or complex homegardens. A classic problem with these complex systems is the seemingly endless major variables which are interconnected. For example, a series of intensive research on the widely spaced *Acacia tortilis* and *Adansonia digitata* in the savannas of East Africa indicates that the biomass of the understorey vegetation is strongly influenced by the average rainfall, grazing intensity, tree population and botanical composition (Belsky *et al.*, 1993). In such situations it seems prudent to begin by examining just one or two key factors. This approach is illustrated by the elegant study of the two trees species (*Vitellaria paradoxa*, *Parkia biglobosa*) in the parklands of Burkina Faso (Kessler, 1992). He observed that the reduction in sorghum yield under the two tree species was largely due to the level of shading even though soil fertility was increased. The dominant role of tree competition was revealed when the trees were cut and subsequent crop yield increased as compared with crop yield reduction under intact trees.

In their recent review of the published literature on ecological interactions in agroforestry systems Anderson *et al.* (1993) highlighted the existing gap between the theoretical basis already developed, and the results of field trials. There is an urgent need to obtain a better understanding of the mechanisms behind the various interactions present in agroforestry systems. They suggested that process-based research has the advantage and potential for extrapolation to other climates, soil types and plant communities. Key research topics in order of priority are belowground studies, e.g. root architecture, root dynamics and competition, nutrient and water cycling and traditional agroforestry practices.

Finally, agroforestry research is no longer confined to tropical regions because in Europe and North America there is an increasing need to reduce the land area in agricultural production by setting aside land for timber production. In Australia, agroforestry is considered as a means to alleviate the problems of a rising water table and salinization, which affect a major proportion of the productive agricultural lands (Eastham *et al.*, 1994). Although temperate and tropical zones differ in climate and tree species the principles and benefits of agroforestry have much in common (Jarvis, 1991).

Acknowledgement

An earlier version of this chapter was published under the citation: Ong, C.K. (1995) The 'dark side' of intercropping: manipulation of soil resources. In: Sinoquet, H. and Cruz, P. (eds) *Proceedings of the International Conference on Ecophysiology of Tropical Intercropping*, 6–11 December 1993, Guadeloupe. Institut National de la Recherche Agronomique, Science Update, pp. 45–56.

References

Akyeampong, E., Duguma, B., Heineman, A., Kamara, C.S., Kiepe, P., Kwesiga, F., Ong, C.K., Otieno, H.J. and Rao, M.R. (1996) A synthesis of ICRAF's research on alley cropping. In: Kang, B.T., Osiname, A.O. and Larbi, A. (eds) *Alley Farming Research and Development: Proceedings of International Alley Farming Conference*. IITA, Ibadan, Nigeria, 14–18 September 1992. IITA, Ibadan, pp. 40–51.

Anderson, L.S. and Sinclair, F.L. (1993) Ecological interactions in agroforestry systems. *Agroforestry Abstracts* 6 (2), 57–91.

Anderson, L.S., Muetzelfeldt, R.I. and Sinclair, F.L. (1993) An integrated research strategy for modelling and experimentation in agroforestry. *Commonwealth Forestry Review* 72, 161–174.

Belsky, A.J., Mwonga, S.M. and Duxbury, J.M. (1993) Effects of widely spaced trees and livestock grazing on understorey environments in tropical savannas. *Agroforestry Systems* 24, 1–20.

Charreau, C. and Vidal, P. (1965) Influence de l'*Acacia albida* Del sur le sol, nutrition minérale et rendements des mils *Pennisetum* au Senegal. *Agronomie Tropicale* 6–7, 600–626.

Corlett, J.E., Black, C.R., Ong, C.K. and Monteith, J.L. (1992) Microclimatic modification in a leucaena/millet alley cropping system II. Light interception and dry matter production. *Agricultural and Forest Meteorology* 60, 73–91.

de Wit, C.T. (1960) On competition. *Verslagen Landbowkundige Onderzoekingen* 66.8.

de Wit, C.T. and van den Bergh, J.P. (1965) Competition between herbage plants. *Netherlands Journal of Agricultural Research* 13, 212–221.

Eastham, J., Scott, P.R. and Steckis, R. (1994) Components of the water balance of tree species under evaluation for agroforestry to control salinity in the wheatbelt of Western Australia. *Agroforestry Systems* 26, 157–170.

Felker, P. (1978) *State of the Art*: Acacia albida *as a Complementary Permanent Intercrop with Annual Crops*. University of California, Riverside, California, USA.

Gichuru, M.P. and Kang, B.T. (1989) *Calliandra calothrysus* (Meissn.) in alley cropping with sequentially cropped maize and cowpea in southwestern Nigeria. *Agroforestry Systems* 9, 191–203.

Grace, J.B. and Tilman, D. (eds) (1990) *Perspectives on Plant Competition*. Academic Press, San Diego, California.

Haggar, J.P. and Beer, J.W. (1993) Effect of maize growth on the interaction between increased nitrogen availability and competition with trees in alley cropping. *Agroforestry Systems* 21, 239–250.

Horsley, S.B. (1991) Allelopathy. In: Avery, M.E., Cannell, M.G.R. and Ong, C.K. (eds) *Biophysical Research in Asian Agroforestry*. Winrock International, USA, pp. 167–183.

Huxley P.A. (1983) Some characteristics of trees to be considered in agroforestry. In: Huxley, P.A. (ed.) *Plant Research and Agroforestry*. ICRAF, Nairobi, Kenya, pp. 3–12.

Huxley, P.A. (1985) The basis of selection, management and evaluation of multipurpose trees – an overview. In: Cannell, M.G.R. and Jackson, J.E. (eds) *Trees as Crop Plants*. Institute of Terrestrial Forestry, Edinburgh, pp. 13–35.

Huxley, P.A., Akunda, E., Pinney, A., Darnhofer, T. and Gatama, D. (1989) Tree–crop interface investigations: preliminary results with *Cassia siamea* and maize. In: Reifsynder, W.S. and Darnhofer, T.O. (eds) *Meteorology and Agroforestry*. ICRAF, Nairobi, Kenya, pp. 361–370.

International Crops Research Institute for the Semi-Arid Tropics (1991) Annual report. ICRISAT, Hyderabad, India.

Jarvis, P.G. (1991) Agroforestry: principles and practices. In: Jarvis, P.G. (ed.) *Proceedings of an International Conference*, 23–28 July 1989, University of Edinburgh.

Kang, B.T. (1993) Alley cropping: past achievements and future directions. *Agroforestry Systems* 23, 141–156.

Kang, B.T., Wilson, G.F. and Sipkens, L. (1981) Alley cropping maize (*Zea mays* L) and leucaena (*Leucaena leucocephala* Lam.) in southern Nigeria. *Plant and Soil* 63, 165–179.

Kang, B.T., Reynolds, L. and Atta-Krah, A.N. (1990) Alley farming. *Advances in Agronomy* 43, 315–359.

Kater, L.J.M., Kante, S. and Budelman, A. (1992) Karité (*Vitellaria paradoxa*) and néré (*Parkia biglobosa*) associated with crops in south Mali. *Agroforestry Systems* 18, 89–105.

Kellman, M. (1979) Soil enrichment by neotropical savanna trees. *Journal of Ecology* 67, 565–577.

Kessler, J.J. (1992) The influence of karité (*Vitellaria paradoxa*) and néré (*Parkia biglobosa*) trees on sorghum production in Burkina Faso. *Agroforestry Systems* 17, 97–118.

Kidd, C.V. and Pimentel, D. (1992) *Integrated Resource Management. Agroforestry for Development*. Academic Press, Inc., San Diego, USA.

Kiepe, P. and Rao, M.R. (1994) Management of agroforestry for the conservation and utilisation of land and water resources. *Outlook on Agriculture* 23, 17–25.

Lal, R. (1989) Agroforestry systems and soil surface management of a tropical alfisol: (1) Soil moisture and crops yields. *Agroforestry Systems* 8, 7–29.

Maclean, R.H., Litsinger, J.A., Moody, K. and Watson, A.K. (1992) The impact of alley cropping *Gliricidia sepium* and *Cassia spectabilis* on upland rice and maize production. *Agroforestry Systems* 20, 213–228.

Malik, R.S. and Sharma, S.K. (1990) Moisture extraction and crop yield as a function of distance from a row of *Eucalyptus tereticornis*. *Agroforestry Systems* 12, 187–195.

Matthews, R.B., Lungu, S., Volk, J., Holden, S.T. and Solberg, K. (1992) The potential of alley cropping in improvement of cultivation systems in the high rainfall areas of Zambia. II. Maize production. *Agroforestry Systems* 17, 241–262.

Monteith, J.L., Ong, C.K. and Corlett, J.E. (1991) Microclimatic interaction in agroforestry systems. *Forest Ecology and Management* 45, 31–44.

Nair, P.K.R. (1984) *Soil Productivity Aspects of Agroforestry*. ICRAF, Nairobi, Kenya.

Ofori, F. and Stern, W.R. (1987) Cereal–legume intercropping systems. *Advances in Agronomy* 41, 41–90.

Ong, C.K. (1991) The interactions of light, water and nutrients in agroforestry systems. In: Avery, M.E., Cannell, M.G.R. and Ong, C.K. (eds) *Application of Biological Research in Asian Agroforestry*. Winrock International, USA, pp. 107–124.

Ong, C.K. (1995) The 'dark side' of intercropping: manipulation of soil resources. In: Sinoquet, H. and Cruz, P. (eds) *Proceedings of the International Conference on Ecophysiology of Intercropping*, 6–11 December 1993, Guadeloupe. INRA, Science Update, pp. 45–46.

Ong, C.K. and Black, C.R. (1994) Complementarity of resource use in intercropping and agroforestry systems. In: Monteith, J.L., Scott, R.K. and Unsworth, M.H. (eds) *Resource Capture by Crops*. Nottingham University Press, Loughborough, pp. 255–278.

Ong, C.K., Odongo, C.W., Marshall, F. and Black, C.R. (1991) Water use of agroforestry systems in semi-arid India. In: Calder, I.R., Hall, R.L. and Adlard, P.G. (eds) *Growth and Water Use of Forest Plantations*. Wiley, Chichester, UK, pp. 347–358.

Ranganathan, R. (1993) Analysis of yield advantage in mixed cropping. PhD thesis, University of Wageningen, The Netherlands.

Rao, M.R., Sharma, M.M. and Ong, C.K. (1990) A study of the potential of hedgerow intercropping in semi-arid India using a 2-way systematic design. *Agroforestry Systems* 11, 243–258.

Rao, M.R., Sharma, M.M. and Ong, C.K. (1991) A tree–crop interface design and its use for evaluating the potential of hedgerow intercropping. *Agroforestry Systems* 13, 143–158.

Reifsnyder, W.S. and Darnhofer, T.O. (1989) *Meteorology and Agroforestry*. ICRAF, Nairobi, Kenya.

Rizvi, S.J.H. (ed.) (1991) *Allelopathy: Basic and Applied Aspects*. Chapman & Hall, London.

Sanchez, P. (1987) Soil productivity and sustainability in agroforestry systems. In: Steppler, H. and Nair, P.K.R. (eds) *Agroforestry, a Decade of Development*. ICRAF, Nairobi, Kenya, pp. 205–226.

Singh, G., Arora, Y.K., Narain, P. and Grewal, S.S. (1990) *Agroforestry Research in India and Other Countries*. Surya Publications, Dehra Dun, India.

Singh, K.S. and Lal, P. (1969) Effect of *Prosopis spicigera* (or *cineraria*) and *Acacia arabica* trees on soil fertility and profile characteristics. *Annals of Arid Zone* 8, 33–36.

Singh, R.P., Ong, C.K. and Saharan, N. (1989) Above and below ground interactions in alley-cropping in semi-arid India. *Agroforestry Systems* 9, 259–274.

Spitters, C.J.T. (1983) An alternate approach to the analysis of mixed cropping experiments 1. Estimation of competition effects. *Netherlands Journal of Agricultural Sciences* 31, 1–11.

Squire, G.R. (1990) *The Physiology of Tropical Crop Production*. CAB International, Oxford, UK.

Szott, L.T. and Kass, D.C.L. (1993) Fertilizers in agroforestry systems. *Agroforestry Systems* 23, 157–176.

Szott, L.T., Fernandes, E.C.M. and Sanchez, P. (1991) Soil–plant interactions in agroforestry systems. *Forest Ecology and Management* 45, 127–152.

Tian, G. and Kang, B.T. (1994) Evaluation of phytotoxic effects of *Gliricidia sepium* (Jacq) Walp prunings on maize and cowpea seedlings. *Agroforesty Systems* 26, 249–254.

Vandenbelt, R.J. and Williams, J.H. (1992) The effect of soil surface temperature on the growth of millet in relation to the effect of *Faidherbia albida* trees. *Agricultural and Forest Meteorology* 60, 93–150.

Wiersum, K.F. (1991) Soil erosion and conservation in agroforestry systems. In: Avery, M.E., Cannell, M.G.R. and Ong, C.K. (eds) *Biophysical Research in Asian Agroforestry*. Winrock International, USA, pp. 209–230.

Willey, R.W. (1979) Intercropping – its importance and research needs: Agronomy and research approaches. *Field Crop Abstracts* 32 (2), 78–85.

Willey, R.W.M., Natajaran, M., Reddy, M.S. and Rao, M.R. (1986) Cropping systems with groundnut: Resource use and productivity. In: *Agrometeorology of Groundnut*. Proceedings of an International Symposium, 21–26 August 1985. ICRISAT Sahelian Center, Niamey, Niger, pp. 193–205.

Young, A. (1989) *Agroforestry for Soil Conservation*. CABI–ICRAF, Nairobi, Kenya.

Young, A. (1991) Soil fertility. In: Avery, M.E., Cannell, M.G.R. and Ong, C.K. (eds) *Biophysical Research in Asian Agroforestry*. Winrock International, USA, pp. 187–207.

Yu, S., Wang, S., Wei, P., Zhu, Z., Lu, X. and Fang, Y. (1991) A study of Paulownia/tea intercropping system – microclimate modification and economic benefits. In: Zhu, Z., Cai, W., Wang, S. and Jiang, Y. (eds) *Agroforestry Systems in China*. CAF-DRC, Singapore, pp. 150–161.

Zhao, L. (1991) Biological control of insect pests and plant diseases in agroforestry systems. In: Avery, M.E., Cannell, M.G.R. and Ong, C.K. (eds) *Biophysical Research in Asian Agroforestry*. Winrock International, USA, pp. 73–90.

2

◆

Mixed Cropping of Annuals and Woody Perennials: An Analytical Approach to Productivity and Management

R. RANGANATHAN[*] AND C.T. DE WIT[†]

Department of Theoretical Production Ecology, Wageningen Agricultural University, PO Box 430, 6700 AK Wageningen, The Netherlands

Introduction

Growing woody perennials with annual crops attempts to provide a strong foundation for conservation oriented farming and meet shortages of fodder and fuelwood. The role of perennials in minimizing leakage of nutrients from the system and recycling them, preventing soil erosion and thereby positively influencing the growth of plants associated with them is the biological premise to agroforestry systems (King, 1979). However, a significant concern is that competition from the perennial to the annual may reduce or possibly override the otherwise positive aspects of such mixed cropping (Verinumbe and Okali, 1985; Singh *et al.*, 1989a, b; Rao *et al.*, 1990; Jama and Getahun, 1991; Yamoah, 1991). This is particularly so in situations where the perennial is of less direct economic value than the annual as farmers seldom consider conservation in itself as a benefit.

This chapter is concerned not so much with the biological premise of agroforestry systems, as with the mathematical and experimental analyses of trade-offs in annual and perennial production. It elaborates on well confirmed theories of competitive interactions between plants and discusses the use of production possibility frontiers, drawn from economics literature.

---

[*] *Present address: ICRISAT Asia Center, Patancheru 502 324, Andhra Pradesh, India.*
[†] *Deceased.*

The production possibility frontier expresses the yield of the annual as a function of the perennial (or vice versa), shows all combinations of maximum yields that can be obtained and allows for the calculation of optimal sowing densities. The trade-off in productivity of one component as a consequence of the other is thereby quantified. With time, the changing morphology and increasing biomass necessitates changes in management to control the dominance of the woody perennial. Management guidelines for maintaining a sustained production of the annual have been derived from studies on crop–weed interactions.

The analytical approaches are supported and illustrated with experimental results from mixed cropping trials of perennial pigeonpea and groundnut in the semi-arid region of India. Some sections of this chapter have already been discussed in other papers and, where this is so, the development of the argument is brief.

Competition and Productivity

The key to increasing productivity in mixed cropping is understanding the nature of interaction between species in the mixture. Plants compete for growth factors such as light, water, nutrients, oxygen and carbon dioxide and the outcome of this competition is, in general, a reduction in plant growth and performance of the species in mixture. An aspect, other than the physiological mechanisms of interaction in mixed crops, is that of population dynamics wherein the effects of competition on productivity are examined without necessarily going into the mechanisms of the interaction. In such studies the effects of interspecific and intraspecific competition are analysed and used to measure yield advantage, if any, achieved through mixed cropping.

Model development

In the analysis of interspecific and intraspecific competition in crop mixtures, de Wit (1960, 1961) used a replacement series design where seed numbers (N_i and N_j) of the two species vary simultaneously in such a way that their sum S remains constant. It was shown that biomass yields (Y_i and Y_j) are often well presented by hyperbolic replacement functions:

$$Y_i = \frac{k_{ij}N_i}{k_{ij}N_i + N_j}M_i \qquad (2.1a)$$

$$Y_j = \frac{k_{ji}N_j}{k_{ji}N_j + N_i}M_j \qquad (2.1b)$$

where M_i and M_j are yields of the sole crops when sown at density S. The parameters k_{ij} and k_{ji} reflect the competitive effect of species j on i and i on j

respectively. De Wit (1960) showed that when the product $k_{ij}k_{ji}$ equals 1, the two species i and j are competing for the same resources at the same time. $k_{ij}k_{ji} > 1$ indicates the species are partly complementary in resource use. The premise of mixed cropping is the spatial and temporal use of resources by crops in the mixture; the crops have different heights and rooting depths, and make their peak demand on resources at different times. Spatial and temporal complementarity is achieved by cropping species with different growth curves.

If species j does not grow at all, its yield is zero and by substituting $(S - N_i)$ for N_j Equation 2.1a can be reduced to the density function:

$$Y_i = \frac{B_i N_i}{B_i N_i - 1} Q_i \qquad (2.2)$$

where Q_i is the maximum yield achieved at high densities and $B_i Q_i$ is the yield of a plant when free from competition from other plants. By combining Equations 2.1 and 2.2, de Wit (1960, 1961) and Spitters (1983) derived the following additive functions:

$$Y_i = \frac{B_i}{B_i N_i + B_{ij} N_j + 1} Q_i \qquad (2.3a)$$

$$Y_j = \frac{B_j}{B_j N_j + B_{ji} N_i + 1} Q_j \qquad (2.3b)$$

where B_{ij} and B_{ji} characterize competitive abilities. It was also shown that for species that are temporally complementary in resource use $0 < B_{ij} < B_j$ and $0 < B_{ji} < B_i$, while $B_{ij} = B_j$ and $B_{ji} = B_i$ in situations where both species compete for the same resources at the same time. From these equations it can be inferred that one plant of species i has the same effect on Y_i as $B_{ij}B_i^{-1}$ plants of species j. Similarly, one plant of species j has the same effect on Y_j as $B_{ji}B_j^{-1}$ plants of species i.

The parameters B_i, B_j, B_{ij} and B_{ji} can take values up to infinity causing some difficulties with the convergence criteria associated with non-linear regression algorithms that may be used in their estimation. Although the yield–density relationship is better visualized in de Wit's notation (as in Equations 2.3a and b), the parameters are more easily estimated using Spitters' (1983) notation. They are:

$$Y_i = \frac{N_i}{b_{i0} + b_{ii} N_i + b_{ij} N_j} \qquad (2.4a)$$

$$Y_j = \frac{N_j}{b_{j0} + b_{jj} N_j + b_{ji} N_i} \qquad (2.4b)$$

It can be shown that the parameters b_{i0}, b_{ii} and b_{ij} from Equation 2.4a are equal to $(B_i Q_i)^{-1}$, Q_i^{-1} and $B_{ij}(B_i Q_i)^{-1}$ respectively.

Such simple mathematical expressions of complex biological processes necessarily introduce some compromise. Yields reach a maximum at finite rather than infinite densities, and at very low densities there is a linear relationship between plant density and yield rather than a hyperbolic one. However, the hyperbolic relationship has been shown by many researchers to be an acceptable description of the biological process of competition (Shinozaki and Kira, 1956; de Wit, 1960, 1961; Willey and Heath, 1969; Spitters, 1983). Exceptions are when one crop profits from the presence of the other or its growth is inhibited by allelopathic effects.

Production possibility frontiers

In mixed cropping one crop cannot be considered independently of the other and measures of yield advantage must express the yield of one crop as a function of the other, so as to determine when more production of one crop and less of the other is advantageous. In addition, such measures must recognize the different requirements of the farmer and incorporate factors other than biological ones which influence his decision to intercrop.

Production functions in economics are analogous to growth curves of crops as they respond biologically to available resources. The production possibility frontier (PPF) has been used as a theoretical tool (Filius, 1982; Raintree, 1983; Tisdell, 1985) to illustrate complementarity or competition between the agricultural and tree components in agroforestry. The PPF as developed by Ranganathan (1992) expresses the yield of a crop as a function of the other and shows maximum combinations of products which can be obtained after consideration of all possible plant density combinations, assuming the most efficient use of available resources. Using Spitters' Equations 2.4a and b as a base for its derivation, Ranganathan (1992) evaluated trade-offs in biological productivity in different intercropping systems such as oats–barley, pigeonpea–sorghum and groundnut–maize.

When N_i and N_j are very large, Equations 2.4a and b can be written in the form:

$$Y_i = \frac{1}{b_{ii} + b_{ij}\dfrac{N_j}{N_i}}$$

$$Y_j = \frac{1}{b_{jj} + b_{ji}\dfrac{N_i}{N_j}}$$

Solving these equations for $N_j N_i^{-1}$ and $N_i N_j^{-1}$ and multiplying the outcomes gives:

$$\left(\frac{1}{b_{ii}Y_i} - 1\right)\left(\frac{1}{b_{jj}Y_j} - 1\right) = C \tag{2.5}$$

where

$$C = \frac{b_{ij}b_{ji}}{b_{ii}b_{jj}}$$

The PPF can be empirically calculated using Equation 2.5 and it represents maximum yield combinations of the two crops obtained after a consideration of all possible plant density combinations.

Field experiments with perennial pigeonpea and groundnut

Pigeonpea, a crop primarily of India, has been successfully intercropped with annuals like sorghum, groundnut and maize. In the southern semi-arid states of India it is extensively intercropped with groundnut and sorghum. Pigeonpea seed is an essential part of the diet while stems, after pod harvest, provide kindling wood. Perennial pigeonpea is a short-lived multipurpose woody species providing grain, fuelwood and green fodder during the dry season. Its deep-rooting and drought-tolerant nature makes it a useful crop in areas of low and uncertain rainfall which characterize much of the semi-arid tropics.

Studies at the International Crops Research Institute for the Semi-Arid Tropics (ICRISAT) with perennial varieties of pigeonpea have shown that they are much like annual varieties in their first year of growth and possess the same slow initial growth that makes annual varieties so suitable for intercropping. Daniel and Ong (1990) demonstrated the possibility of intercropping perennial pigeonpea with annual crops without any serious adverse effects on annual crop yield in the first year.

Field trials were conducted at ICRISAT in Hyderabad, India where long term average rainfall in the rainy season (July–October) is 610 mm, and 148 mm in the post rainy season. Perennial pigeonpea was intercropped with groundnut in four replacement series where groundnut was sown as a sole crop at 8, 16, 32 and 64 plants m^{-2} and pigeonpea at sole crop densities of 1.5, 3, 6 and 12 plants m^{-2}. Accordingly, one pigeonpea plant replaced 5.33 groundnut plants.

At the start of the rainy season in 1989 (Experiment 1), both pigeonpea and groundnut were sown together. Groundnut was harvested 110 days later. After pigeonpea seed harvest, 240 days after sowing, the stand was lopped to a height of 0.5 m and allowed to regrow through the rest of the dry season. At the start of the following rainy season, when pigeonpea entered its second year of growth, it was pruned once again to a height of 0.5 m and 14 days thereafter groundnut was resown into the year-old pigeonpea alleys. Total biomass and marketable yields of groundnut and pigeonpea were measured at every harvest. The relatively small dry season growth of pigeonpea (harvested at the start of the 1990 rainy season) was not included in the analysis. In 1990 a similar trial (Experiment 2) was initiated but discontinued in the following year.

Experimental results and other details are fully described in Ranganathan (1993) so that here only the outcome is presented in Table 2.1 as parameter values and their standard errors. Pigeonpea and groundnut biomass yields obtained from Experiments 1 (years 1 and 2) and 2 (year 1) were used in the calculation of the parameters of Spitters' Equations 2.4a and b. Calculations for the second year data (Experiment 1 – year 2) have been made using the original plant densities of pigeonpea although actual densities were lower because of plant mortality.

The PPFs in Fig. 2.1 were derived for the three data sets. The relatively low pigeonpea yields in Experiment 1 – year 1 (Fig. 2.1a) and the difference in convexity of the first year curves are explained by the delayed sowing of pigeonpea in Experiment 1. Waterlogging caused pigeonpea seedling mortality and resowing at a later date to achieve the planned plant densities. The extremely convex shape of all three curves indicates a large yield advantage (Ranganathan et al., 1991) in intercropping perennial pigeonpea with groundnut. According to Daniel and Ong (1990), perennial pigeonpea is similar to medium-duration annual varieties except for its longer duration to flowering and maturity, lower harvest index, greater ratoonability and deeper rooting habit. Thus it can be inferred that yield

Table 2.1. Estimated parameters of the Spitters' equations.

	b_{i0} (plants 100 g^{-1})*	b_{ii} (m^2 100 g^{-1})*	b_{ij} (m^2 100 g^{-1})*	Adj. R^2
Experiment 1 – year 1				
Pigeonpea	0.151	0.288	0.042	0.90
	(0.089)	(0.030)	(0.012)	
Groundnut	1.127	0.254	0.133	0.94
	(0.303)	(0.197)	(0.112)	
Experiment 2 – year 1				
Pigeonpea	0.044	0.149	0.00001	0.73
	(0.012)	(0.006)	(0.001)	
Groundnut	2.072	0.191	0.739	0.94
	(0.349)	(0.014)	(0.152)	
Experiment 1 – year 2				
Pigeonpea[†]	< 0.044	0.150	0.00001	–
	–	–	–	
Groundnut	2.285	0.244	5.718	0.98
	(0.467)	(0.018)	(0.662)	

* Standard errors of means are shown in parentheses; [†] non-linear algorithm not possible; see Ranganathan (1993) for estimation of parameters.

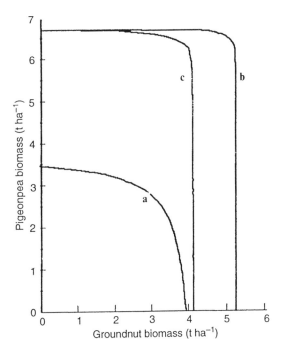

Fig. 2.1. Production possibility frontiers for pigeonpea–groundnut intercrops. a, Experiment 1 – year 1; b, Experiment 2 – year 1 and c, Experiment 1 – year 2.

advantage as shown by the PPFs is due to temporal complementarity of the two species.

The PPF is the envelope of joint production curves, as illustrated in Fig. 2.2. The joint production curves are calculated from Equations 2.4a and b. The curves taking off from the vertical axis show the very gradual decline in pigeonpea yield (Experiment 2 – year 1; pigeonpea density N_p was held constant but at different levels) when groundnut density N_g was increased. Pigeonpea apparently suffered little competition from groundnut. The curves taking off from the groundnut axis (x) represent yields of pigeonpea and groundnut obtained when groundnut density was held constant (at different levels) and pigeonpea density gradually increased. The six curves radiating out from the origin give the yields of pigeonpea and groundnut when densities of the two crops were varied but the ratio $(N_g N_p^{-1})$ kept constant.

Economic evaluation

The importance of an economic evaluation cannot be overestimated because the market value of products provides the farmer with a tool in

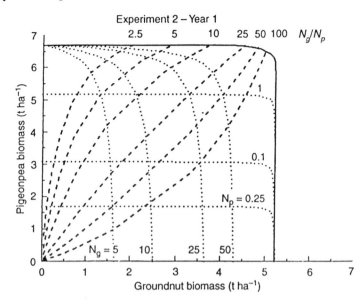

Fig. 2.2. Production possibility frontier for a pigeonpea–groundnut intercrop. Inner curves show yield combinations at different densities, calculated using Spitters' model of competition.

allocating limited resources between competing uses and puts different crops and their products on a comparable basis. The preceding section took total biomass into consideration. In this section on economic analysis, marketable yield rather than total biomass will be considered as it is often the aim of agriculture.

In response to density stress, a plant regulates its allocation of assimilatory products to its various organs thereby affecting the relationship between marketable yield (pods, grain, leaves) and total biomass. This response varies from species to species. When marketable yield and total biomass have a similar response to density, that is, harvest index (HI) remains constant, the PPF for marketable yield can be directly derived from Equation 2.5. But in cases where this is not so, the PPF has to be derived numerically from the relationship between total biomass and marketable yield.

It was shown by Ranganathan (1993) that in mixtures irrespective of whether plants suffer from interspecific or intraspecific competition the relationship between *per-plant* marketable yield and total biomass is well approximated by the same straight line. When the straight line passes through the origin, harvest index remains constant even at high competition stress. However, in many instances the straight line expressing the relationship between per-plant marketable yield and total biomass makes a positive intercept with the x-axis. This reflects a diminishing harvest index with decreasing per-plant yield.

The numerical derivation of the PPF for marketable yield thus requires an estimation of the parameters to Equations 2.4a and b and the linear relationship between per-plant marketable yield and total biomass. Dividing biomass yields per hectare, estimated from Equations 2.4a and b, by the density at which it was obtained gives per-plant biomass. From the linear relationship between per-plant biomass and marketable yield, the corresponding per-plant marketable yield is obtained. Multiplication of these per-plant yields with the density gives marketable yield per hectare for both crops. A plot of the yield combinations gives joint production curves for marketable yield (similar to those in Fig. 2.2) and once again the envelope is the production possibility frontier.

The per-plant marketable yield–biomass relationships for pigeonpea and groundnut in Experiment 2 – year 1 are shown in Fig. 2.3 and the joint production curves with their envelope, the PPF, for marketable yields per hectare in Fig. 2.4. Both per-plant marketable yield–total biomass relationships are linear, but in the case of groundnut the line has a large positive intercept with the per plant biomass axis (x), so that the harvest index decreases in the normal density range with decreasing plant size; first slowly and then rapidly. The positive intercepts are reflected in the joint production functions in Fig. 2.4 turning inward; groundnut yields show a greater decline with increasing density of pigeonpea. The harvest index of pigeonpea in this experiment was less dependent on plant size than

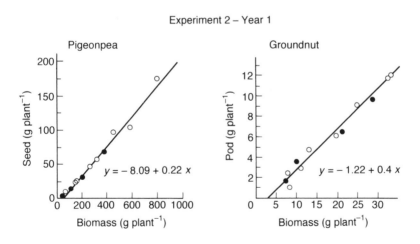

Fig. 2.3. Per-plant marketable yield–biomass relationships for a pigeonpea–groundnut intercrop. Open symbols represent yields in mixture and closed symbols yields in monoculture.

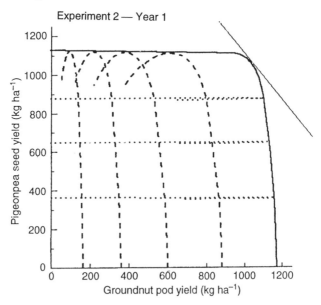

Fig. 2.4. Production possibility frontier for pigeonpea seed and groundnut pod yields. Yields have been calculated from per-plant marketable yield–biomass relationships and Spitters' model of competition.

groundnut, so that the joint production curves obtained with a constant pigeonpea density (dotted lines, Fig. 2.4) show only a slight decrease in seed yield with decreasing biomass (and increasing density).

The economically optimal seed yield combination corresponds to the tangent of the price line to the PPF. A price line reflects a fixed value of production, that is, the total value of the crops expressed as the sum of their constituent prices. The slope of the price line is the negative of the price ratio. For instance, if groundnut pods are priced 20% higher than pigeonpea seeds, the price line would have a slope of − 1.2. Assuming this was so, the price line shown in Fig. 2.4 is tangential to the PPF at the point where groundnut pod and pigeonpea seed yields are close to 1050 and 1025 kg ha^{-1} respectively. Substituting these yields in Equation 2.5 and using parameter values for Experiment 2 – year 1 (in Table 2.1), plant densities that would optimize economic returns to the farmer are 32 and 1 plant(s) m^{-2} of groundnut and pigeonpea. Similar calculations for the data of Experiment 1 – year 2 show groundnut pod and pigeonpea seed yield to be approximately 325 and 1260 kg ha^{-1} and the corresponding plant densities are 25 and 1 plant(s) m^{-2}. From a comparison of yields in the 2 years, it is obvious that in the second year of intercropping groundnut yields are severely depressed due to competition from pigeonpea and even in its first year of intercropping, pigeonpea yields were close to their asymptotic value.

Since pigeonpea and other similar perennials are valued for more than their seed yield, an economic analysis must include all such factors, fodder and fuelwood, and those others which influence farmers' decisions to intercrop. Price per kilogram of seed of the annual is often a decisive factor; it sometimes far exceeds the market price of the produce, on other occasions availability of seeds is low. An economic assessment taking a few such factors is described below.

Figure 2.5 shows the frontier for net returns in the first year of intercropping pigeonpea with groundnut. In this economic assessment pigeonpea fodder and groundnut seed costs have been taken into consideration, in addition to the market prices of pigeonpea seed and groundnut pods. Though groundnut fodder is sold in many parts of the tropics it is assumed to have no economic value for the purpose of this discussion. Pigeonpea fodder has been assigned an arbitrary price of $US 5.34 ton^{-1} ($US 1 = IRs 28.00). This is much lower than the price of other green fodders but in the semi-arid regions of India where pigeonpea is grown, perennial varieties are relatively uncommon and no reliable market exists. It is assumed that groundnut pods have a market price 20% higher than that of pigeonpea seeds and the latter have been assigned a price of

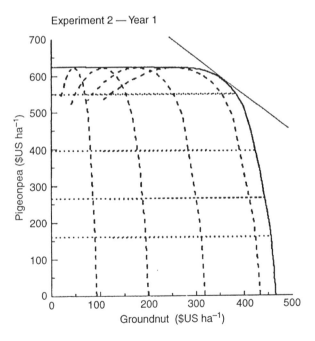

Fig. 2.5. Production possibility frontier for financial returns and price line for a pigeonpea–groundnut intercrop.

$US 0.54 kg^{-1}. The price of groundnut seed is high relative to the price of the produce and has been taken as $US 2.68 kg^{-1}.

Optimum production under such market conditions corresponds to the point where a line with slope -1 is tangential to the frontier. The inner curves show that increasing groundnut density is not economical beyond a certain limit as the cost of seed exceeds the incremental pod yield. Optimum net returns from groundnut and pigeonpea are $US 360 and 600 ha^{-1}, respectively and are achieved at plant densities of 15 and 1 plants m^{-2}. Note that the plant density of pigeonpea is unchanged but reduced, from 32 to 15, in the case of groundnut.

Other factors that influence decisions to intercrop have not been dealt with. Labour costs, for instance, play an important role in decision making where mechanization exists. Mixed cropping imposes design and operational restrictions on the implements used to mechanize crop production. Costs could potentially be so large that it is economically advantageous to grow sole crops and sacrifice the benefits of complementarity, recycling of nutrients and erosion control. Indeed, harvesting difficulties were a main reason why farmers in Western Europe stopped most mixed cropping despite the agronomic advantages. Even where mechanization does not exist, opportunity costs to a farmer may be high. Family labour that may be used in managing the perennial could be more gainfully employed in other farm activities. Conservation benefits obtained from mixed cropping have also not been included in this discussion.

The woody perennial

Dominance

Berendse (1979) studied the coexistence of species, some with the ability to exploit a refugium (resources not available to the others). He showed that in order to coexist, species without the refugium had to be the stronger competitor for common resources. In mixtures of annuals and perennials, the perennial is able to exploit a refugium because of its longer growth period and extensive root system. In the second and/or subsequent years of intercropping it has, therefore, a competitive edge over the annual.

The dominance of a woody perennial over the intercropped annual is visualized in Fig. 2.6 where biomass yields of pigeonpea and groundnut (grown in a replacement series with monocrop densities approaching optimum at 6 plants m^{-2} of groundnut and 32 plants m^{-2} of pigeonpea) are plotted against relative density. Actual yields as observed in the field and those estimated from the yield–density relationship discussed in the earlier sections are presented. The observed yields may deviate, sometimes systematically, from the estimated yields because observations from all four replacement series were used in the determination of parameters.

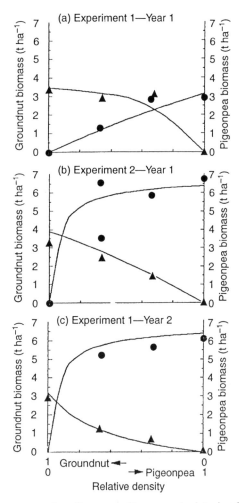

Fig. 2.6. Replacement series diagrams. Curves calculated using Spitters' model of competition, parameters in Table 2.1 (● ▲ actual observations).

In Experiment 1 – year 1, the groundnut crop did not suffer from the presence of pigeonpea, as is reflected by the convexity of yield–density hyperbola. This was due to delayed sowing of much of the pigeonpea. In Experiment 2 – year 1, pigeonpea and groundnut were sown together and groundnut suffered competition from pigeonpea as seen in the near-linear relationship between groundnut relative density and yield. Despite pruning pigeonpea 14 days before and 40 days after planting groundnut, groundnut was dominated by pigeonpea in year 2 of Experiment 1. The value of parameter b_{ij} in Table 2.1, which quantifies the effect of groundnut on pigeonpea yield, confirms these results.

Pigeonpea's dominance in this cropping system may be acceptable because what is lost in annual marketable yield of groundnut is made up by pigeonpea seeds. The direct economic return is, therefore, not very sensitive to the composition of the mixture. However, in many other mixed cropping systems that include woody perennials, the latter do not contribute to food production. In areas where agroforestry is most recommended, food production is the primary objective of subsistence level farming. Thus, a main concern of agroforestry is that competition from the perennial to the annual may override the other benefits obtained from the inclusion of perennials in the system.

Growth

Provided soil fertility is maintained, the yield achieved in time by a pruned woody perennial approaches an equilibrium value that is independent of initial plant density. That this equilibrium yield is directly related to the parameters that govern the density response of the species is demonstrated in this section.

In the case of annuals, seeds that are produced in one year may be resown the following year. When growing conditions are the same, yield in year $n+1$ can be calculated by replacing seed number by yield in year n in Equation 2.2:

$$Y_{n+1} = \frac{BY_n}{BY_n + 1} Q \qquad (2.6)$$

where Y_n and Y_{n+1} are yields (in kg ha^{-1}) in years n and $n+1$ and Q is the maximum yield achieved at high seed densities. The quotient $Y_{n+1}Y_n^{-1}$ approaches the product BQ for low values of Y_n; BQ is accordingly a dimensionless multiplier in this case. If the value of this multiplier is smaller than 1, Y_{n+1} is smaller than Y_n and the yield approaches zero in the course of time. But if BQ is larger than 1, the yield approaches an asymptotic yield Y_a which can be obtained by substituting Y_a for both Y_n and Y_{n+1} in Equation 2.6:

$$Y_a = \frac{BY_a}{BY_a + 1} Q$$

so that

$$Y_a = Q - \frac{1}{B} \qquad (2.7)$$

The asymptotic yield Y_a is always lower than the maximum yield at high plant densities and the more this is so, the smaller the value of B.

For perennials, however, there are no seeds to be resown. But it was shown by van den Bergh (1968) that in grasses the harvested quantity is a good indicator of the quantity of roots and stubble that remain in the field.

This is so because above- and below-ground biomass are positively correlated. For shrubs that are pruned, the amount of stems and leaves that are removed and the amount of stubs and roots that remain on the field for regrowth are likely to be positively related. Although recognizing the need for experimental verification, there is little reason to dispute that Equation 2.6 holds for shrub vegetation that is coppiced every year.

Under the assumption that soil fertility does not change, yield increase of the perennial in subsequent years can be calculated once Q and BQ are estimated from a density experiment that links yields in two successive years. Using Equations 2.2 and 2.6 hypothetical perennial yields in time are calculated assuming Q equals 10 t ha^{-1} and BQ equals 10, 2.5 and 1.5. The resulting curves in Fig. 2.7 show that the asymptotic yield(s) Y_a decreases with decreasing BQ (as in Equation 2.7). The curves also have the familiar S-shape which was shown by de Wit (1960, 1961) to conform to the well-known logistic growth curves of Lotka (1925) and Volterra (1928).

Woody perennials with a low BQ are less competitive to the annual because their regrowth is slow, but to be a good producer of biomass Q should be high. Such perennials may be hard to find because BQ and Q are likely to be positively correlated. This was implicitly shown by van den

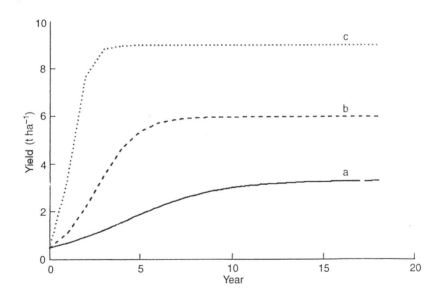

Fig. 2.7. Growth curves for woody perennials that are coppiced every year, calculated with $Q=10$ t ha^{-1} and a: $BQ=1.5$; b: $BQ=2.5$ and c: $BQ=10$.

Bergh (1968) for perennial grasses and explicitly proven by Spitters (1979) for barley.

Figure 2.8 relates the yield of perennial pigeonpea (in Experiment 1) in the second year with the yield measured in the first year of its growth. The value of Y_a appears to be around 5.5 t ha^{-1} although the scatter is large. Even at the lowest density of planting, yields in the second year were already at the asymptotic value (Y_a). Plant densities in the experiment were unfortunately too high to determine the value of BQ. However, from the steepest observed slope of 7 it can be concluded that the value of BQ approaches 10 and the value of Q according to Equation 2.7 is about 6.7 t ha^{-1}. From a simple density experiment of 2 years it is thus possible to estimate the asymptotic yield of a coppiced perennial.

Management

Suggestions have been made on how to overcome the problem of a strongly competitive woody perennial. Daniel and Ong (1990) and Odongo *et al.* (1996) recommend a low perennial plant density. However, in situations where initial planting densities are so small that asymptotic yields as defined by Equation 2.7 are never reached, part of the field would remain fully outside the influence of the woody perennial. This would leave some of the

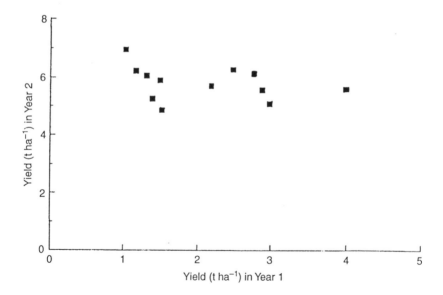

Fig. 2.8. Relationship between pigeonpea yields in the first and second years of growth.

annual as a sole crop and the purpose of the agroforestry system is not fully realized.

Ong (1990) suggests forms of perennial vegetative growth regulation so that there is more complementarity between species in their time dependence for resource sharing. *Faidherbia albida* is an example of a species where resource sharing with the annual is extremely complementary (Miehe, 1986; Poschen, 1986). Mixed cropping systems with *Faidherbia albida* show a unique form of complementarity in that the tree sheds its leaves naturally during the onset of the rainy season and begins leaf production once the understorey crop is well established. This 'competition free' period allows the crop to extend its root system well into the soil profile and to establish a complete crop canopy to capture radiation and water resources (Ong *et al.*, 1992).

One could also experiment with the use of growth retardants on the woody perennial but pruning or coppicing are the alternatives discussed here. Then in the words of Huxley (1983) 'the question is not only what to remove, and how much, but when'.

The effect of management on the annual

The same question was posed in weed research and Cousens (1985) and Kropff and Spitters (1991) dealt with the problem by considering the competitive relationship between weeds and crops. They defined relative yield loss (*YL*) in the crop (*c*) due to the presence of the weed (*w*) as:

$$YL = 1 - \frac{Y_{cw}}{Y_c} \qquad (2.9)$$

where Y_{cw} and Y_c are yields of the crop at the same density but with and without weeds. By substituting Equation 2.3 for Y_{cw} and Equation 2.2 for Y_c, *YL* can be written as:

$$YL = \frac{B_{cw}N_w + 1}{B_c N_c + B_{cw}N_w + 1}$$

Under normal sowing densities this expression simplifies to the hyperbolic expression:

$$YL = \frac{c\dfrac{N_w}{N_c}}{1 + c\dfrac{N_w}{N_c}}$$

where

$$c = \frac{B_{cw}}{B_c}$$

This expression relates the relative yield loss of the crop with the seed densities of the crop and weed. The damage coefficient *c* depends on the competitive abilities of the crop and weed species. However, because of the inherent difficulty of conducting density experiments with weeds and emergence of weeds in flushes, expressions with weed density are not useful for predictive purposes. Kropff and Spitters (1991) characterized the presence of crops and weeds in a mixture by their leaf areas measured at an early stage of growth. Relative yield loss was expressed by:

$$YL = \frac{q\dfrac{LAI_w}{LAI_c}}{1 + q\dfrac{LAI_w}{LAI_c}}$$

or

$$YL = \frac{qL_w}{1 + (q - 1)L_w} \qquad (2.10)$$

where

$$L_w = \frac{LAI_w}{LAI_w + LAI_c}$$

LAI_w and LAI_c are the leaf area indices of the weed and crop at an early stage of growth and q a damage coefficient that differs from c and has to be determined experimentally.

If the perennial is considered to be like a weed in depressing crop yield, establishing q for the mixed crop allows for an estimation of the amount the perennial must be pruned in order to keep crop yield at acceptable levels. The relationship between crop yield loss and relative leaf area of the weed (perennial) for various values of the damage coefficient q is shown in Fig. 2.9. For $q \gg 1$, severe pruning of the perennial is necessary to keep yield loss of the annual down to an acceptable level, whereas light pruning suffices for $q \ll 1$. Experiments with different crop–weed combinations have proved the usefulness of the approach. It was shown that the value of q is large in situations where weed leaf area ratio (leaf area/above-ground weight) is large compared with that of the crop and the weeds overgrow the crops. Relative early emergence of weeds is not reflected in a high value of q, but in a large relative leaf area (L_w) of the weed.

To show the same approach is equally valid for woody perennials, Ranganathan (1993) determined a damage coefficient q for perennial pigeonpea intercropped with groundnut. A 1-year-old stand of perennial pigeonpea was pruned and groundnut was sown within the alleys. Groundnut was also sown as a sole crop to calculate yield loss in intercropped groundnut due to the presence of pigeonpea. At regular intervals both pigeonpea and groundnut were harvested, leaf area indices and biomass determined. Twenty-five days after sowing, groundnut had a leaf area index of 0.23 while that of pigeonpea was 0.6, so that the relative

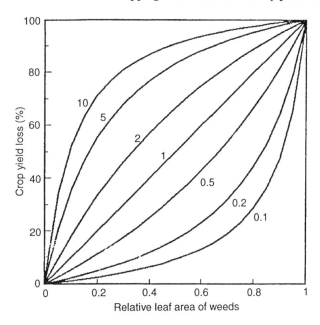

Fig. 2.9. Yield loss functions for different values of q, the damage coefficient (source: Kropff and Spitters, 1991).

leaf area of pigeonpea in the mixture was 0.72. After 60 days, relative leaf area increased marginally to 0.83. Within the period when the crop is established and the onset of competition from the perennial, relative leaf area of the perennial is conservative. Groundnut yield loss was measured at 56% and q was estimated from Equation 2.10 to be 0.37.

The yield loss function for perennial pigeonpea and groundnut is given in Fig. 2.10. The value of the damage coefficient is surprisingly low, when pigeonpea's greater height and similarity in leaf area ratios of pigeonpea and groundnut in the early stages of growth are considered. One explanation may be that pigeonpea was not distributed evenly over the field, but confined to rows that were 1.5 m apart. The alley width appears to be wide enough to prevent early shading of groundnut; and groundnut reached maturity before the greater fraction of pigeonpea biomass was formed. This important aspect of the problem was not further pursued.

To compare predicted yield (from Equation 2.10) with that observed in the field, the relative leaf area of pigeonpea in the mixture was varied in an experiment. Although pigeonpea has been known to recover from pruning and produce yields comparable to unpruned pigeonpea, to reduce the uncertainty of it not surviving corrective pruning, pigeonpea plants were covered with muslin bags for 30 and 60 days. Approximately 50% of the

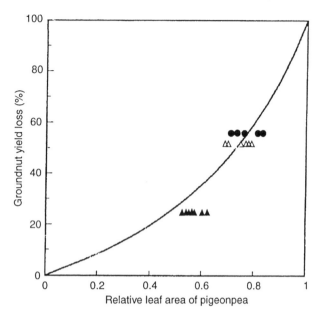

Fig. 2.10. Yield loss function for groundnut–pigeonpea intercrop, $q=0.37$ estimated from an unmanaged stand of pigeonpea (●). Observations from 30-day (△) and 60-day (▲) bag treatments.

incoming light was intercepted by the bags so that the plants continued to grow but at a slower rate.

For the 30-day bag treatment, observed groundnut yield loss was 14% and the relative leaf area of pigeonpea was measured at 0.69 and 0.78 after 25 and 60 days. For the 60-day bag treatment, observed yield loss was only 33% and relative leaf areas were 0.53 and 0.63. It is shown in Fig. 2.10 that these observations and the predicted yield loss were not dissimilar, providing supportive evidence of the applicability of this method to managing perennials. The timing of this pruning is not very critical as long as it is done in the exponential phase of growth of the perennial, and before the annual is unable to recover from the competition that it may briefly suffer before the perennial is pruned.

Effect of management on the perennial

In the previous section the leaf area of the perennial was reduced through corrective pruning in the more or less exponential phase of growth. The deferment time D that it takes pruned shrubs to restore leaf area to its pre-pruning level can be calculated from the expression:

$$L_n = L_m \exp (RD) \tag{2.11}$$

where R is the relative growth rate of the shrub in its exponential phase, L_n and L_m are leaf area indices directly before and after corrective pruning. Accordingly the deferment time D is:

$$D = \frac{\ln L_n - \ln L_m}{R} \qquad (2.12)$$

If M is the biomass produced over the entire period of growth of the perennial and P the length of this period, the linear growth rate G is MP^{-1}. The loss in weight of the woody perennial W_p due to pruning is:

$$W_p = DG \qquad (2.13)$$

Yield loss here depends mainly on the severity of pruning, i.e. ($\ln L_n - \ln L_m$) because G and R themselves are in general proportional.

For pigeonpea yield loss is not compensated for by the extended growth period, because it is governed by daylength (J.C.W. Odongo, Hyderabad, 1991, personal communication). However, some of this loss is saved as the harvest H during the pruning. H can be calculated from:

$$H = \frac{L_n - L_m}{LAR} \qquad (2.14)$$

where LAR is the leaf area ratio calculated as the leaf area of newly formed leaves divided by the weight of these leaves and newly formed twigs. The difference $W_p - H$ is the actual yield foregone due to corrective pruning. If a corrective pruning is applied in the linear phase of growth Equation 2.14 can be generalized by using the expolinear growth function introduced by Goudriaan and Monteith (1990), but not discussed in this chapter.

From the result of the periodic harvest experiment discussed in the previous section it was estimated that pigeonpea's relative growth rate (R) was 0.03 day^{-1}, linear growth rate (G) 0.007 kg m^{-2} day^{-1}, LAR 4 m^2 kg^{-1} and leaf area 0.6 m^2 m^{-2} at 25 days after sowing groundnut. Leaf area of groundnut was 0.3 m^2 m^{-2}. According to Fig. 2.10, leaf area of pigeonpea in the mixture has to be reduced to 0.2 to keep groundnut yield loss at 20%. Using the above information in Equations 2.11, 2.12 and 2.14, deferment time D is 37 days, yield loss of the perennial W_p due to pruning was 2600 kg ha^{-1} and the biomass yielded from pruning (H) is 1000 kg ha^{-1}. These amounts are small compared with the maximum biomass of 12,500 kg ha^{-1} achieved in the season.

Some thoughts on experimental design

It is often the practice in many agroforestry systems to use all or part of the harvested perennial as mulch, which may result in soil fertility and structure changes. Such changes complicate the analysis especially if the perennial is a nitrogen fixer; one crop profits from the presence of the other and the yield–density response is no longer hyperbolic (de Wit *et al.*, 1966). The effect of mulch can be determined by a comparison of unmulched and mulched sole crops of the annual and perennial. This implies that all treatments in the experiment be uniformly mulched, if necessary with clippings of the perennial grown outside the experiment. The combined effects of mulching and mixed cropping can be studied only from such experiments. But experiments of this nature easily expand beyond manageable sizes. However, once the existence of hyperbolic relationships is established or taken for granted, to analyse the effect of mulching only requires five treatments that will allow for the estimation of the six parameters B_i, B_j, B_{ij}, B_{ji}, Q_i and Q_j (from Equations 2.3a and b). The five treatments comprise two monocultures each for the annual and the woody perennial, and one mixture preferably at densities that form part of a replacement series with the high plant density monocrops. Some prior information about the species may be needed to select plant densities such that there are sufficiently large differences in the monocrop yields obtained at high and low plant densities. Likewise, the plant densities in the mixture should be such that the annual has a chance to withstand competition from the perennial. With the establishment of the production possibility frontier and other bio-economic factors, the experiment can then be used to estimate damage coefficients and develop regulatory pruning strategies.

Concluding Remarks

Uncertain weather, pests and diseases that could not be adequately controlled were sometimes the cause of less than satisfactory experimental results, especially so for trials where pigeonpea and groundnut were periodically harvested. Nevertheless a comprehensive analysis of competitive interference between annuals and woody perennials could be made because well-confirmed theories on competition were available and the use of the production possibility frontier proved a valuable analytical tool for evaluating trade-offs in productivity. Sustained food production of the annual is a keystone to successful agroforestry systems and the chapter gives a scientific background to develop guidelines for agronomic management of the woody perennial so that annual production is maintained at acceptable levels in spite of the dominant nature of the perennial.

It may well be argued that agroforestry has many other objectives besides sustaining yield of the annual. But it should be pointed out that in marginal environments where agroforestry is often recommended, the primary concern of the farmer is food production. The perceived benefits of mixed cropping with a perennial, increased production from a unit of land, nutrient recycling, soil conservation, risk alleviation, weed control, integrated pest management, provision of shade to livestock, to name but a few, will be more valued if food production is at least maintained.

Annual pigeonpea in semi-arid environments has proved a very useful companion crop because of its slow initial growth allowing for complementary use of resources. It has been successfully intercropped with not only groundnut but sorghum and maize. The advantage of such intercrops is that there is no significant yield loss in either of the crops due to competition (Willey et al., 1981, 1986) but the disadvantage is, possibly, the recurrent cost of sowing the crop. In intercropping with perennial pigeonpea the cost of regular pruning may be offset by the recurring costs of annual planting and crop establishment but more important are the benefits of additional biomass and utilization of residual moisture in the dry season. It appears that current varieties of perennial pigeonpea are short-lived and not only because of their susceptibility to disease when pruned regularly. This 'self-thinning' nature may be in itself an advantage because it reduces the need for corrective pruning.

References

Berendse, F. (1979) Competition between plant populations with different rooting depths. I. Theoretical considerations. *Oecologia* 43, 19–26.

Cousens, R. (1985) An empirical model relating crop yield to week and crop density and a statistical comparison with other models. *Journal of Agricultural Science* 105, 513–521.

Daniel, J.N. and Ong, C.K. (1990) Perennial pigeonpea: a multipurpose species for agroforestry systems. *Agroforestry Systems* 10, 113–129.

de Wit, C.T. (1960) On competition. *Verslagen Landbouwkundige Onderzoekingen* 66.8, 82 pp.

de Wit, C.T. (1961) Space relationships within populations of one or more species. *Symposium of the Society of Experimental Biology* 15, 314–329.

de Wit, C.T., Tow, P.G. and Ennik, G.C. (1966) Competition between legumes and grasses. *Verslagen Landbouwkundige Onderzoekingen* 687, 60 pp.

Filius, A.M. (1982) Economic aspects of agroforestry. *Agroforestry Systems* 1, 29–39.

Goudriaan, J. and Monteith, J.L. (1990) A mathematical function for crop growth based on light interception and crop expansion. *Annals of Botany* 66, 695–701.

Huxley, P.A. (1983) Phenology of tropical woody perennials and seasonal crop plants with reference to their management in agroforestry systems. In: Huxley, P.A. (ed.) *Plant Research and Agroforestry*. Nairobi, International Council for Research in Agroforestry, pp. 501–525.

Jama, B. and Getahun, A. (1991) Intercropping *Acacia albida* with maize (*Zea mays*) and green gram (*Phaseolus aureus*) at Utwapa, Coast Province, Kenya. *Agroforestry Systems* 14 (3), 193–205.

King, K.F.S. (1979) Concepts of agroforestry. In: Chandler, T. and Spurgeon, D. (eds) *Proceedings of an International Conference on International Cooperation in*

Agroforestry, 16–21 July 1979. International Council for Research in Agroforestry and German Foundation for International Development, Nairobi, Kenya.

Kropff, M.J. and Spitters, C.J.T. (1991) A simple model of crop loss by weed competition from early observations on relative leaf area of the weeds. *Weed Research* 31, 97–105.

Lotka, J. (1925) *Elements of Physical Biology.* Waverly Press, Baltimore.

Miehe, S. (1986) *Acacia albida* and other multipurpose trees on the fur farmlands in the Jebel Marra highlands, W. Darfur, Sudan. *Agroforestry Systems* 4, 89–119.

Odongo, J.C.W., Ong, C.K., Khan, A.A.H. and Sharma, M.M. (1996) Productivity and resource use in a perennial pigeonpea/groundnut agroforestry system (unpublished).

Ong, C.K. (1990) Interactions of light, water, and nutrients in agroforestry systems. In: Avery, M.E., Cannell, M.G.R. and Ong, C.K. (eds) *Biophysical Research for Asian Agroforestry.* Winrock International, New Delhi, pp. 107–124.

Ong, C.K., Odongo, J.C.W., Marshall, F. and Black, C.R. (1992) Water use of agroforestry systems in semi-arid India. In: Calder, I.R. (ed.) *Growth and Water Use of Forestry Plantations.* Wiley, Chichester, UK, pp. 347–358.

Poschen, P. (1986) An evaluation of the *Acacia albida*-based agroforestry practices in the Hararghe highlands of Eastern Ethiopia. *Agroforestry Systems* 4, 129–143.

Raintree, J.B. (1983) Bioeconomic considerations in the design of agroforestry cropping systems. In: Huxley, P.A. (ed.) *Plant Research and Agroforestry.* International Council for Research in Agroforestry, Nairobi, pp. 352–368.

Ranganathan, R. (1992) Production possibility frontiers and estimation of competition effects: the use of *a priori* information on biological processes in intercropping. *Experimental Agriculture* 28 (3), 351–368.

Ranganathan, R. (1993) Analysis of yield advantage in mixed cropping. PhD Thesis, Wageningen Agricultural University, The Netherlands, 90 pp.

Ranganathan, R., Fafchamps, M. and Walker, T.S. (1991) Evaluating biological productivity in intercropping systems with production possibility curves. *Agricultural Systems* 36, 137–157.

Rao, M.R., Sharma, M.M. and Ong, C.K. (1990) A study of the potential of hedgerow intercropping in semi-arid India using a two-way systematic design. *Agroforestry Systems* 11, 243–258.

Shinozaki, K. and Kira, T. (1956) Intraspecific competition among higher plants. VII. Logistic theory of the C-D effect. *Journal of the Institute Polytechnic, Osaka City University, Series D* 7, 35–72.

Singh, R.P., Ong, C.K. and Saharan, N. (1989a) Above and below ground interactions in alley cropping in semi-arid India. *Agroforestry Systems* 9, 259–274.

Singh, R.P., Vandenbeldt, R.J., Hocking, D. and Korwar, G.R. (1989b) Alley cropping in the semi-arid regions of India. In: *Proceedings of a Workshop on Alley Cropping,* 10–14 March 1988. International Institute on Tropical Agriculture, Ibadan, Nigeria, pp. 108–122.

Spitters, C.J.T. (1979) Competition and its consequences for selection in barley breeding. PhD Thesis, Wageningen Agricultural University, The Netherlands, 268 pp.

Spitters, C.J.T. (1983) An alternate approach to the analysis of mixed cropping experiments. 1. Estimation of competition effects. *Netherlands Journal of Agricultural Sciences* 31, 1–11.

Tisdell, C.A. (1985) Conserving and planting trees on farms. Lessons from Australian cases. *Review of Marketing and Agricultural Economics* 53 (3), 185–194.

van den Bergh, J.P. (1968) An analysis of yields of grasses in mixed and pure stands. *Verslagen Landbowkundige Onderzoekingen* 714, 1–71.

Verinumbe, I. and Okali, D.V.V. (1985) The influence of coppiced teak (*Tectona grandis* L.F.) regrowth and roots on intercropped maize (*Zea mays* L.). *Agroforestry Systems* 3, 381–386.

Volterra, V. (1928) Variations and fluctuations of the number of individuals in animal species living together. In: Chapman, R.N. (ed.) *Animal Ecology*. New York.

Willey, R.W. and Heath, S.B. (1969) The quantitative relationship between plant population and crop yield. *Advances in Agronomy* 21, 281–321.

Willey, R.W., Rao, M.R. and Natarajan, M. (1981) Traditional cropping systems with pigeonpea and their improvement. In: *Proceedings of the International Workshop on Pigeonpea*, Vol. 1, 15–19 December 1980. International Crops Research Institute for the Semi-Arid Tropics, ICRISAT Center, India, pp. 11–25.

Willey, R.W., Natarajan, M., Reddy, M.S. and Rao, M.R. (1986) Cropping systems with groundnut: resource use and productivity. In: *Proceedings of the International Symposium on Agrometeorology of Groundnut*, 21–26 August 1986. International Crops Research Institute for the Semi-Arid Tropics, Niamey, Niger, pp. 193–206.

Yamoah, C. (1991) Choosing suitable intercrops prior to pruning Sesbania hedgerows in an alley configuration. *Agroforestry Systems* 13 (1), 87–94.

3

◆

Mulch and Shade Model for Optimum Alley-cropping Design Depending on Soil Fertility

M. van Noordwijk

International Centre for Research in Agroforestry, ICRAF-SE Asia, PO Box 161, Bogor 16001, Indonesia

Agroforestry or Woodlots Plus Cropped Fields

In many popular texts on agroforestry, the simple fact that tree and crop products can both contribute to a viable farming enterprise seems to be sufficient reason to use agroforestry systems. According to the definition used by ICRAF, however, agroforestry systems are not simply farming systems where both trees and crops or animals give useful products to the farmer, but systems where tree and crop (and/or animal) production interact (Lundgren and Raintree, 1982; Nair, 1993). Understanding and predicting such interaction should thus be at the heart of an agroforestry research programme.

Figure 3.1 gives a tentative classification of agroforestry systems, based on the degree of spatial and temporal overlap of the tree and crop components. Systems in the lower left corner do not fall under the definition of agroforestry, as here trees and crops do not interact. Systems in the upper right corner are generally not viable as competition between the tree and crop component will be too severe. In the upper left corner we find (improved) fallow systems, where a crop and a tree phase alternate on the same land. From a biophysical point of view, such systems are fairly simple and can be successful, but farmers will hesitate to put efforts into improving the fallow phase; it is thus understandable that developments in the direction of relay-establishment of the fallow vegetation are sought, which move the system towards the centre of the graph. Alley cropping was

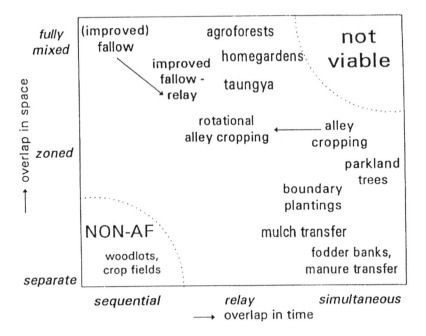

Fig. 3.1. Classification of agricultural systems based on both trees and crops, with regard to the degree of overlap in time (*x*-axis) and space (*y*-axis).

first developed as a fully simultaneous spatially zoned system, but recently interest in 'rotational alley cropping' has also moved the system towards the centre of the graph. Agroforestry systems with full-grown trees are either based on low tree densities (parklands, boundary plantings) or on relay systems, with a short crop and long tree phase (taungya, homegarden, agroforests).

The total yield of an agroforestry system in any given year can be described as the sum of the crop yield, the yield of tree products, the yield of animal products and the change in land quality, which reflects the concerns over the long term sustainability of the system. If we restrict ourselves to agroforestry systems without an animal production component, we obtain:

$$Y_{tot} = E_c Y_c + E_t Y_t + E_L \Delta L \qquad (3.1)$$

where:

Y_{tot} = total yield [\$ ha^{-1}]
E_c = price per unit crop yield [\$ kg^{-1}]
Y_c = crop yield [kg ha^{-1}]
E_t = price per unit yield of tree products [\$ kg^{-1}]

Y_t = yield of tree products (or 'net present value' of future productivity) [kg ha^{-1}]

E_L = price per unit change in land quality [\$ X^{-1}]

ΔL = change in land quality for future production in units X to be further specified [X ha^{-1}].

From this equation we can explore under which conditions a maximization of total yield will lead to a choice for an agroforestry system, with both a tree and a crop component, and under which conditions pure tree or crop production will be preferred.

In the most simple case we may describe all tree–crop interactions as linear functions of the relative tree area f_t. For crop yield Y_c we may formulate:

$$Y_c = (1 - f_t)(Y_{0c} + f_t F - f_t C_{tc}) \tag{3.2}$$

where:

f_t = relative tree area (for an agroforestry system: $0 < f_t < 1$)

Y_{0c} = crop yield in the absence of trees [kg ha^{-1}]

F = positive effect of trees on crop yield, e.g. due to (short term) soil fertility improvement, per unit relative tree density [kg ha^{-1}]

C_{tc} = crop yield decrease due to competition by the tree, per unit relative tree density [kg ha^{-1}].

For the yield of tree products we may consider a negative interaction only:

$$Y_t = f_t(Y_{0t} - (1 - f_t)C_{ct}) \tag{3.3}$$

where:

Y_{0t} = yield of tree products in the absence of crops [kg ha^{-1}]

C_{ct} = decrease in yield of tree products due to competition by the crop, per unit relative crop density [kg ha^{-1}].

The change in land quality for future production, ΔL, may be negative for a pure crop system and may become more positive with increasing relative tree density (unless a substantial part of tree products is exported from the field):

$$\Delta L = (1 - f_t)\Delta L_c + f_t \Delta L_t = \Delta L_c + f_t(\Delta L_t - \Delta L_c) \tag{3.4}$$

where:

ΔL_c = (normally negative) change in land quality for future production while under monoculture crop [X ha^{-1}]

ΔL_t = (normally positive) change in land quality for future production while under tree cover [X ha^{-1}].

Two contrasting views exist on trees: in agroforestry trees are generally seen as 'soil improvers', especially where fast-growing N_2 fixing trees are used, while in plantation forestry there is serious concern about soil depletion due to short rotation forestry, especially where fast-growing trees are used (Sanchez *et al.*, 1985; Bruijnzeel, 1992). The different perceptions are partly due to different conditions (poor soils used for plantation forestry) and management practices (more damage may be done to the soil while harvesting the timber than by the nutrient export as such).

If sustainability is considered a hard constraint ($\Delta L > 0$), then:

$$f_t \frac{-\Delta L_c}{\Delta L_t - \Delta L_c} \tag{3.5}$$

Alternatively, the costs of land degradation may be considered to be outweighed by direct benefits and be restored later as happens in shifting cultivation or fallow rotation systems.

Agroforestry systems are defined as systems that combine trees and crops (thus $0 < f_t < 1$) and where tree–crop interactions occur. An optimum tree density $f_{t,opt}$ may be found for $dY_{tot}/df_t = 0$, if $d^2Y_{tot}/(df_t)^2 < 0$. Only if this optimum tree density is between 0 and 1, agroforestry practices are the best choice.

$$\frac{dY_{tot}}{df_t} = E_c(F - C_{tc} - Y_{0c}) + E_t(Y_{0t} - C_{ct}) + E_L(\Delta L_t - \Delta L_c)$$
$$- 2f_t(E_c(F - C_{tc}) - E_tC_{ct}) \tag{3.6}$$

The requirement $d^2Y_{tot}/(df_t)^2 < 0$ leads to:

$$E_cF > E_cC_{tc} + E_tC_{ct} \tag{3.7}$$

which shows that the positive interaction term on the left hand should be larger than the negative one on the right hand; otherwise it is better to have crops and trees on separate plots. Yet, it is possible to compensate a negative interaction term with a larger positive other term. A positive overall interaction can be obtained for systems where neither the crop nor the tree component shows an absolute benefit. For $f_{t,opt}$ we then obtain:

$$f_{t,opt} = \frac{-E_cY_{0c} + E_tY_{0t} + E_L(\Delta L_t - \Delta L_c)}{2(E_c(F - C_{tc}) - E_tC_{ct})} + 0.5 \tag{3.8}$$

which can be rewritten as:

$$f_{t,opt} = \frac{X - 1 + L}{2(I_{tc} + X I_{ct})} + 0.5 \tag{3.9}$$

where:

$X = (Y_{0t}E_t)/(Y_{0c}E_c)$ is the ratio of financial returns on a pure tree and a pure crop system

$I_{tc} = (F - C_{tc})/Y_{0c}$ is the scaled net tree crop interaction
$I_{ct} = (-C_{ct})/Y_{0t}$ is the scaled net crop tree interaction
$L = E_l(\Delta L_t - \Delta L_c)/(Y_{0c}E_c)$ is the scaled relative importance of changes in land quality.

The constraint $0 < f_{t,opt} < 1$ then leads to:

$$\frac{1 - L - I_{tc}}{1 + I_{ct}} < X < \frac{1 - L + I_{tc}}{1 - I_{ct}} \qquad (3.10)$$

Outside the constraints (Equation 3.10) one would prefer either a pure tree system ($f_t = 1$) or a pure crop system ($f_t = 0$), depending on the values of $E_c Y_{0c}$ and ($E_t Y_{0t} + E_l L$). The equations also show that the choice for an agroforestry or a more simple system not only depends on the biophysically determined parameters, but also on the 'value' assigned to the various possible products (trees, crops and land).

Figure 3.2 shows a general demarcation of the domain for agroforestry based on the economic value of tree and crop production (the tree products should be discounted for the length of the harvest cycle) and the strength of the interaction term. The larger the positive interaction on crop production ($F - C_{tc}$), the larger the scope for agroforestry (i.e. the range of price ratios which lead to $0 < f_{t,opt} < 1$). For realistic estimates of the interaction term, the tree products need to have some direct value to the farmer to justify agroforestry. If trees have no direct value, the interaction term ($F - C_{tc}$)/Y_{0c} has to be 1.0 or more, i.e. the net positive effect of trees on crop yield per unit tree area has to exceed the monocultural crop yield per unit area.

With these equations one can describe approximately stationary systems, as approximated in alley cropping where the normal growth of the tree component is checked by regular pruning. For most other agroforestry systems, however, the tree–crop interactions change from year to year. When the tree component is more valuable than a pure crop, there may still be scope for crops in the first year(s) when the trees are still small. The approximately stationary situations of 'alley cropping', however, may offer good opportunities for further investigation. To get one step further, we will attempt to formulate the tree–crop interaction term based on external resources limiting crop production, such as light and nutrients.

Hedgerow Tree–Crop Interactions

In hedgerow intercropping systems, trees and food crops are interacting in various ways. As both positive and negative interactions occur, site-specific optimization of the system may be required. The most important interactions probably are:

1. Mulch production from the hedgerows, increasing the supply of N and other nutrients to the food crops,

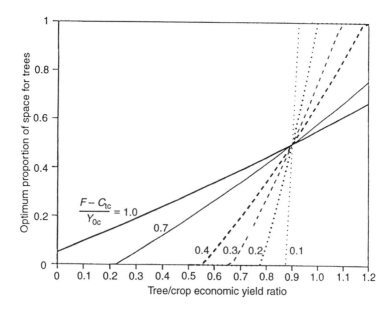

Fig. 3.2. Optimum allocation of land to trees, $f_{t,opt}$, as a function of the economic yield ratio of tree and crop products ($Y_{0t}E_t/(Y_{0c}E_c)$) and the relative interaction term $F - C_{tc}/Y_{0c}$, for $C_{tc}/Y_{0t} = 0.1$ and $E_1(\Delta L_t - \Delta L_c)/(E_c Y_{0c}) = 0.1$ (based on Equation 3.8).

2. Shading by the hedgerows, reducing light intensity at the crop level,
3. Competition between tree and crop roots for water and/or nutrients in the topsoil,
4. Nitrogen supply by tree roots to crop roots, either due to root death following hedgerow pruning or by direct transfer if nodulated roots are in close contact with crop roots,
5. Effects on weeds, pests and diseases,
6. Long term benefits by erosion control and maintenance of soil organic matter.

Disregarding the long term effects through the soil and the more direct effect (positive or negative) through weeds, pests and diseases, interactions 1 and 4 are positive, 2 and 3 are negative. Little quantitative information about the below-ground interactions 3 and 4 is available as yet, the overall effect may be positive or negative. Wherever water availability is limiting crop growth (roughly in zones with less than 150 mm of rainfall per month in the growing season, based on 4 mm day^{-1} of evapotranspiration and 20% losses by runoff and deep infiltration), however, negative effects will normally predominate.

For some time, tree–crop interactions have been studied primarily by focusing on the tree–crop interface. For below-ground interactions, however, no clear 'interface' exists as tree roots may extend 30 m from the stem or more (see Chapter 7).

As a first step we will here consider only the above-ground interactions 1 and 2 and derive a simple model for *mulch* and *shade*, which can be used for optimization of hedgerow intercropping in conditions where water is not limiting. In hedgerow intercropping the following choices can be made:

1. Tree species,
2. Distance between hedgerows,
3. Pruning regime (height and frequency),
4. Crop, cultivar, crop population density and plant spacing,
5. Additional fertilizer input level.

As these factors are interacting and field experiments involving all factors are too large to be manageable, a simple model describing the interaction may help to predict optimum patterns beforehand so as to test relevant ones in a field experiment. In the following we will use the distance between hedgerows as the parameter to be optimized, given characteristics of the tree, the pruning regime, the crop and soil fertility level. The conventional control for hedgerow intercropping experiments, crops without trees, fits in this scheme as it represents an infinitely large distance between hedgerows. A point of warning is appropriate here, as in practical experiments below-ground interactions may not have been effectively excluded in so called 'control' plots; the control plot may therefore give a lower yield and the positive effects of alley cropping may be overestimated. On most smallholder farms a similar mining of adjacent (or neighbours') plots by trees can be expected (Coe, 1994).

Shade and Mulch Model

Crop growth as influenced by N-uptake and shade

The relationship between dry matter production and N uptake can often be described by the rising branch of a quadratic function. Under severe N limitation dry matter production increases almost linearly with increasing uptake; when the maximum production is approached the N concentration in the crop as a whole is about twice the minimum value. A simple model was formulated by van Noordwijk and Wadman (1992) and van Noordwijk and Scholten (1994):

$$Y_c = f_p U \left(2 - \frac{f_p U}{Y_{m,l}} \right) \tag{3.11}$$

and

$$U = f_u (N_s + f_a N_a) \tag{3.12}$$

where:

Y_c = dry matter yield of crop [Mg ha^{-1}]

f_p = crop efficiency in producing yield per unit N taken up from the soil [kg yield per kg N uptake]. f_p equals the harvest index divided by the required N concentration in biomass (van Noordwijk and Scholten, 1994)

U = amount of N taken up by the crop [kg ha^{-1}],

f_a = application efficiency of added nutrient sources, e.g., prunings; increase in available soil N per unit applied N [kg kg^{-1}]

f_u = uptake efficiency [kg N uptake per kg N available]

N_a = amount of N added [kg ha^{-1}]

N_s = amount of N available in the soil, without any (organic) fertilizer [kg ha^{-1}]

$Y_{m,l}$ = maximum yield, not limited by N shortage, under current light conditions [Mg ha^{-1}].

The quadratic function reaches a maximum value for:

$$N_a = \frac{Y_{m,l} - N_s \, f_u \, f_p}{f_a \, f_u \, f_p} \tag{3.13}$$

In the present model we will assume that internal regulation of N uptake is complete, so N uptake will not be more than this value required for maximum crop production, but this assumption is not important as we do not consider supra-optimal N supply.

Under reduced light intensity the crop cannot reach the maximum yield value. In the humid tropics with relatively short days and under overcast sky, we may assume that for light demanding crops any further reduction in light leads to a proportional (linear) decrease in maximum (potential) yield level, Y_{max}:

$$Y_{m,l} = f_j Y_{max} \tag{3.14}$$

with:

Y_{max} = maximum yield potential without shading by trees,

f_j = fraction of light *not* intercepted by trees above the crop canopy.

Figure 3.3 shows the assumed relation between N uptake and crop yield for some numerical examples. The circles on the line for $f_j = 0$, $f_j = 0.25$ and $f_j = 0.5$ suggest the possibility of optimizing crop yields by hedgerow intercropping. Decreasing the distance in between hedgerows may lead to increased N uptake as well as increasing shade (f_j). As N

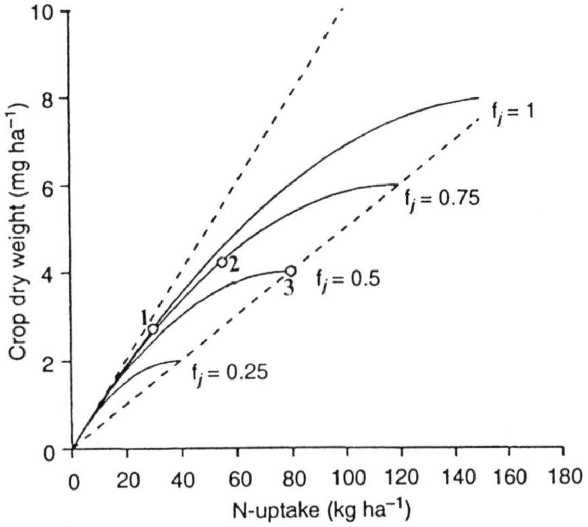

Fig. 3.3. Assumed relation between N uptake and crop yield according to Equations 3.10 and 3.13, for a value of Y_{max} of 8 Mg ha^{-1} and for f_p of 20 kg kg^{-1}. The circles on the line for $f_j=0$, $f_j=0.25$ and $f_j=0.5$ indicate the possibility of optimizing crop yields by hedgerow intercropping.

limitation becomes less severe but shading increases, crop yields may show an optimum response curve to distance between hedgerows.

Mulch production by the hedgerow trees

The amount of N supplied to the soil in prunings per crop is equal to:

$$N_a = HP_fD_pN_p = \frac{10^4 P_f D_p N_p}{D_h} \tag{3.15}$$

with:

H = length of hedgerows per ha [m ha^{-1}] = $10^4/D_h$
D_h = distance between two hedgerows [m]
P_f = frequency of pruning per crop growing season
D_p = dry weight of prunings per cutting cycle per m of hedgerow [kg m^{-1}]
N_p = N concentration in prunings [kg kg^{-1}].

For the biomass production D_p we can formulate:

$$D_p = \frac{LAI_p w}{LAR_p} = \frac{LAI_p w}{LWR_p SLA} \tag{3.16}$$

with:

LAI_p = leaf area index in the hedgerow tree canopy at pruning [$m^2\ m^{-2}$]
w = width of hedgerow tree canopy at pruning [m]
LAR_p = leaf area ratio = leaf area per unit (pruned) shoot dry weight [$m^2\ kg^{-1}$]
LWR_p = leaf weight ratio = leaf dry weight as fraction of (pruned) shoot dry weight [$kg\ kg^{-1}$]
SLA = specific leaf area = leaf surface area per unit leaf dry weight [$m^2\ kg^{-1}$].

Combining Equations 3.14 and 3.15 gives:

$$f_a N_a = \frac{H w f_a LAI_p N_p P_f}{LWR_p SLA} = HM \tag{3.17}$$

where:

$$M = \frac{w f_a LAI_p N_p P_f}{LWR_p SLA} \tag{3.18}$$

The factor M combines several tree characteristics to obtain the amount of N available to the crop from prunings per metre of hedgerow. It is relatively high for trees with a wide crown, a high leaf area index, a high N content, frequent pruning, thick leaves (low SLA), many twigs and branches (low LWR) and a well-timed N release (high f_a).

Crop shading by hedgerow trees

For the shading by the hedgerows we may simplify the description by distinguishing between the crops directly underneath the canopy of the hedgerow trees (an area wH) and the crops outside the tree canopy (an area $10^4 - wH$). If all light came under an angle of $90°$ the shade factor would be:

$$f_j = 1 - \frac{H w (1 - \varepsilon^{x LAI_p})}{10^4} = 1 - SH \tag{3.19}$$

with:

$\varepsilon =$ light interception fraction by foliage at $LAI = 1$
$x =$ fraction of hedgerow tree canopy above the crop canopy.

If light comes from a different angle, the shade intensity in the zone w may be less and in the remaining zone it will be more. As a first approximation, Equation 3.18 may still hold. The factor S indicates the equivalent space [ha m^{-1}] occupied by the hedgerow canopy at complete light interception, per unit hedgerow length. It is high for wide crown canopies and high values of ε and LAI_p.

Combined effects

Combining Equations 3.10, 3.12, 3.16 and 3.18 we obtain crop yield Y as a function of hedgerow length H:

$$Y = f_p f_u \left[(2N_s + MH) + \frac{f_p f_u (N_s + MH)^2}{Y_m(SH - 1)} \right] \tag{3.20}$$

By differentiating Equation 3.19 with respect to H, we can derive that yield Y is maximum when $H = H_{opt}$:

$$H_{opt} = \frac{1}{S} \left(1 - \sqrt{1 - \frac{N_m - N_s - 0.5SN_s^2/M}{N_m + 0.5S/M}} \right) \tag{3.21}$$

where N_m is the available N level required for obtaining maximum yield:

$$N_m = \frac{Y_{max}}{f_p f_u} \tag{3.22}$$

The optimum hedgerow length is inversely proportional to the S parameter, increases less than proportionally with increasing nitrogen production per unit hedgerow length, M, and decreases with increasing soil N supply, N_s. For high values of N_s the best solution is one without hedgerow trees ($H_{opt} = 0$). The critical value for $N_{s,crit}$ where H_{opt} becomes 0 is:

$$N_{s,crit} = \frac{M}{S} \left(\sqrt{\frac{2SN_m}{M} + 1} - 1 \right) \tag{3.23}$$

$N_{s,crit}$ is approximately proportional to the $M:S$ ratio.

The yield advantage due to alley cropping at optimum tree spacing, compared with a situation without trees is:

$$Y_{opt} - Y_0 = f_p H_{opt}\left(2M - \frac{SN_s^2 + 2Mf_p N_s + M^2 H_{opt}}{N_m(1 - SH_{opt})}\right) \tag{3.24}$$

The yield advantage apparently decreases with increasing N_s.

Results

Figure 3.4 shows some examples of calculations. With increasing N_s, i.e., N available in the soil for the crop from other sources than the prunings, the crop yields obviously increase. For values of N_s which are insufficient for obtaining the maximum yield, crop yields show an optimum response curve to the distance between hedgerows. With decreasing N_s the optimum becomes more pronounced and occurs at smaller distances between the hedgerows.

From curves as shown in Fig. 3.4 we can establish an optimum length of hedgerows per hectare, as a function of N_s. Figure 3.5 gives some

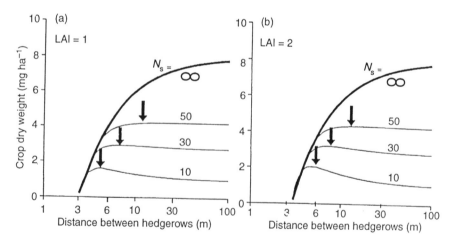

Fig. 3.4. Calculated crop yields as a function of distance between hedgerows and N_s, for a set of data representing maize production; N_s represents background soil fertility; (a) gives results for a rather open hedgerow canopy (LAI = 1); (b) results for a more dense canopy (LAI = 2). Arrows indicate optimal hedgerow spacing for each level of N_s (kg ha^{-1}).

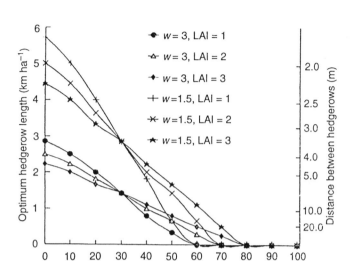

Fig. 3.5. Optimal length of hedgerow per hectare as influenced by N_s and characteristics of the hedgerow tree canopy.

numerical examples, for hedgerow canopies of different width and density (LAI_p).

Figure 3.6 shows crop yields at optimum hedgerow spacing as a function of background nitrogen supply, reflected in N_s. This graph shows that on poor soils, with low N_s, hedgerow cropping at optimum hedgerow spacing can give a considerable yield increase. At yield levels more than about 50% of non-N limited yields, however, no benefit from hedgerow intercropping may be expected from our simple model.

Figure 3.7 identifies the domain where at least some forms of alley cropping, with a near-optimum tree spacing, may increase crop production. Based on Equation 3.20 the upper limit of the soil N supply divided by crop N demand N_m can be related to the $M : S$ ratio of the tree, which gives the N supply per unit fully shaded area. The higher the $M : S$ ratio of a tree, the better are its prospects for alley cropping. If one wants alley cropping to work in a range where the control plots allow a crop N uptake near 50% of the maximum, the $M : S$ ratio has to be 50–125 kg N ha^{-1} shaded, for N_m in the range 200–500 kg ha^{-1}.

Table 3.1 gives the required parameter values for several tree species used in alley-cropping trials on an acid soil in Sumatra, Indonesia. The values are based on Hairiah *et al.* (1992) and some new observations. In

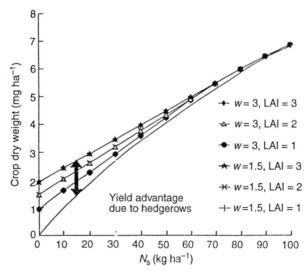

Fig. 3.6. Crop yields at optimum hedgerow spacing as a function of background nitrogen supply, reflected in N_s. A reference line for crop yields without hedgerows is given as well.

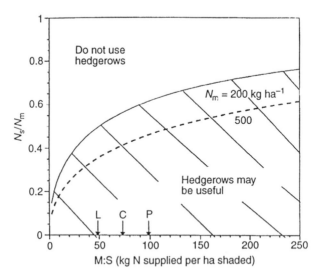

Fig. 3.7. Domain where at least some versions of hedgerow intercropping will give a yield advantage, as determined by the $M:S$ ratio of the tree and the relative fertility of the site (N_s/N_m), based on Equation 3.20. The letters L, C and P indicate the approximate $M:S$ ratio of *Leucaena leucocephala*, *Calliandra calothyrsus* and *Peltophorum dasyrrachis*.

Table 3.1. Parameter values for a number of hedgerow trees used in alley-cropping experiments on an acid soil in Sumatra (Hairiah et al., 1992; van Noordwijk et al., 1995); the last two columns give total dry weight of prunings (DW_{tot}) and total N contents (N_{tot}) measured at a 4-m hedgerow spacing and about 5 pruning events per year.

Tree species	w (m)	LAI_p	LWR	SLA (m²kg⁻²-1)	D_p (kg m⁻¹)	N_p (kg kg⁻¹)	f_a	x	DW_{tot} (Mg ha⁻¹ year⁻¹)	N_{tot} (kg ha⁻¹ year⁻¹)
Leucaena leucocephala	3.5	2.5	0.7	24	0.53	0.024	0.6	0.7	8	190
Gliricidia sepium	3.0	2.3	0.6	22	0.53	0.029	0.6	0.9	8	230
Calliandra calothyrsus	3.7	3.0	0.7	20	0.86	0.030	0.5	0.7	12	360
Peltophorum dasyrrachis	1.5	3.0	0.6	12	0.53	0.021	0.5	0.4	8	170
Erythrina orientalis	2	1.6	0.7	20	0.27	0.028	0.6	0.5	4	110

Figure 3.8 and Table 3.2 the predicted optimum hedgerow spacing is given, for three levels of inherent soil N supply. The calculations show a three-fold variation in both the M and S parameters between the five tree species, and a two-fold variation in the $M:S$ ratio (50–97 kg N ha^{-1} shaded). The soil N supply above which alley cropping never has an advantage (whatever the tree spacing) only varies from 100 to 123 kg ha^{-1}. The optimum tree spacing at N_s below the critical value depends strongly on the tree species, however. The tree spacing of 4 m used in the experiment from which the tree parameters were derived was approximately optimal at a N_s level of 60 kg ha^{-1} for *Peltophorum* and *Erythrina*, but was too narrow for the three other trees even at $N_s = 30$ kg ha^{-1}. The calculations thus show that a comparison of different tree species at a single hedgerow spacing cannot be extrapolated to other hedgerow spacings, as the spacing used can be below optimum for some of the trees, about right or above optimum for others. The table also shows that in situations where the inherent soil N supply is changing over time, any hedgerow spacing chosen has to be a compromise; a spacing which was good initially will become too narrow if soil fertility is improving and too wide if the soil fertility is declining. Unfortunately, hedgerow spacing cannot be easily adjusted, unless the trees are grown in pots (which would jeopardize any nutrient recycling functions of the trees). The best practical way to modify the effective tree density may be to vary pruning height and switch to ground-level pruning if trees are too competitive while soil fertility is adequate.

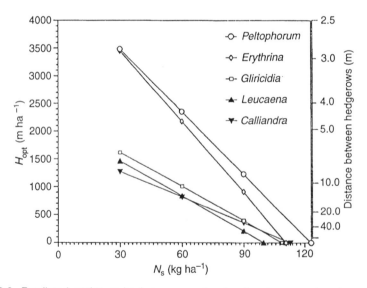

Fig. 3.8. Predicted optimum hedgerow spacing for five tree species, based on the parameters in Table 3.1, as a function of inherent soil fertility.

Table 3.2. Calculated optimum hedgerow spacing for the tree parameters of Table 3.1 for a crop with a maximum N uptake of 200 kg ha^{-1}; $\varepsilon = 0.25$, $P_f = 2$ prunings per crop.

Tree	M (kg m^{-1})	S (ha m^{-1})	M/S (kg ha^{-1})	$N_{s,crit}$ (kg ha^{-1})	$N_s = 30$ kg ha^{-1}		$N_s = 60$ kg ha^{-1}		$N_s = 90$ kg ha^{-1}	
					H_{opt} (m ha^{-1})	D_h (m)	H_{opt} (m ha^{-1})	D_h (m)	H_{opt} (m ha^{-1})	D_h (m)
Leucaena	0.0159	3.2	50	100	1460	6.8	833	12	208	48
Gliricidia	0.0184	2.8	66	109	1616	6.2	1005	9.9	395	25
Calliandra	0.0258	3.5	74	113	1271	7.9	813	12	355	28
Peltophorum	0.0117	1.2	97	123	3484	2.9	2355	4.3	1226	8.2
Erythrina	0.0091	1.3	70	111	3451	2.9	2179	4.6	907	11

Discussion

Apparently, hedgerow intercropping, where trees have to be a source of N to the crop, is restricted to situations with a low soil N supply from other sources and crops that can respond to considerably higher N supply than is available. At the economic level, the labour costs for obtaining N this way have to be evaluated against the costs of obtaining N from other sources.

The description given by this model assumes that the prunings are spread equally over the land available, including the area reserved for hedgerow trees. In practice the prunings will be concentrated in the crop zone, but the trees will partly compete with the crops for the N mineralized; as a first approximation the current description therefore may be acceptable. As the crops in the middle of the alley will receive more light, it would make sense to concentrate the pruning only in this middle strip, and accept that the crops closer to the hedgerows have a limiting level of both N and light. The equations given are rather pessimistic on the possibilities of hedgerow intercropping, as they do not consider this possibility of strategic mulch placement (neither do most of the experiments reported in the literature...). Although the model suggests a limited 'niche' for hedgerow intercropping, considerable scope remains for selecting hedgerow trees which are most suitable.

For the situation described, the best hedgerow tree is one with a high $M:S$ ratio, which can be based on a combination of a narrow but compact hedgerow canopy, thick leaves, the major part of the tree canopy not exceeding that of the crop, a high N content and a suitable N release pattern from the prunings, coinciding with crop demand. The need for fine tuning of the N release pattern increases with decreasing residence time for mineral N in the crop root zone, because of shallow rooting and/or high rainfall infiltration surplus over evapotranspiration (van Noordwijk *et al.*, 1992).

The evaluation given may be too pessimistic on the scope for alley cropping for several situations:

1. *Other litter inputs*: Spontaneous litterfall from the trees (turnover of leaf biomass) will add to the mulch supply, without causing further shade. With the intensive pruning regimes required to check the tree growth during the cropping period, however, litterfall will be low for most trees.
2. *Temporal segregation*: If part of the growing season can be reserved for tree growth, litterfall and an increased pruned biomass can be obtained without shading a crop. Off-season biomass production by the tree is probably the major way for agroforestry to increase total resource capture.
3. *Shade-tolerant crops*: The prospects for alley cropping greatly improve if the crop is light-saturated under full sunlight. This opens the option of 'free light interception' by an upper tree canopy, resulting in mulch supply to the

crop without shade costs. In that case sparse open canopies are better than dense hedgerows. This situation is more likely to exist under the clear skies of the semi-arid tropics than under the usually overcast skies of the humid tropics, and more so for C3 than for C4 crops.

Essentially, the approach can also be used for evaluating the optimum tree density for sparse upper storey trees with little or no pruning. The example of tea gardens in Java shows, however, that shade costs of *in situ* N production are considerable. Tea gardens used to have a complete tree cover of trees such as *Paraserianthes falcataria* as a source of N and organic matter, before inorganic fertilizer became abundantly available, but now nearly all trees are removed. Cheaper fertilizer combined with less shade-tolerant new tea varieties induced the change from an agroforestry to a pure crop system. Data on the N response of tea in NE India in the presence or absence of shade trees (Fig. 3.9) are in line with the shade and mulch model presented here: without N fertilizer the presence of *P. falcataria* trees could induce a 50% yield increase. That same yield increase, however, can be obtained with 45 kg ha^{-1} of N in the absence of trees. At a fertilizer rate of 90 kg ha^{-1} trees have no effect on tea yield, but they reduce the maximum attainable tea yield by 250 kg ha^{-1}. If the direct value of the shade trees (e.g. by timber production) plus the saving of about 90 kg N fertilizer per year would offset this yield loss of 250 kg tea per year, it would become worthwhile to reintroduce the trees.

The scope for systems with simultaneous trees and crops has clear limitations. A partial temporal separation may be needed. If we uncouple

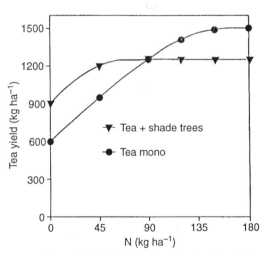

Fig. 3.9. Response of tea yield to N fertilizer application in the presence or absence of *Paraserianthes falcataria* as N$_2$ fixing tree in NE India (B.C. Barbora, personal communication).

the mulch production from the amount of shade cast, e.g. by having a period of the year devoted to tree growth and thus mulch production and a period in which the trees are set back severely and the crops are growing, the potential for *in situ* mulch production will increase. For such a system, however, we need either trees that have the right phenology, such as the famous *Faidherbia albida*, or trees which survive a heavy pruning, e.g. at a low height, take 2 or 3 months to recover, but can still have a good biomass production rate afterwards. If the terms B and BQ in the models of Ranganathan and de Wit (Chapter 2, this volume) are correlated as they suggest, it will be difficult to find such trees. Yet it will be worthwhile to look for the exceptions, i.e. trees which recover slowly from a setback by pruning but have a high biomass production rate once recovered, and solve the shade and mulch dilemma by partly temporal separation of tree and crop growth.

Acknowledgements

Thanks are due to Radha Ranganathan and Keith Shepperd for constructive remarks on an earlier draft.

References

Bruijnzeel, L.A. (1992) Sustainability of fast-growing plantation forests in the humid tropics with particular reference to nutrients. In: Jordan, C.F., Gajaseni, J. and Watanabe, H. (eds) *Taungya: Forest Plantations with Agriculture in Southeast Asia*. CAB International, Wallingford, pp. 51–67.

Coe, R. (1994) Through the looking glass: ten common problems in alley-cropping research. *Agroforestry Today* 6 (1), 9–11.

Hairiah, K., Van Noordwijk, M., Santoso, B. and Syekhfani, M. (1992) Biomass production and root distribution of eight trees and their potential for hedgerow intercropping on an ultisol in Lampung. *AGRIVITA* 15, 54–68.

Lundgren, B.O. and Raintree, J.B. (1982) Sustainable agroforestry. In: Nestel, B. (ed.) *Agricultural Research for Development: Potential and Challenges in Asia*. ISNAR, The Hague, The Netherlands, pp. 37–49.

Nair, P.K. (1993) *An Introduction to Agroforestry*. Kluwer Academic, Dordrecht, 499 pp.

Sanchez, P.A., Palm, C.A., Davey, C.B., Szott, L.T. and Russell, C.E. (1985) Tree crops as soil improvers in the humid tropics? In: Cannell, M.G.R. and Jackson, J.E. (eds) *Attributes of Trees as Crops*. Institute of Terrestrial Ecology, Huntingdon, pp. 327–358.

van Noordwijk, M. and Scholten, J.H.M. (1994) Effects of fertilizer price on feasibility of efficiency improvement: case study for an urea injector for lowland rice. *Fertilizer Research* 39, 1–9.

van Noordwijk, M. and Wadman, W. (1992) Effects of spatial variability of nitrogen supply on environmentally acceptable nitrogen fertilizer application rates to arable crops. *Netherlands Journal of Agricultural Research* 40, 51–72.

van Noordwijk, M., Widianto, Sitompul, S.M., Hairiah, K. and Guritino, B. (1992) Nitrogen management under high rainfall conditions for shallow rooted crops: principles and hypotheses. *AGRIVITA* 15, 10–18.

van Noordwijk, M., Sitompul, S.M., Hairiah, K., Listyarini, E. and Syekhfani, M.S. (1995) Nitrogen supply from rotational or spatially zoned inclusion of leguminosae for sustainable maize production on an acid soil in Indonesia. In: Date, R.A., Grundon, N.J., Rayment, G.E. and Probert, M.E. (eds) *Plant–Soil Interactions at Low pH: Principles and Management.* Kluwer, Dordrecht, pp. 779–784.

Appendix: Parameters and Symbols Used

C_{ct} = decrease in yield of tree products due to competition by the tree, per unit relative crop density [kg ha^{-1}]

C_{tc} = crop yield decrease due to competition by the tree, per unit relative tree density [kg ha^{-1}]

D_h = distance between two hedgerows [m]

D_p = dry weight of prunings per cutting cycle per m of hedgerow [kg m^{-1}]

ΔL = change in land quality for future production, units X to be specified [X ha^{-1}]

ΔL_c = (normally negative) change in land quality under monoculture crop [X ha^{-1}]

ΔL_t = (normally positive) change in land quality under tree cover [X ha^{-1}]

E_c = price per unit crop yield [\$ kg^{-1}]

E_t = price per unit yield of tree products [\$ kg^{-1}]

E_L = price per unit change in land quality [\$ X^{-1}]

ε = light interception fraction by foliage at LAI = 1

F = positive effect on crop yield due to soil fertility improvement, per unit relative tree density [kg ha^{-1}]

f_a = application efficiency of added nutrient sources, e.g. prunings; increase in available soil N per unit applied N [kg kg^{-1}]

f_j = fraction of light not intercepted by trees available to the crop

f_p = crop efficiency in producing yield per unit N taken up from the soil [kg yield per kg N uptake]

f_t = relative tree area (for an agroforestry system: $0 < f_t < 1$)

$f_{t,opt} = f_t$ which maximizes crop yield

f_u = uptake efficiency [kg uptake per kg N available]

H = length of hedgerows per ha [m ha^{-1}] = $10^4/D_h$

$H_{opt} = H$ which maximizes yield

I_{ct} = scaled net crop–tree interaction

I_{tc} = scaled net tree–crop interaction

L = scaled relative importance of changes in land quality

LAI_p = leaf area index in the hedgerow tree canopy at pruning [m^2 m^{-2}]

LAR_p = leaf area ratio = leaf area per unit (pruned) shoot dry weight [m^2 kg^{-1}]

LWR_p = leaf weight ratio = leaf dry weight as fraction of (pruned) shoot dry weight [kg kg^{-1}]

M = effective N supplied to crop via mulch, per m of hedgerow [kg m^{-1}]

N_a = amount of N added [kg ha^{-1}]

N_m = amount of available N required for obtaining maximum yield Y_{max} [kg ha^{-1}]

N_p = N concentration in prunings [kg kg^{-1}]

$N_{s,crit}$ = critical soil N level above which alley cropping gives no yield benefit [kg ha^{-1}]

N_s = amount of N available in the soil, without any (organic) fertilizer [kg ha^{-1}]

P_f = frequency of pruning per crop growing season

S = effective crop area shaded by per m hedgerow [ha m^{-1}]

SLA = specific leaf area = leaf surface area per unit leaf dry weight [m^2 kg^{-1}]

U = N uptake by crop [kg ha^{-1}]

w = width of hedgerow tree canopy at pruning [m]

x = fraction of hedgerow tree canopy above the crop canopy

Y_c = dry matter yield of crop [Mg ha^{-1}]

Y_{0c} = crop yield in the absence of trees [kg ha^{-1}]

Y_{0t} = yield of tree products in the absence of crops [kg ha^{-1}]

$Y_{m,l}$ = maximum yield, not limited by N shortage, under current light conditions [Mg ha^{-1}]

Y_{max} = maximum yield without shading by trees [Mg ha^{-1}]

Y_{tot} = total yield of agroforestry system [\$ ha^{-1}]

Y_t = yield of tree products (or 'net present value' of future productivity) [kg ha^{-1}]

4

♦

Principles of Resource Capture and Utilization of Light and Water

C.K. ONG,[1] C.R. BLACK,[2] F.M. MARSHALL[2] AND J.E. CORLETT[3]

[1] International Centre for Research in Agroforestry, ICRAF House, PO Box 30677, Gigiri, Nairobi, Kenya; [2] Department of Physiology and Environmental Science, University of Nottingham, Sutton Bonington Campus, Loughborough, Leicestershire LE12 5RD, UK; [3] Horticulture Research International, Wellesbourne, Warwick CV35 9EF, UK

Introduction

> The most serious challenge to plant physiologists involved in intercropping and agroforestry research is how to translate retrospective understanding into actual productivity increases.
>
> (Fukai and Trenbath, 1993)

Physiologists and ecologists have long argued that a sound understanding of the processes and mechanisms involved in resource capture and use and their interactions with the environment is essential for the development of more reliable and productive systems (Trenbath, 1976; Willey, 1979). It is therefore surprising that there is still considerable doubt concerning the impact of such understanding after the impressive accumulation of knowledge of intercropping in recent years (Fukai and Trenbath, 1993). A collection of reviews on the physiological and agronomic aspects of intercropping appeared recently in a special issue of *Field Crops Research* (Volume 34, 1993).

The purpose of this chapter is to review the principles of resource capture, drawing heavily on the intercropping literature to illustrate the difficulties involved. Appropriate methods of measurement and analysis are outlined and their application to agroforestry research is considered.

Special attention is devoted to consideration of the key differences between intercropping and agroforestry systems, major opportunities for translating an improved understanding of these systems into increased productivity or improved sustainability, and the main 'bottlenecks' to future progress in research.

Competition and complementarity: definition

Competition between species in mixed stands (*inter*specific competition) differs from that between plants within monocultures (*intra*specific competition) in that the component species of intercrops may impose different demands on the available resources. The intensity of competition is greatest when site requirements are similar, to the point where species with overlapping niches may be unable to coexist within the same community. Vandermeer (1989) suggested that competition may be more severe between similar species than between species with contrasting growth habits. However, the opportunity for complementarity of resource use between species is restricted by the fact that all plants are competing for the same, usually finite, resources (light, CO_2, water, nutrients). Thus, there is extensive overlap between species in their resource requirements, with the partial exception of legumes which are capable of fixing atmospheric nitrogen.

Plants growing in monocultures are usually all of the same genotype and their growth and development proceed synchronously. They are therefore of a similar size and have equal access to available resources. Although interplant competition is intense, this does not affect survival and reproductive success; indeed, there is good evidence that biomass accumulation and yield are unaffected by a wide range of populations above a specific threshold. Since each plant inevitably captures a decreasing proportion of the available resources as planting density increases, the initially linear relation between population and biomass becomes curvilinear as competition intensifies above and below ground. The maximum attainable biomass for individual species depends primarily on the availability of light, water and nutrients. To increase productivity further, crops must either capture more of these resources or use them more efficiently.

The components of intercrops or agroforestry systems often differ greatly in size, with the result that the growth of the smaller understorey species may be inhibited by shading, and possibly also by competition for water and nutrients. Competition for light is the primary limitation when water and nutrients are freely available. However, in many tropical systems, water (e.g. semi-arid regions) or nutrient availability (e.g. acidic, leached or degraded soils) rather than light is the major limiting factor. It is not always straightforward to establish which is the primary limitation when more than one factor is marginal. For example, a species which establishes an early advantage in light capture through more rapid initial shoot growth may also exhibit greater root growth and hence resource capture because of the increased availability of photosynthate. This may in turn further improve shoot growth and light interception to the detriment of the less competitive species in the mixture.

When resources are not limiting, densely planted monocultures usually provide the most efficient resource capture systems. However, where one (or more) resource is limiting, it may be possible to improve productivity by using species mixtures if the component species capture more of the available resources or use them more efficiently for growth. In such instances, mixtures may provide a greater yield than the combined yield of the corresponding sole crops. Such yield advantages have been widely reported for intercrops (e.g. Ofori and Stern, 1987), and are often expressed in terms of Land Equivalent Ratio (LER) or Crop Performance Ratio (CPR) (Willey, 1979; Harris et al., 1987; Azam-Ali et al., 1990). The area–time equivalency ratio (ATER; Hiebsch and McCollum, 1987) is an adaptation of LER that takes account of the land that is left unused after harvesting the shorter duration sole crop. However, great care must be taken to ensure that the sole crops against which the performance of intercrops is assessed are themselves grown at their optimal densities. When this precaution is observed and LERs exceed 1, it has often been assumed that the productivity of the mixture is superior to that of sole-cropping, and that complementarity has occurred.

Principles of resource capture

The basic principle underlying the concept of resource capture is that complementary or competitive interactions between species depend on their ability to capture and use the most limiting essential growth resources effectively (Monteith, 1981a). Capture (c) of the limiting resource (e.g. light, water or nutrients) depends on the number, surface area, distribution and effectiveness of the individual elements within the canopy or root system of the species or mixture involved. The use of the acquired resources depends on the conversion coefficient (e) of the species involved and environmental influences such as temperature extremes or drought. Biomass production (B) over a time period (t) may be expressed as:

$$B = \bar{c}.\bar{e}.t \tag{4.1}$$

where B has units of g m^{-2} of land area; the mean values of c and e ($\bar{c}$ and $\bar{e}$ respectively) have units of MJ m^{-2} and g MJ^{-1} m^{-2} respectively when the limiting resource is light, and mm and g mm^{-1} H$_2$O when the limiting resource is water; t has dimensions of days.

The major advantage of expressing productivity in these terms is to emphasize the apparent conservativeness of e. For example, conversion coefficients expressed in terms of total intercepted shortwave radiation are generally around 1.0 g MJ^{-1} for C3 species and 2.0 g MJ^{-1} for C4 species. The corresponding conversion coefficients for water (e_w) are dependent on atmospheric saturation deficit (D) such that e_w is proportional to $1/D$, where D has units of kPa. Conservative values for e_w of around 4 kg mm^{-1} kPa^{-1} for C3 species and 8 kg mm^{-1} kPa^{-1} for C4 species have been reported (Squire, 1990; see later for more detailed discussion).

Choice of systems for study

Despite the apparent simplicity of the principles involved, surprisingly few attempts have been made to quantify resource capture in intercropping or agroforestry systems, largely because of the technical difficulties and expense involved in intensive studies of light and water use until recent advances in microprocessor technology became available. A further complication arises from the numerous combinations of species, populations, spacing and fertilizer treatments that are possible. Unlike agronomic trials, it is essential to limit the number of treatments examined to the absolute minimum required to quantify resource capture and utilization in intercrops or agroforestry systems adequately. A final problem is posed by the choice of treatments or systems in which resource capture should be examined. For example, is there a unique relationship between species A and species B in terms of resource utilization which is independent of environmental conditions or management practices? In other words, will mixtures of species A and species B always increase productivity? If not, there is little practical advantage in understanding the mechanisms of resource use if these cannot be extrapolated to other sites and other seasons.

There are several pitfalls to avoid once a particular combination of species has been identified for investigation. To illustrate these and the types of interaction that may occur, two hypothetical examples illustrating competition or complementarity are examined below. In both cases, the x and y axes represent the biomass of the sole stands of each species, while the hypotenuse represents that of the mixtures (Fig. 4.1). Such graphs are known as production possibility curves or competition productivity frontiers and provide a visual representation of all combinations of treatments involving both species (see Chapter 2). Vandermeer (1989) made the point that such curves are applicable to isoresource situations where 'biomass' is a measure of productivity, i.e. output expressed in terms of resource input.

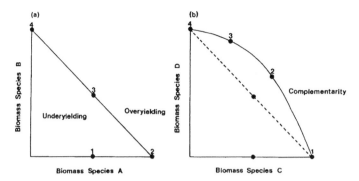

Fig. 4.1. Competition and complementarity expressed as hypothetical production possibility frontiers. The *x* and *y* axes represent biomass production by pure stands of each species, and mixtures are shown on the hypotenuse; see text for explanation.

Example 1

This example (Fig. 4.1a) represents a combination of species that share the same finite quantity of potentially limiting growth resources in such a way that the best yields of A and B are identical. Increasing the proportion of one species decreases the quantity of resources available to the other (treatments 2 and 4) and so the combined resource capture remains the same and productivity stays within the hypotenuse line. The potential pitfall here is that comparison of resource capture by treatments 1 and 3 leads to the conclusion that species A (treatment 1) is using less resources than the mixture (treatment 3).

In mixtures of competing species there is no advantage in changing the proportion of each species as far as resource capture is concerned. An excellent example of this type of interaction is the *Leucaena leucocephala*–sorghum system described by Rao *et al.* (1992) for a semi-arid site at Hyderabad, India. It is essential to understand why the competitive interactions in such systems are so extreme in order to avoid their repetition elsewhere. Recent studies at another semi-arid site at Machakos, Kenya suggest that the roots of *Leucaena* and maize exploit water from the same soil horizons during the rainy season, with the result that the intercropped maize captures less water than sole maize (Howard *et al.*, 1995). Results from *Leucaena*-based systems elsewhere in the tropics also indicate that its competitive nature and vigorous growth usually reduce the yield of associated crops.

A reduction in the yield of the crop component is not in itself sufficient to demonstrate that a particular tree–crop combination is underyielding, i.e. exhibiting a reduction in production. Over- or underyielding occurs when the combined production of either biomass or harvested materials is greater or less than that of sole stands grown on the same area of land. Overyielding and underyielding resulting from competition (which mainly affects crop performance) usually occur concurrently because the trees and crops both require water, light, space and nutrients. Overyielding can therefore only be established in relation to the appropriate sole stands.

Example 2

The second example (Fig. 4.1b) provides a graphical representation of overyielding or complementarity in resource use, where treatments 1 and 4 represent the most productive sole stands of species C and D. Treatments 2 and 3 represent mixtures of species C and D, where the biomass of species D is reduced relatively more than species C in treatment 2 but is much less affected in treatment 3. Both mixtures are nevertheless overyielding. Experimental evidence for this type of response comes from both intercropping (sorghum–pigeonpea, pigeonpea–groundnut) and agroforestry (*Cassia spectabilis*–cowpea) systems. In terms of assessing resource capture, the appropriate sole stands are treatments 1 and 4, and the use of sub-optimal populations or management of either sole stand will create a bias and result in overestimation of the advantage offered by the mixtures. Such pitfalls are common because researchers often use identical plant populations in all treatments for convenience of statistical analysis. Another common pitfall is to manage the sole and intercropped trees in the same way, for example by pruning, even though this is intended to reduce competition with the associated crop. This also creates a spurious bias in favour of the mixture.

Mechanisms responsible for overyielding

Provided such precautions are taken, the mechanisms involved in overyielding can be examined in situations where the capture of light or water is the major limiting factor. For instance, in the humid tropics where water and nutrients are not limiting for cowpea, a rapidly growing crop, the canopy can only intercept 45–50% of the incident radiation during the 80 day crop duration, equivalent to about 800 MJ m^{-2} of interception (Fig. 4.2). Assuming a mean seasonal conversion of intercepted radiation to dry matter (e) of 0.5 g MJ^{-1}, then 4 t ha^{-1} of biomass would be produced. However, if cowpea were intercropped with a tree with a similar timecourse of canopy development and e value, the total capture of light would be expected to increase even in the presence of any shade-induced effects on the canopy development and specific leaf area of the cowpea canopy. This increase in light capture results from the more rapid expansion of the combined tree–crop canopy, and is most advantageous when one of the component species is slow to develop its canopy, in this case the tree. An additional potential benefit is that the long term e value for the mixed canopy may be higher than in sole cowpea, in which photosynthesis is light-

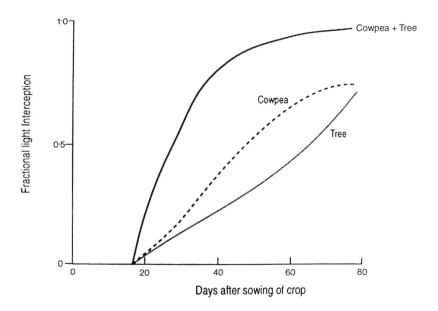

Fig. 4.2. Seasonal trends in fractional interception of radiation by sole cowpea, sole trees and cowpea + tree treatments to illustrate spatial complementarity (ICRAF, 1994).

saturated in full sunlight (16–20 MJ m^{-2} day^{-1}). The influence of shade on light and water use efficiencies is considered in greater detail later in this chapter.

Mixed cropping systems offer considerable scope for improving seasonal water use in the semi-arid and arid tropics. For example, most annual cropping systems use only 30–35% of the total rainfall because much of this is lost by soil evaporation (S) from sowing until the canopy closes, as runoff (r), or is left as residual moisture at final harvest (R). There is evidence that intercrops may make better use of available soil moisture than shorter duration sole crops and it may be postulated that agroforestry may be more effective still in increasing the proportion of rainfall used for transpiration (T) and hence overall productivity. The capture of water by an agroforestry system can be written as:

$$T_t + T_c = P - I - r - R - D - S \tag{4.2}$$

where T_t and T_c denote transpiration by the tree and crop components and P, I and D represent precipitation, interception losses and deep drainage beyond the rooting zone. All values are expressed in millimetres of water.

The limited available evidence on water use by agroforestry systems uniformly supports the hypothesis that substantial increases in water use are possible, although the relative importance of reductions in r, R and S in achieving this are unknown. Water use by the trees has rarely been examined but light interception and conversion by the crop component are much better understood; e values for the latter may be expected to increase despite reduced interception because of microclimatic changes induced by the trees. Measurements of water use are considered further later in this chapter and in Chapter 6.

Calculations of resource capture

Agronomic methods of expressing yield advantage have dominated the analysis of resource capture in mixtures (Natarajan and Willey, 1980b; Marshall and Willey, 1983; Morris and Garrity, 1993). However, these have serious shortcomings because they overestimate the advantages of inter-cropping and often lead to the awkward situation where apparent intercrop advantages of 30–60% are achieved in the absence of any difference in resource capture by the mixture and sole crops. This is largely an artefact of expressing resource capture as a function of the density of the intercrop relative to the sole crop. For example, if the quantities of water transpired by the intercrop and sole crops of the intercropped species A and B are respectively C_{ic}, C_a and C_b, then the relative change in water use resulting from intercropping (ΔC) may be described as:

$$\Delta C = (C_{ic}/(P_a C_a + P_b C_b)) - 1 \qquad (4.3)$$

where P_a is the proportion of species A in the intercrop relative to its own sole crop density and P_b is the corresponding value for species B (after Morris and Garrity, 1993).

However, this calculation is incorrect for most of the examples cited in the literature because it assumes that resource capture is linearly related to the population of plants in the sole crop. However, since the relationship between resource capture and plant population is often curvilinear, the apparent intercrop advantage of 30–40% is an artefact of the supposed linearity (Ong and Black, 1994). When the results are recalculated using the curvilinear relationship, it is possible to reconcile the reduced intercropping advantage with relatively constant absolute values for resource capture. A typical example is the study of sorghum–groundnut intercrop reported by Harris *et al.* (1987). To avoid such pitfalls, it is better to express resource capture in terms of absolute values in the intercrop and the corresponding sole stands. In addition, it is important in resource capture studies that at least one component of the intercrop is maintained at its full sole stand population wherever possible.

Relevance of Intercropping Research to Agroforestry

It has been argued that, since intercropping and agroforestry both involve mixtures of species, they should share many common processes including competition, environmental modification, transfer of nitrogen to non-legume associates and, of primary interest here, resource utilization. As pointed out in Chapter 1, there are also major differences between the two systems which will greatly influence the extent of the interactions involved. The first of these is that the woody perennial component of agroforestry systems has a well-established root system, at least after the initial establishment period. Thus, the woody species already has a substantial advantage in its access to below-ground resources when the crop component is sown. This is also true for the capture of light unless the tree canopy is managed (e.g. by lopping or pruning) before sowing. Second, because of its size and age, the woody component has a considerable advantage in sequestering resources from a large area and in enhancing soil physical and chemical properties under its canopy (Kessler, 1992; Belsky *et al.*, 1993). This is well illustrated by the effects of isolated trees on the understorey environment and vegetation described by Belsky *et al.*. (1993) in the semi-arid rangelands of East Africa, where the productivity of understorey vegetation under *Acacia tortilis* and *Adansonia digitata* (baobab) was 50 and 20% higher than outside the tree canopy. In addition, organic matter content and total N, P, K and Ca concentrations were significantly higher and soil bulk density significantly lower under the tree canopies than in the open. Although several explanations have been offered for this phenomenon of microsite enrichment, the interactions involved are highly complex. There is nothing comparable in intercropping, particularly on this scale. Finally, it has been suggested that the woody species in agroforestry systems are capable of better spatial and temporal utilization of the available growth resources than annual species.

There is a paucity of published information concerning the extent and consequences of such interactions in agroforestry systems. None the less, it should be possible to extract important lessons from the vast body of information available on intercropping.

Evidence for complementarity in resource capture

Spatial complementarity
It was stated earlier that overyielding in intercropping systems may occur even when the crop yield is reduced by the associated tree species, depending on the management and population of the two species. This is clearly a problem if the farmer's objective is to maintain a full yield of the crop component relative to the corresponding sole stand. However, if a partic-

ular association (e.g. between species C and D in Fig. 4.1b) is always advantageous (i.e. complementary in resource capture), then farmers may modify the proportion of each species to suit their own requirements without any sacrifice. In this instance the capture of the major limiting growth resources by the mixture always exceeds that by the corresponding sole stands irrespective of the proportions of species C and D. This situation applies when two (or more) species complement each other in capturing growth resources. Complementarity is two-way when a full yield of one component or the other can be achieved simply by varying the proportion of the desired species (Fig. 4.3a).

Examples of two-way complementarity in intercrops reported in the literature include maize–cowpea, maize–sorghum and pigeonpea–groundnut, although recent re-analysis of light capture has cast doubt on the earlier interpretations (Keating and Carberry, 1993). This study demonstrated that intercrops rarely capture more light than the best sole crops and that some previous comparisons were not valid because the appropriate sequential cropping systems were not used as controls. Instead, the observed complementarity was often explicable in terms of an increased conversion efficiency (*e*) for radiation.

Trenbath (1974, 1986) was amongst the first to suggest that a combination of tall C4 species and short C3 species may increase the conversion coefficient for intercepted radiation. Studies at ICRISAT by Marshall and Willey (1983) provided the first unequivocal demonstration that *e* was increased by 30% in a millet–groundnut intercrop, almost identical to the yield advantage. Later workers at ICRISAT confirmed this classic observation (Harris *et al.*, 1987; Ong *et al.*, 1992) and attention has

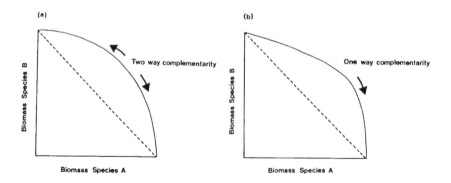

Fig. 4.3. Two types of complementarity expressed as production possibility frontiers.

subsequently turned to how these approaches may be applied to agroforestry systems. Sadly, little experimental work of this nature has been conducted outside the ICRISAT studies and research currently in progress at ICRAF.

The most detailed comparable study of an agroforestry system involved a *Leucaena leucocephala*–pearl millet alley-cropping system at ICRISAT (Ong *et al.*, 1991; Corlett *et al.*, 1992a, b). Experimental details are provided elsewhere in this chapter. Unlike the millet–groundnut intercropping study of Marshall and Willey (1983), the *Leucaena* hedges (C3 legume) were pruned to a height of 0.65–0.70 m before sowing and remained above or at the same height as the C4 pearl millet throughout the season. There is some evidence that *e* was increased in the agroforestry system in 1986 (Table 4.1) but this was insufficient to offset the reduced light capture by the pearl millet resulting from competition from the *Leucaena*. This is a good example of one-way complementarity in which the taller species is clearly dominant in capturing growth resources (Fig. 4.3b). This type of complementarity is difficult to modify by spacing or pruning and so provides less flexibility for farmers who wish to maintain the contribution of the crop component. This problem is less serious if the tree component is capable of producing high value products (e.g. fruit) which offset the loss in crop yield (Kessler, 1992). Several examples of one-way

Table 4.1. Total intercepted radiation, above-ground biomass and conversion efficiency (*e*) in a *Leucaena leucocephala*–pearl millet alley-cropping system, Hyderabad, India (data from Corlett, 1989). See text for details of radiation measurements and calculations of radiation interception.

	Intercepted radiation (MJ m^{-2})		Biomass (t ha^{-1})		Conversion coefficient (g MJ^{-1})	
	1986	1987	1986	1987	1986	1987
Rainy season (July–Aug)						
Sole millet	580	504	4.7	5.0	0.81	0.99
Alley millet	290	148	3.1	0.9	1.07	0.61
Sole *Leucaena leucocephala*	520	861	4.0	7.1	0.77	0.82
Alley *Leucaena leucocephala*	520	780	4.0	6.4	0.77	0.82
Total alley system	810	928	7.1	7.3	0.88	0.79
Dry season (Sept 1986–June 1987)						
Sole *Leucaena leucocephala*	1270		1.5		0.12	
Alley *Leucaena leucocephala*	1160		1.7		0.15	
Annual value (July 1986–June 1987)						
Total alley system	1970		8.8		0.45	

complementarity have been reported for maize–pigeonpea, millet–ground-nut and pigeonpea–groundnut intercrops, in which the first-named dominant component exhibits more rapid canopy development.

Temporal complementarity

The yield advantages provided by mixtures cannot always be explained by more effective use of growth resources at specific times. Indeed, there are substantial opportunities for temporal complementarity if species make their major demands on available resources at different times, thereby reducing the possibility of competition. For example, the usually advantageous intercropping of the slow-growing pigeonpea with fast-grow-ing cereals such as maize or sorghum produces only limited increases in total light or water use (Natarajan and Willey, 1980b; Sivakumar and Virmani, 1984) but nevertheless provides intercrop advantages of 40–70%. This study has often been wrongly cited as an example of greater resource capture because the calculations were based on the concept of Land Equivalent Ratio (Willey *et al.*, 1986). The real explanation for the advan-tage of this system is the considerable increase (70%) in the harvest index of the intercropped pigeonpea; similar improvements in harvest index have been reported for cassava–soybean (Tsay *et al.*, 1987). This improvement occurs because excessive vegetative growth of the sole crop was suppressed by the dominant associated species, resulting in greater partitioning of assimilate to grain production. Such improvements in partitioning are more spectacular in crops with an inherently low harvest index or an indeter-minate growth habit, both of which are features of crops commonly grown by subsistence farmers.

As far as we know, there are no published reports of spatial complementarity and improvements in assimilate partitioning in agro-forestry systems, possibly because previous studies have focused largely on the crop rather than the tree component. However, unpublished evidence from several perennial pigeonpea-based systems indicates that partitioning to grain yield in pigeonpea was unaffected by intercropping (Odongo *et al.*, 1996; Table 4.2).

Conservation of growth resources

Unlike radiation, growth resources such as water and nutrients can be stored and/or lost before the vegetation can absorb them. Agroforestry technologies for soil and water conservation offer substantial potential advantages over intercropping, especially on hill slopes (Young, 1989). In the semi-arid tropics, soil evaporation is the main cause of water loss (with the possible exception of steeply inclined sites), but there have been few quantitative studies to establish the effectiveness of agroforestry in reducing such losses. An analysis of this potential benefit is presented by Wallace in

Table 4.2. Harvest index of perennial pigeonpea in sole and intercropped systems (J.C.W. Odongo, unpublished).

	Harvest index (%)	
	ICP 8094	ICP 8084
Sole pigeonpea	10.0	10.0
Pigeonpea + maize	10.0	13.3
Pigeonpea + sorghum	10.6	13.8
Pigeonpea + groundnut	9.3	10.5
Pigeonpea + cowpea	10.0	10.7

Chapter 6 and a theoretical framework describing losses by leaching is discussed in Chapter 9 by van Noordwijk *et al.*

Principles of Light Interception and Water Use

The principles underlying the capture of light and water by mixtures will be reviewed before considering the experimental approaches that may be used and the practical difficulties encountered in quantifying resource capture by the components of mixed communities. As indicated earlier, the capture of light depends on two factors, first the fraction of the incident photo-synthetically active radiation (PAR) that is intercepted by each species, and second the efficiency of conversion of the intercepted radiation by photosynthesis. Each of these stages in dry matter production will be discussed briefly; fuller reviews are provided by Squire (1990) and Keating and Carberry (1993).

Light interception – sole crops

For simplicity, intercepted radiation (S_i) is normally taken as the difference between the quantity of solar radiation incident upon the canopy (S) and that transmitted to the soil (S_t). However, this method of determining S_i makes no allowance for the fraction of incident radiation that is reflected from the canopy surface (perhaps 5–20% of S depending on leaf angle, surface characteristics and moisture content). Few field studies of light interception by field crops have attempted to correct for this source of error.

Mean values for total shortwave solar radiation (400–3000 nm) during the cropping season range from about 12 MJ m^{-2} day^{-1} in cloudy upland regions to 24 MJ m^{-2} day^{-1} in some semi-arid areas. Light interception by crops growing in different climatic regions may therefore be best compared by using the ratio of S_i/S to describe fractional interception (f), rather than the absolute values for intercepted radiation. Fractional interception for a specific canopy is hardly affected by the absolute values of S, but shows great seasonal variation depending on developmental stage, leaf area index, plant height, canopy structure and plant water status.

When water is freely available, fractional interception is related to leaf area index (L) in many sole crops by the expression:

$$f = 1 - \exp(-kL) \qquad (4.4)$$

where k is an extinction coefficient for the canopy which is dependent on leaf angle and distribution. Fractional interception therefore increases as k and L increase. In crops such as cereals where the stems, leaf sheaths and panicles provide a significant photosynthetic area, it is important to substitute total green area index (G) for L in Equation 4.4.

Light penetration into the canopy is greater when the leaves are erect (low k value) than when they are horizontally oriented (high k value; Fig. 4.4a). The optimum arrangement for effective light interception is a progressive change from erect leaves near the top of the canopy to more horizontal leaves deeper in the canopy. Values of k differ greatly between species with different canopy structures (Fig. 4.4a), but are generally conservative within specific genotypes over a range of conditions. The k values for total shortwave radiation range from 0.3 to 0.45 in crops with erect leaves (e.g. cereals) to about 0.8 in species with evenly distributed horizontal leaves (e.g. groundnut, cassava). Nevertheless, the variation in k

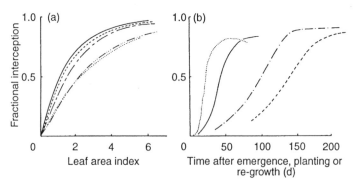

Fig. 4.4. (a) Leaf area index and fractional interception of total solar radiation, as influenced by the extinction coefficient. The k values used for each species are shown in brackets. For groundnut (0.60) ——, cassava (0.58) -----, pigeonpea (0.52) — -, oil palm (0.34) — ··· and pearl millet (0.30) ······ (Squire, 1990). (b) Seasonal changes in fractional interception of total solar radiation. For short-season sorghum (······), groundnut (——), cassava (— · —) and a ratoon stand of sugar cane (-----) (Squire, 1990).

between genotypes within morphologically variable species such as rice may be as great as that between species (Squire, 1990).

The amount of radiation intercepted by a crop depends on the quantity of solar energy received, canopy duration and fractional interception. Seasonal timecourses for fractional interception vary greatly (Fig. 4.4b) depending on canopy characteristics such as growth rate, size and duration. The f values increase much more rapidly in cereals such as sorghum than in legumes like groundnut, reflecting their differing rates of leaf initiation and expansion. The slower increases in sugar cane and cassava reflect their lower initial planting densities. The variation in f values between crops is generally smaller than those in the green area index, partly because k values are often larger in species with slowly expanding canopies; maximum f values therefore frequently differ little between crops grown under non-limiting conditions.

Mean f values calculated over the life cycle of the crop $(\bar{f})$ are generally smallest in short-duration cereals ($c.$ 0.5) and legumes ($c.$ 0.15) and largest ($c.$ 0.9) in perennial species, mainly because of the differing durations of ground cover (Squire, 1990). Similarly, f values are typically higher in long-duration genotypes than in short-duration cultivars of the same species. A primary goal of agroforestry is to increase f values and hence overall dry matter production; this contrasts with intercropping where improvements in productivity are often attributable to increased e values, as discussed earlier.

Light interception – mixed cropping systems

The relationship between f and canopy characteristics defined by Equation 4.4 assumes that the canopy is a homogeneous arrangement of randomly distributed leaves, and that there is no spatial variation in light transmission. Clearly these conditions are not met in intercrops or agroforestry systems because of the extensive horizontal and vertical variation in canopy structure introduced by different species combinations and differences in planting date and arrangement, leaf size, shape and orientation and plant height. Canopy architecture is also constantly changing in mixed cropping systems because of the differing growth rates and durations of the component species. For example, low-lying legumes growing adjacent to taller cereals are initially subjected to greater competition from neighbouring plants than in the corresponding sole crops because of the faster growth of the cereal, but subsequently experience less competition than in the sole crop for much of their reproductive phase due to the earlier harvest of the cereal component. Light interception by the remaining legume component is not identical to the sole crop because the 'missing' rows are no longer intercepting radiation, and may even differ from a sole crop planted in the same discontinuous arrangement because of the residual effects of the taller, more competitive cereal on the canopy architecture of the legume (Azam-Ali, 1995).

The principles underlying temporal complementarity of light capture in intercrops are now well understood and largely self-evident, but those involved in spatial complementarity are much more complex and require a more rigorous theoretical framework than in monocultures to account for the spatial heterogeneity and clumping of leaves within the canopy. The best documented example of temporal complementarity for light in intercrops involves mixtures of short duration cereals and longer duration legumes. In such systems, light interception typically tracks the sole crop of the faster growing component initially, falls sharply when the short-duration component is harvested and then increases again as the canopy of the slower growing, long-duration legume continues to enlarge (Fig. 4.5). Since light interception by the shorter duration intercrop component is little affected by competition but overall crop duration is increased, cumulative interception by the intercrop is greater than in the sole cereal but similar to the sole legume. For example, f values for cereal, legume and intercrops were respectively 0.39, 0.53 and 0.45 for pearl millet–groundnut and 0.36, 0.51 and 0.53 for sorghum–pigeonpea systems (data recalculated from Reddy and Willey (1981) and Natarajan and Willey (1980b) by Squire (1990)). However, seasonal interception has also been reported to be greater in sorghum–pigeonpea intercrops than in either of the sole crops (Natarajan and Willey, 1980b, 1985). In most of the earlier studies where the productivity of intercrops was correlated with the capture and use of growth resources such as light (Reddy *et al.*, 1980; Sivakumar and Virmani,

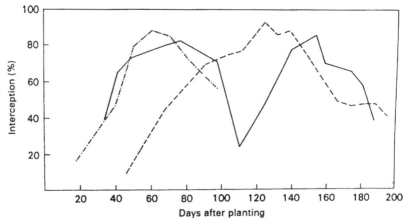

Fig. 4.5. Seasonal interception of photosynthetically active radiation (PAR) in sole maize (— · —), sole pigeonpea (-----) and a maize–pigeonpea intercrop (——) (Sivakumar and Virmani, 1980).

1980) or water (Hulugalle and Lal, 1986), no attempt was made to quantify the contributions of the component species, largely because of the practical difficulties involved. However, Marshall and Willey (1983) and Azam-Ali *et al.* (1990) were successful in partitioning radiation interception within millet–groundnut and sorghum–groundnut intercrops. In both cases the cereal component of the intercrop intercepted 70–112% more photosynthetically active radiation (PAR) calculated on a per-row basis than the equivalent sole crop, while the groundnut intercepted 15–22% less than the sole crop.

In 'addition intercrops' (i.e. where one component is planted at its normal sole crop density and a second component is added so that overall planting density is increased), canopy structure can be considered as horizontally homogeneous but vertically heterogeneous. The vertical variation in canopy structure depends on the foliar characteristics of the species involved and the extent of any interactions between them. A modified form of Beer's Law satisfactorily describes the light attenuation profiles within such systems, except during the early stages of growth before canopy closure. An additional constraint is that the k values derived for sole crops may not be applicable in intercrops.

Light interception in addition intercrops may be described by stratifying the canopy into several horizontal layers such that:

$$S_l = S(1 - e^{-(k_{lA}L_{lA} + k_{lB}L_{lB})}) \tag{4.5}$$

where S denotes the quantity of solar radiation reaching the surface of canopy layer l, S_l is the quantity of radiation intercepted by layer l and k_{lA}, k_{lB} and L_{lA} and L_{lB} represent the extinction coefficients and leaf (or green) area indices for layer l in species A and B. Interception by species A in layer l (S_{lA}) may be calculated as (after Keating and Carberry, 1993):

$$S_{lA} = S_l \frac{k_{lA}L_{lA}}{k_{lA}L_{lA} + k_{lB}L_{lB}} \tag{4.6}$$

Similar calculations may be made for species B, and total daily interception for the intercrop and by species A and B may be obtained by summing the appropriate interception values for all canopy layers.

Equation 4.6 indicates that light interception by addition intercrops will be greater than in the sole crops provided the k values for the component species are similar but green area index is increased. Similarly, the addition of an understorey component with a high k value to a crop with a low k value would increase the overall extinction coefficient, thereby increasing interception. However, such beneficial interactions may be difficult to achieve when resources are non-limiting because sole crops grown at high densities are capable of attaining high leaf area indices and almost complete radiation capture.

Analysis of light attenuation in replacement intercrops (i.e. where some rows of one crop are replaced with another without necessarily altering total plant density) is more complex because the canopy is both horizontally and vertically heterogeneous, except possibly when the areas of the individual components are large enough to form 'mini-sole' crops (Keating and Carberry, 1993). This complexity is increased even further in agroforestry systems where the tree component may be widely dispersed and much larger than associated crops. Various models have been developed to describe the interception of radiation by horizontally heterogeneous canopies. For instance, Trenbath (1974) proposed a model which took account of varying solar angle and foliage distribution, while Sinoquet and Andrieu (1993) described the 'gap frequency concept' which is funda-mental to statistical models of radiation penetration and interception by discontinuous canopies; this concept assumes that the probability (P) of a canopy intercepting radiation is given by $1 - P_0$, where P_0 represents the gap frequency or probability of a beam of light reaching a specific level in the canopy without being intercepted. Nilson (1971) discussed the influence of canopy characteristics such as leaf area index, leaf angle distribution and leaf distribution, and also used a Markov chain model to provide a theoretical description of light penetration through non-randomly distributed foliage.

Thus the practical usefulness of a unique value for k is often limited in heterogeneous canopies, particularly when the non-random distribution of foliar elements occurs because the intercrop components are planted in rows or clumps, or never achieve complete ground cover due to stress. The influence of drought on extinction coefficients within specific crops is discussed by Azam-Ali *et al.* (1993). There is little information concerning the extent to which extinction coefficients for the same species differ between sole and intercrops, and few reliable estimates of k values for either intercrops or their component species. However, Sivakumar and Virmani (1984) observed no significant difference in k between sole maize (0.64), sole pigeonpea (0.69) and a maize–pigeonpea intercrop (0.66) in which maize contributed two-thirds of the leaf area involved. However, data obtained by Harris *et al.* (1987) suggest that the k value for sorghum was lower than that for the understorey groundnut, but that the combined value for the intercrop was close to that for sole groundnut, which formed the major component of the intercrop.

Conversion efficiency

Dry matter production in any cropping system is often linearly related to the quantity of radiation absorbed by its canopy, in the absence of other limiting factors (Monteith, 1981a). The slope of this relation provides a measure of the light conversion coefficient (e), which is usually expressed as the quantity of above-ground dry matter produced per unit of intercepted radiation (g MJ^{-1}). This relation holds because the net photosynthetic rate (P_n) of individual leaves increases linearly with irradiance up to a level where they become light saturated (Fig. 4.6a); whole canopy assimilation tends to exhibit a similar response, but with saturation occurring at higher irradiances because of the influence of canopy architecture and mutual shading on the irradiance incident upon individual leaves (Biscoe et al., 1975). Dry matter production is proportional to mean canopy photosynthetic rate because respiratory and photorespiratory losses of carbon generally comprise a conservative fraction of assimilation within individual species (Squire, 1990). The higher light-saturated photosynthetic rates of C4 species as compared to C3 plants are reflected by differences in both conversion coefficients and maximum growth rates.

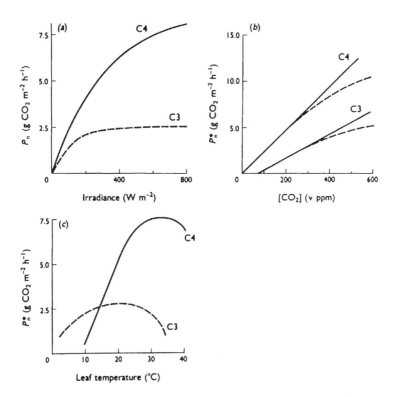

Fig. 4.6. Comparative response of leaf photosynthesis of C3 and C4 species to (a) irradiance at ambient [CO_2] and optimum temperature; (b) [CO_2] at high irradiance and (c) temperature at high irradiance and ambient [CO_2] (Loomis and Connor, 1992).

When water and nutrient supplies are not limiting, dry matter production (W) by a crop stand may be formalized as:

$$W = \overline{S}\,\overline{f}\,\overline{e}\,t \qquad (4.7)$$

where $\overline{S}$ represents mean daily solar radiation (MJ m^{-2} day^{-1}), $\overline{f}$ is the seasonal mean fractional interception, $\overline{e}$ is the seasonal mean conversion coefficient (g MJ^{-1}) and t is the canopy duration in days (Corlett *et al.*, 1992b). The conversion coefficient ($\overline{e}$) is therefore a key factor in determining the quantity of dry matter produced.

Conversion coefficients for intercepted radiation may be calculated on various biological and time scales. For example, the initial slope of the asymptotic relation between the net photosynthetic rate (P_n) for individual leaves and intercepted radiation provides an estimate of instantaneous quantum efficiency. This provides a basis for comparing the intrinsic efficiency of the photosynthetic systems in different species. An important consequence of the curvilinear light response curve for P_n is that quantum efficiency calculated for the whole canopy is greatest when the incident radiation is distributed evenly across a large leaf area so that relatively few leaves are light saturated and the majority are operating on the initial linear phase of the light response curve. Canopy architecture is important in determining the distribution of radiation within the canopy and therefore influences whole-canopy conversion coefficients.

In theory, C4 species should have a lower quantum efficiency than C3 species because of the greater energy costs of the additional carboxylation steps and intercellular transfers of intermediate metabolites (Osmond *et al.*, 1982). However, the quantum efficiencies of C3 species are adversely affected by photorespiration, to an extent dependent on temperature, $CO_2 : O_2$ ratio and drought. Quantum efficiencies are therefore similar in C3 and C4 species at 20–25°C, but are greater in C3 plants at lower temperatures. Thus the photosynthetic temperature response curves of C3 and C4 species differ greatly, with the latter usually declining sharply at temperatures below 15–20°C (Fig. 4.6c). Photosynthesis increases with increasing atmospheric CO_2 concentration within the range 0–800 ppm, but is consistently higher in C4 species (Fig. 4.6b).

Stress factors such as extremes of temperature, salinity or drought and acclimation to shade may predispose the photosynthetic system to damage by excess PAR which cannot be used for photosynthesis. Photoinhibition occurs when the rate of damage to photosystem II (PSII) exceeds the rate of repair (Baker, 1991), and may be observed during drought stress or when shade plants are exposed to high PAR fluxes (Evans *et al.*, 1988). Down-regulation of PSII, involving thermal dissipation of absorbed energy, may help to minimize damage. However, down-regulation and injury both reduce quantum efficiency under low light conditions and may also decrease net photosynthesis at high irradiances. This, in conjunction with the relatively low PAR fluxes required to saturate photosynthesis in C3 species (Fig. 4.6a), may account for the reported improvements in conversion coefficient when C3 crops such as groundnut are intercropped; the shading effect of the taller component reduces the interception of excess radiation by the legume, thereby avoiding photoinhibitory effects and coincidentally improving water use ratio since less energy has to be dissipated by transpiration.

Species differ greatly in their photosynthetic responses to temperature (Fig. 4.6c). Most C4 species are adapted to high temperatures, but are susceptible to chilling injury at temperatures below 10–12°C. However, a limited number of C4 species are adapted to temperate environments. Most C3 species are capable of tolerating temperatures down to 0–5°C, although some are also susceptible to chilling injury (e.g. banana and coffee), while others such as groundnut, cotton and sunflower perform well at temperatures of 30–40°C. In general, the temperature optimum for photosynthesis is broader than that for growth, emphasizing the importance of other factors that influence assimilate production and use. The shape of the relationships between photosynthesis and light (Fig. 4.6a) and temperature (Fig. 4.6c) confers a growth advantage on C4 species at high PAR fluxes and temperatures above 20–25°C; this behaviour reflects the greater quantum efficiency and maximum photosynthesis rates in C4 species under these conditions (Loomis and Connor, 1992).

In the longer term, conversion coefficients may be calculated from cumulative values for dry matter production and light interception. These values may be based either on periodic estimates of biomass and intercepted radiation, or on a single analysis at final harvest. Particularly in the latter case, the estimates are susceptible to error because dry matter accumulation often slows and may even become negative towards the end of the growing season due to senescence. A further major source of error in long term estimates of e is that the contribution of roots is rarely included; root biomass may contribute only a small fraction of the total (10–20%) in annual crops growing on moist soil, but may reach 50% in herbaceous species grown under dry conditions, or in woody perennials. Clearly the exclusion of root biomass from the calculation introduces major uncertainties into estimates of the impact of intercropping or agroforestry on conversion efficiency.

A further consideration is that the energy content per unit dry weight varies between different plant tissues, particularly those containing large quantities of energy-rich compounds such as lipids. This has important implications where the component species differ in their lipid fraction (e.g. oilseeds and legumes vs. cereals) or where intercropping alters the harvest index of one or both species. In such cases, total biomass may need to be corrected for the varying energy equivalents of the organs involved. Failure to account for differences in energy equivalents would only slightly underestimate the true conversion coefficient in cereals, but would underestimate *e* by up to 30% in oil crops (Squire, 1990). In species such as legumes and oil palms, the dry mass of lipid may be converted to an equivalent mass of carbohydrate in order to derive energy-corrected stand dry weights and conversion coefficients. Alternatively, biological productivity may be expressed in terms of energy content, which has the potential advantage that the interception of solar energy and storage of chemical energy in biomass both have the same units (MJ).

Water use – sole crops

Water use by crops can be treated in an analogous manner to light by breaking its utilization down into 'capture' and 'conversion efficiency' components. As with light, the quantity of dry matter produced (W) depends on the quantity of water captured (mm) and the 'efficiency' (g mm^{-1} H$_2$O) with which it is used in dry matter production. The ratio of dry matter produced to water transpired is known as the water use ratio (e_w). Dry matter production may be expressed as:

$$W = e_w \Sigma E_t \tag{4.8}$$

where ΣE_t represents cumulative transpiration. As for light, the quantity of dry matter produced is often linearly related to the quantity of water transpired, indicating that e_w is conservative (Azam-Ali, 1983; Cooper *et al.*, 1987). Equation 4.8 is based on the close interdependence of CO_2 and water vapour exchange during photosynthesis and transpiration; whenever the stomata open to permit entry of CO_2 into the leaves, water is lost by transpiration.

The leaves of most crop species have evolved to provide effective carbon assimilation at the expense of potentially rapid water losses. Thus the leaves of mesophytic species typically have a large surface area to maximize light interception and an effective gas exchange system to permit entry of CO_2 to the chloroplasts. These features also favour rapid transpiration whenever the stomata are open and solar radiation and evaporative demand are high. The pathways for CO_2 and water vapour exchange differ in length because evaporation occurs mainly from the cell walls, while the CO_2 must penetrate to the chloroplasts of the mesophyll cells. The stomata present on the surfaces of most green organs are therefore ideally located to regulate gaseous exchange.

The air in the intercellular species within the leaf is always close to saturation, even in water stressed plants (water potential of -2 to -3 MPa), and so the driving force for transpiration depends predominantly on temperature and atmospheric humidity. Thus, the rate of transpiration from a leaf (E_l) is determined by:

$$E_l = (v_i - v_a)/r' \qquad (4.9)$$

where v_i and v_a are the intercellular and atmospheric concentrations of water vapour and r' is the diffusive resistance to water movement from the leaf. Similarly, the net uptake of CO_2 (A_l) can be expressed as:

$$A_l = (C_a - C_i)/r \qquad (4.10)$$

where C_a and C_i are the atmospheric and intercellular CO_2 concentrations and r is the diffusive resistance to CO_2 uptake. Water use ratio (e_w) may therefore be calculated as:

$$e_w = A_l/E_l = (C_a - C_i)/\beta(v_i - v_a) \qquad (4.11)$$

where β is the ratio of the diffusive resistances for CO_2 and water vapour ($\sim 1.6 : 1$). ($v_i - v_a$) is much more variable than ($C_a - C_i$) and so the ratio A_l/E_l is also variable. When air and leaf temperature are similar, ($v_i - v_a$) is proportional to the atmospheric saturation vapour pressure deficit (D) (Brown et al., 1987) and in such cases A_l/E_l should be inversely proportional to D (El-Sharkawy and Cock, 1984). Saturation deficit may therefore be expected to influence dry matter production by crops because of its influence on transpiration and hence on the ratio A_l/E_l. Results for numerous species indicate that e_w is inversely proportional to saturation deficit (e.g. groundnut, Ong et al., 1987), and that $e_w D$ may not be conservative in droughted or nutrient deficient plants. For example, in a study of groundnut growing in drying soil, e_w decreased as D increased, but $e_w D$ also decreased from 5.6 to 3.1 g kPa^{-1} kg^{-1} (Ong et al., 1987). This effect may have been caused by stress-induced stomatal closure and inhibition of photosynthesis.

Control of transpiration
Actively growing crops that provide full ground cover and are well supplied with water exert minimal control over transpiration and transpire at rates determined by the evaporative demand of the atmosphere. The maximum rate of water use for prescribed environmental conditions has been termed potential evaporation (E_0); E_0 defines the upper limit to the actual evapotranspiration (E_t), which is smaller than E_0 when ground cover is incomplete, the soil surface is dry or water stress enforces stomatal closure. Although atmospheric conditions predominantly determine E_0, it is the attributes of the vegetation that limit transpiration and determine E_t in water-limiting environments. These attributes may be divided into canopy characteristics which influence the movement of water from within the leaf to the atmosphere and rooting characteristics which influence the supply of water to the plant. These attributes are not independent since the rates of absorption and transpiration are intimately linked.

Control of water use by the canopy can be subdivided into short and long term regulatory mechanisms, i.e. those which provide protection against transient or more prolonged periods of water stress. Short term responses operate over timescales of minutes or hours to reduce transpiration per unit area. These include stomatal closure to reduce diffusive losses of water vapour and leaf movements or rolling to reduce radiation interception and hence leaf temperature and the saturated vapour pressure within the transpiring leaf. However, it is important to recognize that partial stomatal closure may be ineffective in reducing transpiration because of concomitant increases in leaf temperature and hence $(v_i - v_a)$ in Equation 4.9 (Black *et al.*, 1985). Longer term adjustments in transpiring area are achieved over periods of days or weeks through premature senescence, usually of the older leaves, or reductions in the production and/or expansion of new leaves. Regardless of how they are achieved, decreases in transpiration are usually accompanied by reductions in assimilation and growth.

Transpiration from crops is controlled by the aerodynamic and canopy conductances (g_a and g_c respectively). Canopy conductance (g_c) takes account of both the physiological and morphological characteristics of the canopy since it is usually calculated as the sum of the product of leaf area index (L) and leaf conductance (g_l) for the various layers of foliage within the canopy. The contributions of leaf sheaths, pods and panicles must be included in such calculations since they may contribute up to 30% of g_c and canopy transpiration, and 15% of the total water transpired between sowing and maturity (e.g. rice, Batchelor and Roberts, 1983). There is good evidence that transpiration is often controlled primarily through regulation of the transpiring area rather than leaf conductance during progressive drought (e.g. groundnut, Black *et al.*, 1985). The main determinant of transpiration from open, stressed or senescent canopies is g_c, but the aerodynamic conductance (g_a), which is dependent on windspeed, may be important in tall, dense canopies, particularly at low windspeeds. Depending on aerodynamic conductance, canopy conductances of 1.5–2.0 cm s^{-1} are required before most crops can transpire at near the potential rate; to achieve this, a leaf area index of 3–4 and a mean leaf conductance averaged over the canopy of about 0.6 cm s^{-1} are required. The value of g_c is usually much less than 1.5 cm s^{-1} under water-limiting conditions because of reductions in both L and g_l.

The root system comprises numerous elements which are normally distributed in a non-random manner within the soil because the spatial variation in the rooting environment is much greater than in the atmosphere. Factors that influence rooting depth and density include the content and distribution of water, nutrients and oxygen, and the presence of compacted or stony layers. Crops growing in semi-arid areas must often rely on stored water for at least part of their life cycle and may experience potential evaporation rates of up to 8–10 mm day^{-1}; E_0 values of 5–6 mm day^{-1} are common in more humid areas (Loomis and Connor, 1992). The quantity of water available to meet these demands depends on rainfall, soil storage capacity and the ability of the root system to exploit the available reserves effectively. The first two factors are outside the plant's control, but root systems exhibit extensive plasticity in their distribution and functional efficiency, both between and within species.

The maximum quantity of water potentially available to the crop is determined by the depth of the soil profile and its extractable water content. The available water is usually defined as the quantity of water held at tensions between field capacity and the permanent wilting point (usually taken as 10–1500 kPa). However, it should be recognized that the permanent wilting point, beyond which the remaining water is unavailable to plants, is not a unique value applicable to all species, but may vary according to the osmotic content of the root tissues; thus species capable of extensive osmotic adjustment can dry the soil further. The volumetric water content between field capacity and permanent wilting point represents the available water and may range from 7% in very light sandy soils to 40% in soils with a high organic or clay content; values for loams and clay loams are typically 20–25%.

Absorption of water occurs down the water potential gradient established between the roots and the surrounding soil by transpirational water losses from the canopy. In very broad terms, the flux of water (F) may be described as (Gardner, 1965):

$$F = (\Psi_s - \Psi_r)/(R_s + R_r) \tag{4.12}$$

where Ψ_s and Ψ_r denote soil and root water potentials and R_s and R_r represent the hydraulic resistances within the radial flow pathway from the soil to the xylem. However, it must be recognized that in drying soils Ψ_s may vary greatly between different sections of the rooting zone or with increasing distance from the root surface. Similar variation in Ψ_r may also occur under such conditions. In moist soils, R_r is approximately 100 times larger than R_s, but the latter increases by a factor of approximately 10,000 as the soil dries from field capacity to the permanent wilting point because of the exponential decrease in soil hydraulic conductivity over this range. Thus the rate of absorption and hence plant water status depend critically on both soil moisture content and rooting depth and density, since the latter determine the total volume of soil that is colonized. R_r depends on the total surface area and the radial conductance of roots to water movement, and is normally expressed per unit length rather than surface area (Loomis and Connor, 1992). The most effective means for plants to maintain absorption and hence transpiration on drying soils is through sustained root proliferation, since this confers the dual benefits of minimizing both R_r by increasing total root length and R_s by tapping previously uncolonized regions of soil which still have a high moisture content and hydraulic conductivity. Deep, densely branched root systems may be more able to sustain absorption over extended periods, but are achieved only at the cost of increased assimilate investment to support their growth and maintenance respiration. Additional metabolic costs are incurred when inorganic and, more especially, highly reduced organic solutes such as proline and betaine are accumulated and sustain absorption through osmotic adjustment (Loomis, 1985).

Grasses and cereals possess a dual root system which enables them to cope with varying conditions of water supply. Seminal roots with a low hydraulic conductivity provide slow access to water throughout the profile, while nodal or adventitious roots proliferate rapidly in the surface horizons following rewetting; for example, new adventitious root primordia may be detected in teff (*Eragrostis teff*) within 48 h of rewetting the soil surface. However, the development of high root densities is not always desirable since this may cause more rapid depletion of available soil water, whereas a sparser root system would enforce a more conservative use of soil moisture reserves by reducing stomatal conductance and/or leaf area. Indeed, Passioura (1983) suggested that selection for root systems with a high hydraulic resistance would improve conservation of water supplies during the vegetative period for subsequent use during reproductive development.

The balance between transpiration and absorption depends on both atmospheric and soil moisture conditions. The root systems of well-watered crops generally extract water at rates close to the prevailing potential evaporation (E_0). In monsoon climates with E_0 values of about 5 mm day^{-1}, this requires mean absorption rates of 1.7–2.5 g H_2O m^{-1} root day^{-1} for groundnut or millet crops with rooting densities of 2–3 km m^{-2} of land area (Gregory and Reddy, 1982). In these and similar studies of rice (Yoshida and Hasegawa, 1982) and cassava (Aresta and Fukai, 1984), the rate of absorption was probably limited by E_0 through its effects on plant water status and hence ($\Psi_s - \Psi_r$) in Equation 4.12, rather than by rooting or soil characteristics. The converse applies in dry soil where absorption depends much more on the size and distribution of the root system and soil moisture content than on atmospheric conditions. For example, Squire (1990) reported that transpiration from groundnut or millet growing on drying soil was only 2–3 mm day^{-1}, even though total root lengths were greater than in constantly moist soil; thus, millet growing on stored moisture at Hyderabad required a mean absorption rate of 1 g m^{-1} day^{-1} over the total root length of 3 km m^{-2} to sustain a transpiration rate of 3.1 mm day^{-1}. Equivalent values for millet grown in Niger were 0.65 g m^{-1} day^{-1} over a root length of 5.9 km m^{-2} to support a transpiration rate of 3.8 mm day^{-1}. Thus, mean absorption rates were only 25–50% of those in moist soil, primarily because of the greatly reduced soil hydraulic conductivity.

Limits to the extraction of stored water

Crops vary greatly in their maximum potential rooting depth, which ranges from perhaps 70 cm in short-duration cereals to 1.5–3.0 m in longer duration legumes or cereals growing on deep friable soil. Rooting depths may be much greater in woody perennials growing in arid or semi-arid environments, for example 53 m in *Prosopis* sp. growing in Arizona (Phillips, 1963) and 60 m in *Zizyphus lotus* growing in southern Morocco (Le Houéron, 1972), although these extreme rooting depths are not typical of most agroforestry systems. The quantity of water available to crops in a drying soil depends on the rate at which roots extend through the profile. Measurements of soil moisture content indicate that the progress of 'extraction' or 'drying' fronts downwards through the profile is closely related to that of the 'rooting front'. The velocity of the extraction front is greater in light sandy soils than in heavy clay soils, with values typically ranging from 10 to 40 mm day^{-1}, but reaching 70 mm day^{-1} in rapidly growing species such as millet (Squire *et al.*, 1984a).

In annual species, absorption begins as soon as or shortly after the roots reach specific horizons, increases to a maximum a few days later as roots proliferate and then decreases over the ensuing 1–2 weeks as hydraulic conductivity falls, unless the soil is rewetted. The depth of maximum extraction therefore lags behind the extraction front. Extraction of water from drying soils is limited by two factors: absorption is initially slow when roots reach specific horizons because of the low rooting density; and by the time the roots have proliferated sufficiently to achieve maximum extraction, soil water content may have decreased to the point where it restricts absorption. Thus, the principal limitation on water uptake by annual crops growing on stored moisture is the rate at which roots extend and proliferate. However, a very different situation applies for woody perennial species which already possess established root systems at the onset of the rainy season, and are therefore well placed to exploit the available moisture rapidly. This may be an important aspect of detrimental tree–crop interactions in agroforestry systems where the roots of both components are distributed within the same soil volume and hence competing for the same resource pool.

The temporal and spatial patterns of water use differ between determinate and indeterminate species or genotypes; the latter usually extract more water over greater rooting depths, particularly in perennials with established root systems. Cotton has been reported to extract *c.* 500 mm of water from a rooting depth of 3.5 m over a 110 day period, whereas maize extracted *c.* 50 mm over a 70 day period, little of which was drawn from below 1 m (Squire, 1990). The period during which the rooting and extraction fronts continue to move downwards in the soil profile may be even briefer (30–40 days) in short-duration cereals, since determinate crops usually show marked reductions in root activity and absorption after flowering, when priority switches to reproductive growth.

Interdependence of roots and shoots

In the longer term, the rates of soil water extraction and transpiration are often equal, a balance which may be defined as:

$$RI = \frac{v_i - v_a}{r_s + r_a} L \tag{4.13}$$

where R and I are respectively total root length per unit ground area and mean absorption rate per unit root length, $(v_i - v_a)$ is the leaf to air vapour pressure difference averaged over the canopy, r_s and r_a are suitably averaged stomatal and boundary layer resistances and L is leaf area index. In some instances, it is obvious which of the factors in Equation 4.13 is rate-limiting for transpiration. For example, in well-watered crops growing in humid conditions, $(v_i - v_a)$ is small and r_s is usually non-limiting, with the result that transpiration is determined by the rate of leaf area expansion. Since the ratio $R:L$ is often conservative, the extraction rate (I) is probably also conservative. Similarly, in dry soils transpiration may be limited because R and I are both constrained.

However, in other cases it is difficult to establish whether transpiration is limited primarily by above- or below-ground factors. For example, during grain filling in determinate species, the foliage and roots may both be senescing, affecting L and R, while their functional efficiency expressed as stomatal resistance or absorption rates (r_s and I in Equation 4.13) may also be declining; in such cases, it is extremely difficult to disentangle the causes and effects underlying the concomitant changes in water use. Plants exhibit great adaptive plasticity in their responses to drought. Thus an increase in saturation deficit (increased atmospheric drought) is often at least partially offset by reductions in L or g_l or both (Squire *et al.*, 1984a), particularly when soil moisture is also limiting. Similarly, plants growing under water-limited conditions frequently exhibit more rapid root growth, greater rooting depths and greater extraction of water at depth. These adaptive responses serve to reduce canopy conductance and increase root:shoot ratio, thereby improving the balance between supply and demand for water.

Water use – mixed cropping systems

As with light interception, intercropping and especially agroforestry systems offer the opportunity for spatial and temporal complementarity of water use, resulting in an improved exploitation of available moisture relative to sole crops. However, the opportunity for significant complementarity is likely to be limited unless the component species differ appreciably in their rooting patterns or duration; the former would permit exploitation of different soil reserves, while the latter would allow continued abstraction of residual moisture following the harvest of the shorter duration component.

Several field experiments suggest that total water use by intercrops may be little different from sole crops. Table 4.3 summarizes data for water use and water use ratio from ten studies spanning a wide range of intercrops and environments. Despite the substantial variation in seasonal rainfall from 84–140 mm (Rees, 1986a) to 575 mm (Natarajan and Willey, 1980a), water use by the intercrops ranged from 6% below to 7% above the corresponding sole crops in all except the last two examples shown. In the pigeonpea–sorghum system examined by Natarajan and Willey (1980a, b), there was no difference in water use between the intercrop and sole crops when the shorter duration sorghum was harvested, but a further 154 and 168 mm of water was extracted by the sole and intercropped pigeonpea before final harvest 10 weeks later. Thus, the later-maturing pigeonpea used residual water within the profile and late-season rains which would otherwise have remained untapped. Although total water use by the intercrop (585 mm) greatly exceeded that of sole sorghum (434 mm), there was no advantage over the longer duration sole pigeonpea (584 mm). In a similar study, Reddy and Willey (1981) showed that water use by a pearl millet–groundnut intercrop was 10% greater than in sole groundnut and 34% greater than in sole millet; a major contributory factor was that leaf area index exceeded 2 for only about 20 days in the 82 day sole millet crop, but for about 50 days in the 105 day intercrop.

Thus, much of the available evidence suggests that water use is similar in sole and intercrops, particularly when water losses from land left fallow following the harvest of the shorter duration species are taken into account (Morris and Garrity, 1993). However, there is some evidence that substantial differences may occur when species with contrasting root distributions are used. For example, in a study of water use during short drying cycles 2 months after planting rice and pigeonpea, Jena and Misra (1988) recorded mean extraction rates of 2.8, 7.9 and 4.2 mm day^{-1} for sole rice, sole pigeonpea and a rice–pigeonpea intercrop. The sole pigeonpea extracted water from depths below 1 m at a mean rate of 2.5 mm day^{-1}, whereas drainage through this plane occurred at a rate of 1.4 mm day^{-1} under sole rice in one of the two seasons reported. The relatively low daily water use of the intercrop, which contained the full density of pigeonpea, may have occurred because its growth was initially suppressed by competition from the rapidly growing rice. However, the observation that its yield was reduced by only 11% suggests that the intercropped pigeonpea was able to exploit residual moisture at depth within the profile following the rice harvest to effect an almost complete recovery. Thus, if season-long water use had been determined, the value for sole rice would probably have been much lower than in either sole

Table 4.3. Water use (WU) and water use ratio (e_w) of intercrops and changes in water use and water use ratio (ΔWU and Δe_w) relative to sole crops (after Morris and Garrity, 1993).

Reference	Location	Species	Concurrency* (%)	Crop proportion†	Yield measure	WU (mm)	ΔWU (%)	e_w (kg mm⁻¹)	Δe_w (%)
Rees (1986a, b)	Botswana	Sorghum + cowpea	70	93 : 22	Grain	125 to 198	−2 to 6	0.6 to 2.9	−25 to 0
Singh et al. (1988)	India	Pearl millet + cowpea	100	100 : 133 100 : 100	Dry matter	129 to 204 517 to 571	1 to 7 6 to 7	0.2 to 2.2 8.0 to 13.0	−42 to 3 −8 to 55
Morris et al. (1990)	Philippines	Sorghum + cowpea	71	50 : 50	Grain	177 to 271	2 to 5	12.0 to 12.6	25 to 53
Hulugalle and Lal (1986)	Nigeria	Maize + cowpea	100	50 : 50, 50 : 30	Grain	248 to 554	−5 to 0	1.6 to 4.2	4 to 99
Kushwaha and De (1987)	India	Mustard + chickpea	86	33 : 67, 67 : 33	Grain	199 to 222	−6 to 0	9.8 to 16.4	4 to 39
Suwanarit et al. (1984)	Thailand	Maize + mungbean	74	100 : 100	Dry matter	362	4	20.5	78
Singh (1985)	India	Pearl millet + cluster bean	Not indicated	100 : 30, 100 : 60	Grain	210 to 435	5 to 6	6.7 to 7.1	17 to 31
Natarajan and Willey (1980a, b)	India	Pigeonpea + sorghum	50	100 : 100	Grain	263 to 474 585(168)‡	0 to 5 15	7.3 to 7.6 8.3	22 to 33 38
Reddy and Willey (1981)	India	Pearl millet + groundnut	78	75 : 25	Dry matter	406	15	21.4	18

* Percentage of time that the earlier-harvested species was in the field with the later-harvested species; † crop proportions relative to respective sole crop plant densities; ‡ WU of the intercrop after sorghum harvest.

pigeonpea or the intercrop because of the drainage losses under rice and its shorter duration.

Although intercropping is frequently advocated for drought-prone environments, significant interactions between management factors (e.g. irrigation) and cropping methods (e.g. sole vs. intercropping) are unlikely when the growth of the component species is largely concurrent and water use by the sole and intercrops is similar. For example, planting density (Natarajan and Wiley, 1980a, b; Rees, 1986b; Kushwaha and De, 1987) and arrangement (alternative rows vs. mixtures of species within rows; Singh *et al.*, 1988) have both been shown to have little effect on water use. Similarly, while beneficial effects of irrigation have been documented, no significant interaction between irrigation and cropping practice has been reported (Singh, 1985; Mandal *et al.*, 1986; Singh *et al.*, 1988). A similar conclusion applies for intercrops containing similar species mixtures, particularly when the changes involve the minority component, as in mixtures of millet with soybean or pigeonpea (Singh *et al.*, 1988), or pigeonpea with cowpea or mungbean (Hegde and Saraf, 1979). In the latter study, seasonal water use was only slightly increased when the cowpea or mungbean was replaced by blackgram, which matured 15–18 days later. Thus, the effect of altering the subsidiary species is likely to be small unless the attributes of the component species differ appreciably, as in the study of pigeonpea and rice reported by Jena and Misra (1988).

There is limited evidence that the ratio of water use to available soil moisture may decline as water availability increases. For example, Singh *et al.* (1988) showed that this ratio decreased as irrigation increased, although 84% of the available water was used even in the wettest treatment; total water use was 47% greater in the wettest treatment than in the driest. A similar correlation appears to exist under rainfed conditions since Singh and Russell (1981) reported that maize–pigeonpea intercrops extracted 58% and 75% of the available soil moisture during the period of concurrent growth in seasons when available soil water was 868 mm and 573 mm; the absolute quantity of water extracted was nevertheless greater when rainfall was high (503 mm vs. 429 mm).

Water use ratio

The importance for crop production in water-limited environments of maximizing both the quantity of water available and the proportion that is used for transpiration has already been addressed. However, the 'efficiency' with which the transpired water is used in dry matter production is also a critical factor. Water use ratio (e_w in Equation 4.11), like the conversion coefficient for intercepted radiation, may be calculated on various time-scales ranging from instantaneous measurements of the ratio of net CO_2 fixation to transpirational water loss to season-long estimates based on measurements of dry matter accumulation and crop water use. Season-long estimates are invariably much lower than short term values derived from gas-exchange measurements under favourable conditions because of respiratory carbon losses, which may account for up to 50% of the photosynthate produced in herbaceous species, and the effects of adverse environmental (e.g. drought, nutrient deficiency) and biological factors (e.g. pests, diseases, weeds). Below-ground biomass is rarely considered when calculating e_w and above-ground biomass may decline as the crop approaches maturity due to senescence, leading to underestimation of the true value.

As with the conversion coefficient for light, e_w may be expressed as the ratio of biomass, yield or their glucose equivalents to total water use. The first two options are most commonly used because of their relative simplicity, but the latter is important when comparing species with differing chemical composition, such as grain and oil crops. The value of e_w should be calculated on the basis of *transpired* water rather than evapotranspiration (E_t) since water evaporated from the soil plays no part in dry matter production. Soil evaporation dominates E_t early in the season but transpiration becomes increasingly important as the canopy enlarges: soil evaporation may contribute 30–70% of the total seasonal water use depending on the rate of canopy development and the maximum leaf area attained. As indicated earlier, dry matter production is often linearly related to the quantity of water transpired (e.g. Tanner, 1981; Connor *et al.*, 1985), although the slope of the relationship, or water use ratio, varies according to environmental factors such as saturation deficit (D) which influence transpiration.

Water use ratios for tropical C4 cereals are generally a little more than double those for C3 species under comparable conditions, similar to that expected from the difference in $C_i : C_a$ ratios (Equation 4.11) typically found between C4 and C3 species. For example, reported e_w values for maize and sorghum were 2.2 times greater than in C3 species (Angus *et al.*, 1983), and similar differences in $e_w D$ between pearl millet and groundnut have been recorded (Squire, 1990). Under similar atmospheric conditions (mean saturation deficit of 2.0–2.5 kPa), season-long e_w values of 3.9 and 4.6 g kg^{-1} have been reported for millet (Squire *et al.*, 1984a) and 1.5, 1.9 and 2.0 g kg^{-1} for groundnut (Ong *et al.*, 1987; Mathews *et al.*, 1988; Azam-Ali *et al.*, 1989). However, C4 species do not always have higher water use ratios than C3 crops, since similar values have been reported for drought-tolerant C3 species such as cowpea and cotton and relatively drought-sensitive C4 maize and sorghum cultivars (Rees, 1986b). It is often difficult to make direct comparisons between e_w values obtained in different studies because of the impact of environmental factors such as saturation deficit which influence transpiration. For instance, Squire (1990) reported that season-long e_w values for groundnut decreased from 5.2 to 1.5 g kg^{-1}

as mean daytime saturation deficit increased from 1 to 2 kPa, in the absence of any change in other environmental factors. Similarly, e_w values for millet are reported to vary between 2.1 and 6.4 g kg^{-1} within the saturation deficit range 4.0–1.4 kPa (Azam-Ali et al., 1984; Squire et al., 1984a). According to Squire (1990), the influence of saturation deficit on e_w is one of the most important factors restricting productivity in dryland areas since dry matter production decreases by a factor of at least two as saturation deficit increases from 1 to 4 kPa.

Loomis and Connor (1992) calculated water use ratios corrected for differences in relative humidity experienced during the growing period by a wide range of species in an approach analogous to that proposed by Monteith (1986), who suggested that the product of water use ratio and saturation deficit ($e_w D$) is frequently conservative. Loomis and Connor (1992) demonstrated that the adjusted e_w values for C4 species were at least double those of a range of C3 cereals, legumes and non-leguminous species. Among the C3 species, the adjusted values were lower in legumes than in non-legumes, reflecting the higher metabolic costs incurred by the host in supporting the nitrogen fixing symbiont. An important point to emerge was that the corrected e_w values showed little benefit of an intensive breeding programme sustained over a 50-year period when calculated on the basis of total biomass, but did when computed against economic yield because of the improved partitioning of biomass to the grain.

The e_w values shown in Table 4.3 are difficult to compare directly because biomass production was assessed differently in different studies; some recorded total above-ground biomass, others only final grain yield. However, the Δe_w values (change in e_w relative to corresponding sole crops) clearly demonstrate that the water use ratio was greatly increased by intercropping in most studies. Many of the observed increases exceed the value of 18% reported for a millet–groundnut intercrop by Reddy and Willey (1981), which was attributed to an increase in the proportion of the available water used for transpiration relative to the sole crops. This may be a common feature of intercrops where the inclusion of a rapidly growing component, usually a cereal, leads to more rapid canopy development and a consequent reduction in soil evaporation. The detrimental effect of intercropping reported by Rees (1986a, b) occurred under harsh conditions of low rainfall and was attributed to extensive plant mortality induced by excessive early extraction of water from a limited supply.

The effects of intercropping on Δe_w summarized in Table 4.3 varied greatly between studies, even when identical species mixtures were used. For example, experiments with sorghum–cowpea intercrops have provided very different Δe_w ranges of -42% to $+3\%$ (Rees, 1986a, b) and 25% to 53% (Morris et al., 1990); in both cases the calculations were based on grain yields. A major difference between the two studies was that seasonal evapotranspiration was 700 mm in the former and 420 mm in the latter.

The effects of factors such as crop proportion, planting density and row spacing on water use ratio have proved inconsistent. Natarajan and Willey (1980a, b) demonstrated that variation in the planting density of either pigeonpea or sorghum affected the e_w values for intercrops by altering pigeonpea yields. Similarly, e_w was positively correlated with the planting density of chickpea in both sole and intercrops containing 7.5–30 chickpea plants m^{-2}, but was unaffected by the planting density of mustard (Kushwaha and De, 1987). The effects of crop proportion (Natarajan and Willey, 1986; Kushwaha and De, 1987) and intimacy (Singh, 1985; Hulugalle and Lal, 1986; Singh *et al.*, 1988) are generally small.

There is some evidence that the generally positive influence of intercropping on water use ratio may depend on water availability. For example, Stewart (1983) obtained basically linear correlations between grain yield and the quantity of water applied when sole and intercrops of maize and bean (*Phaseolus vulgaris*) growing under semi-arid conditions were irrigated. The response of each species was smaller when intercropped than in the sole crop (14.6 vs. 16.3 and 2.7 vs. 10.4 kg grain yield mm^{-1} water applied for maize and bean respectively). Re-analysis of the data to establish the effects of intercropping on e_w relative to the sole crops produced values of -49, -1, $+17$ and $+30\%$ for irrigation supplements of 235, 335, 435 and 535 mm (Morris and Garrity, 1993). These data indicate that the intercrops showed a greater response to irrigation than the sole crops, possibly because they depleted available soil moisture more rapidly when only a limited quantity of irrigation was applied, with the result that insufficient water remained to support reproductive growth. Natarajan and Willey (1986) also used line-sprinklers to supplement rainfall and observed that grain yields were essentially linearly related to the quantity of water received when sorghum, millet and groundnut were grown either as sole crops or replacement intercrops; the response to irrigation was greatest in sorghum and least in pearl millet. Grain yield per millimetre of applied water (rainfall + irrigation) approximately doubled as irrigation was increased from 297 to 584 mm in sorghum–groundnut (1 : 2), sorghum–millet (1 : 1) and millet–groundnut (1 : 1) intercrops. The value of e_w was invariably higher in the intercrops than in the sole crops; in the sorghum–groundnut intercrop e_w increased relative to the sole crops as water availability decreased, but in the two intercrops involving millet the changes were smaller and less consistently related to water availability.

None of the studies summarized in Table 4.3 provides information concerning the division of resources between the component species of intercrops, or whether the water use and e_w values of either component were altered relative to the sole crop, possibly because of microclimatic modification. The mean values shown in Table 4.3 may therefore obscure substantial changes caused by interactions between the crop components. To elucidate the nature and extent of such changes requires separate measurements of the capture and utilization of light and water by each component. A major obstacle to such progress in both intercropping and agroforestry systems has been the technical difficulties involved in making the necessary measurements. These are discussed in greater detail below.

The preceding discussion has demonstrated that the beneficial effects of intercropping on water economy generally originate from improvements in e_w rather than seasonal water use. Several factors may be involved. First, intercropping may increase the proportion of available soil moisture used for transpiration because the more rapid canopy development reduces soil evaporation. However, as indicated earlier, enhanced canopy development during early vegetative growth may also adversely affect long term e_w values and productivity because of premature depletion of available moisture. It is therefore essential to strike an appropriate balance between these opposing influences on the seasonal water budget. Second, a more vigorous crop component with an inherently high e_w value may capture a greater proportion of the available water, thereby increasing its contribution to the final yield and the overall water use ratio of the system. In many intercrops, the dominant and subordinate species are respectively C4 and C3 crops with relatively high and low e_w values. Third, intercropping may bring microclimatic benefits to at least one of the species involved. Equations 4.9 and 4.11 demonstrate the importance of the leaf to air vapour pressure gradient ($v_i - v_a$) in determining transpiration and hence e_w. The windbreak and shading effect of the taller component of intercropping and agroforestry systems tends to reduce air temperature and windspeed and increase the atmospheric humidity experienced by the shorter component, thereby decreasing evaporative demand (see also Chapter 5). The shorter component of intercrops, though not necessarily of agroforestry systems, is often a C3 species with a low light-saturated photosynthetic rate, and so the reduced PAR fluxes caused by partial shading may not reduce assimilation. This does not apply to C4 species with their much greater light-saturated photosynthetic rates, in which even minimal shading is likely to reduce assimilation. Thus, the e_w values of understorey C3 species may be improved by reductions in transpiration in the absence of effects on photosynthesis. These comments are much less applicable to agroforestry systems because the dominant overstorey species is invariably C3, with its lower light-saturated photosynthetic rate and e_w, the beneficial microclimatic changes experienced by associated crops may be largely eliminated by the wider tree spacing, and the understorey component is often a C4 cereal which does not respond favourably to shading.

Water use – agroforestry systems

The preceding sections have clearly demonstrated that seasonal water use is often not increased by intercropping unless the component species differ appreciably. In annual systems, resource capture and biomass production are confined to the growing period of the crops involved, and the land lies fallow and unproductive for much of the year, particularly in areas of uni-modal rainfall. However, appreciable residual water frequently remains in the soil after harvest, and significant off-season rainfall may go unused in annual cropping systems. For example, work at ICRISAT Centre, India showed that substantial quantities of available water were left in the 45–90 cm horizons when short-duration sole sorghum was harvested in Oct-ober, and that available water remained even after the longer duration sole pigeonpea was harvested some 3 months later (Ong *et al.*, 1992). At Hyderabad, 20% or 152 mm of the annual rainfall is received outside the normal cropping season. Agroforestry offers considerable potential for exploiting these residual water supplies in the surface horizons, deeper re-serves beyond the rooting depth of annual crops and the off-season rain-fall. Additional benefits may accrue if the proportion of annual rainfall lost by runoff, deep percolation and soil evaporation can be reduced. The scope for improvement in this area is considerable. For example, traditional sorghum–pigeonpea intercrops grown on the alfisols of the Deccan plateau in India have been reported to use only 41% of the annual rainfall, while the remainder is lost as runoff (26%) or deep drainage (33%) (Ong *et al.*, 1992). Similarly, Cooper *et al.* (1983) and Wallace (1991) have reported that soil evaporation may account for 30–60% of the rainfall in semi-arid areas of the Middle East and West Africa. Thus any reduction in soil evaporation, runoff or deep drainage resulting respectively from increased shading and reduced soil surface temperature, the physical barriers pre-sented by the tree component (e.g. contour-planted hedgerows on hill-slopes) or increased abstraction during the rainy season would increase the proportion of the annual rainfall retained for transpiration and hence biomass accumulation. The close correlation between dry matter accumu-lation and the quantity of water transpired is well established, although the underlying physical and physiological principles have only been established relatively recently. A possible disadvantage is that interception losses, i.e. the evaporation of rainfall intercepted by the canopy, may range from 10 to 30% for agroforestry systems (Ong and Black, 1994).

The hypothesis that annual crops cannot fully utilize available water and that agroforestry systems may improve productivity by using a larger proportion of the annual rainfall has gained support in recent years. For example, Ong *et al.* (1992) reported that hedgerow plantings of sole *Leucaena* extracted more of the available soil moisture than either sole crops or intercrops of sorghum and pigeonpea. Widely spaced alley crops (4.4 m between hedges) extracted even more water than sole *Leucaena*, indicating that the agroforestry systems were more effective in utilizing available moisture than annual crops. However, in this and other experiments in the semi-arid tropics, severe competition between the extremely vigorous

Leucaena and associated annual crops reduced crop yields by 50–80% (Rao *et al.*, 1990; Corlett *et al.*, 1992b). Subsequent research at ICRISAT and ICRAF has concentrated on water use in agroforestry systems which are less competitive with crops, for example dispersed plantings of deeper rooting or less vigorous trees than *Leucaena* which are both efficient in their water use and compatible with dryland crops. Several systems involving species such as perennial pigeonpea, *Faidherbia albida* and *Grevillea robusta* show promise. For instance, measurements of transpiration using the heat balance approach (see below) showed that the total annual transpiration of perennial pigeonpea grown in an agroforestry system with groundnut was 887 mm or 84% of the annual rainfall, double the water use of the most productive intercrop system (Ong and Black, 1994). Almost half (47%) of the total transpiration occurred between January and June, when only 211 mm of rain fell, indicating that 205 mm was extracted from soil reserves. In contrast, an excess of 420 mm received during the rainy season between July and November was lost as runoff and deep drainage. Separate measurements in a similar system indicated that sole pigeonpea was much less effective in preventing runoff (up to 30% of rainfall during high intensity storms) than either sole groundnut or the pigeonpea–groundnut mixture (runoff ≤ 5%).

These results emphasize the potential of agroforestry for increasing rainfall utilization and productivity. However, the influence of the planting density and arrangement of trees on community water use and the productivity of its components is complex, not least because the optimum arrangement varies depending on climatic and soil conditions and the species combination involved, and changes with time as the trees grow larger and increasingly competitive. Thus an arrangement which increases overall water use and productivity without greatly reducing crop yield in the first year or two may cause massive yield reductions in subsequent years, implying that a reduction in the tree population may be advantageous. Decisions on optimum planting arrangements will hinge on whether the key objective is to maximize productivity, irrespective of the effect on crop yield (e.g. by increasing fodder production during the dry season when demand is most acute), or to maintain a full yield from the staple or economically important crop even though a higher tree population would increase the overall intercropping advantage.

Measurement and Interpretation of Light Interception and Use in Agroforestry Systems

Despite the considerable potential benefits of complementarity in agro-forestry systems, few quantitative attempts have been made to analyse their productivity in terms of resource capture and use. The vast majority of the existing knowledge concerning resource partitioning in mixed cropping systems comes from intercropping, and is often based on circumstantial evidence or empirical yield studies in which the utilization of only a single resource such as light has been measured. Such studies have produced conflicting evidence concerning the source of any intercrop advantage for two major reasons: differing criteria have been used to assess the performance of the intercrops; and separate quantification of resource capture and use by the components of mixed systems is physically, technically and intellectually extremely challenging. A fundamental understanding of how agroforestry systems utilize available resources is nevertheless crucial for establishing species combinations, planting arrangements, tree densities and management strategies suitable for different locations.

In monocultures consisting of uniform plants, the experimental approaches and instrumentation required to quantify light interception, water use and resource use efficiency are now well established and relatively straightforward in their application. The number of sampling positions required is confined by the high degree of spatial homogeneity within the stand. However, the problems involved in monitoring sufficient, appropriately sited locations to characterize the spatial and temporal variation in mixed communities are much greater. The scale of the spatial heterogeneity is relatively modest and capable of resolution in annual row intercrops (e.g. Marshall and Willey, 1983; Azam-Ali *et al.*, 1990), but is much greater in both horizontal and vertical planes in agroforestry systems, particularly those involving large overstorey trees. An additional dimension adding to the complexity of resource use studies in agroforestry systems is the extended timescale and constantly changing relationship between the tree and crop components. In annual intercrops the measurement period is normally confined to a period of 100–150 days, whereas the tree component of agroforestry systems is present and usually actively using water and light throughout the year. Tree size and the extent of their interaction with associated crops may alter rapidly during the cropping cycle and from year to year; for example, in a current experiment at ICRAF's Machakos field station in Kenya, *Grevillea robusta* trees being grown as poles have reached a height of 4–5 m in under 3 years. Thus, the nature of the interactions and the partitioning of resources between trees and crops are constantly changing, with the result that rigorous and highly intensive studies are required to obtain a sound quantitative understanding of the processes involved.

The complexity of the interactions depends on the design, composition and maturity of the agroforestry system involved. In the first cropping season after planting the trees, the system bears comparison to conventional intercropping systems because the trees have not yet achieved their full competitive vigour and stature. In more mature systems, the nature and extent of the interactions will vary according to planting arrangement. In alley-cropping systems, a concave gradient of decreasing competition between the tree and crop component may be observed towards the centre of the alley, which can be characterized, at least in principle, by measuring each of the relevant variables along the transect between adjacent hedges. The situation is more complex with dispersed arrangements, particularly when the trees are not planted on a regular grid pattern. For example, in the dispersed pigeonpea treatment shown in Fig. 4.10b, differing gradients of competition extend in three directions between the trees forming the corners of each grid; competitive interactions will be least in the centre of the grid and most intense close to the trees.

Vertical gradients in resource sharing may also be important, depending on the relative heights of the tree and crop components. The latter is inevitably shorter and has a much smaller leaf area and rooting system than the tree component early in the growing season, and is therefore at a competitive disadvantage relative to the trees with their established root system and larger canopy. During the early stages of crop growth, it may be possible to treat the tree and crop canopies as vertically discrete strata intercepting light, and a similar situation may prevail throughout the season in systems with tall overstorey trees. In such cases, separate quantification of light interception and conversion by each component is relatively straightforward using arrays of sensors located above and below the tree and crop canopies. These arrays should ideally include instruments to determine reflection from the soil and canopy surfaces, although these are often omitted for technical reasons. This omission introduces a degree of uncertainty into the estimates of intercepted radiation and the derived values for conversion coefficients. However, in alley-cropping systems which are kept relatively short by pruning, the canopies of the component species increasingly intermingle as the crop grows taller and/or the side branches of the hedgerow extend, rendering separate assessment of light interception extremely difficult. This is inevitably the case where the tree and crop components are of a similar stature and have a high degree of intimacy.

An additional complication is that shading affects the radiation environment experienced by understorey crops by altering not only the quantity of radiation received but also its spectral composition. Previous studies of the partitioning of light in mixed communities have often relied on measurements of total shortwave radiation and assumed that the photosynthetically active radiation (PAR) forms a constant fraction of the shortwave spectrum (usually taken as 45–50%). However, this assumption is less valid for heavily shaded understorey species whose main sources of light are diffuse radiation, which has penetrated unintercepted through gaps in the overstorey canopy, and transmitted radiation which has previously been intercepted and transmitted, perhaps several times, by the foliage of the tree canopy. The former is relatively rich in PAR and so is used with apparently greater efficiency for dry matter production (Sinclair *et al.*, 1992), while the latter may be severely depleted in the photosynthetically most effective red and blue wavelengths. Thus, detailed measurements of PAR interception are essential for a full understanding of the influence of the dominant component in any mixed community on the light use, growth and productivity of subordinate species.

Proper characterization of the impact of the extensive spatial and temporal variation in canopy architecture on resource capture in agroforestry systems has major implications in terms of capital outlay for instrumentation, technical support staff and the computing facilities necessary to process extensive databases. For this reason, even the most heavily instrumented experiments with good technical support must compromise between what may be experimentally ideal and what is feasible within the available technical and human resources. Two field experiments conducted at ICRISAT Centre, India will be used as case studies to illustrate some of the practical and theoretical difficulties involved in quantifying resource capture in agroforestry systems in relation to the growth and productivity of component species. Some of the data obtained will be used to illustrate the influence of agroforestry on resource capture and utilization.

Case Study 1 – Leucaena–millet alley cropping

Pearl millet (*Pennisetum typhoides*) and *Leucaena* (*Leucaena leucocephala*) were grown for 2 years as alley crops and sole crops on a medium-shallow alfisol. The alleys were 2.8 or 3.3 m wide and the hedges contained two rows of *Leucaena* 50 cm apart with an inter-tree spacing of 25 cm. The alleys contained five or six rows of millet depending on the treatment involved, 47 cm apart with a spacing of 15 cm between plants. Vertical polythene barriers were installed to a depth of 50 cm in one treatment to separate the root systems and assist in distinguishing between above- and below-ground interactions. The millet was grown during the rainy season between July and mid-September. Total annual rainfall was below average in 1986 but above average in 1987 at 712 mm and 879 mm respectively, of which 50% and 38% was received during the millet growing season.

The *Leucaena* was 11 months old and well established when the first millet crop was sown. The trees were lopped to a height of 65–70 cm immediately before sowing and after harvesting the millet, and again 29 days after sowing in 1986. In 1987, only the side branches were lopped during the rainy season to examine the effects on millet of increased shading. The growth and development of millet were analysed weekly, while dry matter production in *Leucaena* was assessed from the loppings and stem dry weight increments for designated sampling trees. Windspeed, saturation deficit, soil and leaf temperatures, light interception and soil water content were monitored routinely throughout the experiment. Full details are given by Corlett (1989) and Corlett *et al.* (1992a, b).

Figure 4.7 illustrates the differing strategies regarding instrument placement that were adopted during the two rainy seasons examined. The objectives in 1986 were to compare the microclimatic conditions experienced by sole and alley-cropped millet and establish the extent of the variation in light interception and temperature across the alleys. Measurements were concentrated on two of the three replicates to keep the number of instruments and datalogger channels required within manageable limits. In 1987, temperature and light interception were measured in all three replicates to allow statistical examination of the relationships between microclimate and growth and development; a reduced number of instruments was used in each replicate to remain within the total number of datalogger channels (128) available. In both years, saturation deficit and windspeed were measured in a single replicate of each sole stand and the alley crop. These differing approaches reflect the difficult decisions that must be made regarding the numbers of treatments, replicates and sites that can be examined; the choice is often between studying a limited number of treatments in detail or a larger number less comprehensively. In this study we adopted the former approach wherever possible to obtain a detailed mechanistic understanding of the interactions involved.

Solar radiation was measured above and below the tree and millet canopies using tube solarimeters (Green and Deuchar, 1985), which were cleaned regularly and calibrated against a Kipp solarimeter before and after each season. More detailed guidance on the installation and use of tube solarimeters is given by Monteith (1993). Incident radiation was measured using a pair of solarimeters located at a height of 3 m outside the experimental area. In 1986, transmitted radiation was measured using seven solarimeters per plot in sole *Leucaena* (SL) and the alley crop (LM) and one per plot in sole millet (SM). One tube was placed at an angle of 27° under each millet and *Leucaena* row in LM so that it spanned half of the inter-row area on either side (Fig. 4.7); this approach enabled row-by-row values for light interception to be determined. Tubes were placed at the same locations in SL and the single tube per replicate in SM was again oriented at an angle of 27°. Thermocouples were used to measure soil temperature at a depth of 5 cm and the temperature of the youngest fully expanded leaf at locations close to the tube solarimeters for each of the five millet rows.

In 1987, the tubes were oriented at 90° to the crop rows so that light interception could be monitored across the entire alley in LM5 and SL using only three tubes, as opposed to the seven required in 1986 (Fig. 4.7). Two additional tubes located above the millet but below the unpruned *Leucaena* canopy in LM5 were used to measure radiation incident on the

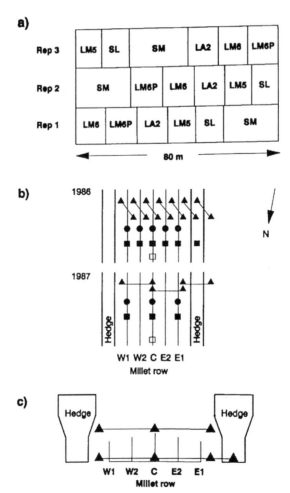

Fig. 4.7. Experimental design in a *Leucaena*–millet alley crop, ICRISAT. (a) SM and SL indicate sole millet and *Leucaena* plots, LM5 and LM6 represent alley crops containing five or six rows of millet and LM6P denotes an alley crop containing six millet rows with vertical polythene barriers separating the root systems of *Leucaena* and millet. Plots marked LA2 were not used. (b) Instrumentation in 1986 and 1987; tube solarimeter (▲——▲); two leaf thermocouples (●); soil thermocouples at 5 cm (■) and 20 cm (▲) depths; anemometer and psychrometer mast (□). (c) Solarimeter placement in 1987 (Corlett *et al.*, 1992a).

millet; these were raised as the millet grew taller. A single tube was again placed under the millet in each replicate of SM. Soil and leaf temperatures were also monitored at fewer sites within each treatment than in 1986 (Fig. 4.7). The reduced number of instruments in LM and SL naturally reduced the resolution provided by the row-by-row analysis adopted in 1986, but nevertheless permitted light interception by the tree and crop components to be separated. A major benefit of the reduced sampling intensity in LM and SL was that it released instruments to monitor a third replicate per treatment, thereby increasing the statistical certainty of the data obtained.

Two approaches were used to estimate light interception by the tree and crop components of the alley system. *Method 1* assumed that the conversion coefficient for intercepted radiation was conservative in *Leucaena* and that the value calculated for SL could be extrapolated to the alley system. Interception by the alley millet could therefore be calculated as the difference in interception between the alley and sole *Leucaena* systems. This approach assumes that: (i) the greater interception under the hedges in LM than in SL was attributable to interception by millet; (ii) light interception measured in the alleys of LM represented the sum of interception by *Leucaena* and millet; and (iii) the conversion coefficient for *Leucaena* was conservative. The first two assumptions are justifiable provided light interception by weeds is unimportant. The third was not tested but appears reasonable because the trees were all of the same cultivar, were managed in the same way and experienced similar environmental conditions. The data for light interception by the alley millet and *Leucaena* shown in Table 4.1 were calculated in this way.

More sophisticated approaches were also adopted in both years (*Method 2*). In 1986 solarimeters were positioned not above the millet but below the *Leucaena* canopy, and so interception by millet was calculated on a row-by-row basis by assuming that the differences between LM and SL in interception at specific locations across the alley were attributable to the millet occupying that row position in LM. Daily estimates obtained in this way were accumulated to obtain seasonal interception by millet at each row position. This approach assumes: (i) that the hedges in SL and LM were identical in shape, leaf area and extinction coefficient; and (ii) that the five solarimeters in the alleys of LM each measured interception by the millet row under which they were located, together with any interception by the *Leucaena*. The first assumption is supported by field observations that the hedges were similar in shape and above-ground dry matter production in 1986; they were therefore assumed to intercept equal quantities of light. The second assumption is valid when solar elevation is high and the canopies of the individual millet rows are discrete; when solar angle is small and the canopies intermingle, the resultant errors will only be important when the adjacent rows differ significantly in leaf area index. The estimates of light interception by individual millet rows are therefore most reliable for the central three rows and least reliable for the rows adjacent to the hedges.

In 1987, the placement of additional solarimeters above the millet canopy permitted a more direct approach. In making the calculations it was assumed that there was: (i) no intermingling of the *Leucaena* and millet canopies; and (ii) no net gain or loss of solar radiation across the vertical plane separating the two species below the surface of the millet canopy. The first assumption is reasonable since the side branches of *Leucaena* were removed before planting the millet, most regrowth branches developed at the top of the lopped stem and the side branches were lopped again at 43 days after sowing (DAS). The errors introduced by the second assumption are difficult to assess but, as transmission through the *Leucaena* canopy was generally lower than that through the millet canopy, the *Leucaena* would have made a net contribution to light interception between the above-millet and below-millet solarimeters, causing interception by millet to be overestimated. This error would have increased in size during the season but is unlikely to have been larger than the residual error caused by crop variability. Based on these assumptions, daily interception (y) by millet in LM was calculated as:

$$y = S\,(f_b - f_a) \tag{4.14}$$

where S denotes the incident solar radiation, f_b is $1 - t_b$ and f_a is $1 - t_a$, where t_a and t_b represent fractional transmitted solar radiation measured by the solarimeters above and below the millet canopy. As in 1986, the daily values were accumulated to determine total seasonal interception.

The values for seasonal interception by the alley millet calculated by the different approaches show good agreement. In 1986, estimates of 290 and 260 MJ m^{-2} were obtained using Methods 1 and 2, a difference which arose mainly because measurements from solarimeters located under the *Leucaena* were included in Method 1 but not in Method 2. The corresponding values in 1987 were 148 and 128 MJ m^{-2}. These differences in estimated interception would introduce uncertainties of 12–15% into the calculated conversion coefficients.

Figure 4.8a shows fractional light interception in the sole stands over a 10-month period. The sole *Leucaena* achieved a maximum f value of 0.6, as compared with 0.7 in pearl millet, but because the *Leucaena* retained some leaves during the dry season, it intercepted more than twice as much radiation as sole millet over the measurement period (Table 4.1). Figure 4.8b shows the total interception by the agroforestry system as measured, and as predicted on the basis of interception by the sole crops. These calculations indicate that interception by the alley-cropped millet was only 40% of that in sole millet, as opposed to the 'expected' value of 70% based on its population (5 rows millet : 2 rows *Leucaena*). Nevertheless, the alley system greatly increased total interception (810 MJ m^{-2}) and produced 2.4–3.1 t ha^{-1} more dry matter than the sole stands during the 1986 rainy season. Light interception and dry matter accumulation were again greatest in the alley system in 1987, although the advantage over sole *Leucaena* was much smaller. Both variables were much lower in the alley millet than in the sole crop, reflecting the severe shading imposed by the increasingly vigorous *Leucaena*; interception by the alley millet contributed only 14% of the total for the alley system, as opposed to 40% in 1986.

Conversion coefficients (e) were generally low during the 1986 rainy season, presumably because of the severe drought resulting from rainfall 36% below the long term average. For example, a value of 1.6 g MJ^{-1} was obtained when the same millet cultivar (BK 560) was grown during the rainy season at ICRISAT (ODA, 1987) and the maximum recorded value for millet of 2.5 g MJ^{-1} was maintained throughout the life of the glass-

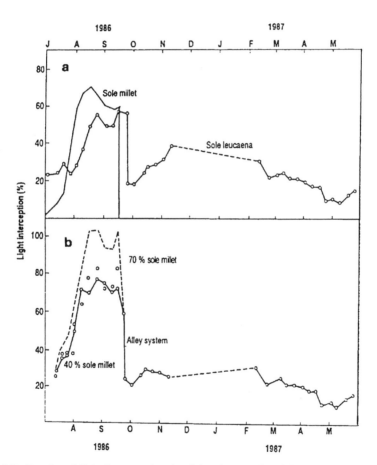

Fig. 4.8. Fractional light interception by (a) sole stands of *Leucaena* and pearl millet and (b) the alley system (——) and estimates based on interception by its components, assuming that millet intercepted either 70% (-----) or 40% (○) of the sole millet value. The broken lines between November and February represent periods when no measurements were made. See text for details (Monteith *et al.*, 1991).

house crops examined by Squire *et al.* (1984b). The conversion coefficient for alley-cropped millet was 32% greater than in the sole crop, possibly because the reduced light-saturating photon flux density for photosynthesis associated with drought was attained less frequently under conditions of partial shade. However, the overall *e* value for the alley system was only 9% greater than in sole millet. During the rainy season the *e* value for *Leucaena* was lower than expected for a C3 species in the absence of stress, but was surprisingly close to that for sole millet. There was a sharp decrease in *e* in *Leucaena* during the dry season to $0.12-0.15$ g MJ^{-1} at a time when evaporative demand was greatest $(D \approx 4.0$ kPa$)$ and leaf senescence was extensive. The *e* value for sole millet was much greater in 1987 than in 1986, whereas the converse applied for the alley millet, in which *e* was 38% lower than in the sole stand. There was a corresponding decrease in the overall value for the alley system because the conversion efficiency of *Leucaena* was only slightly greater than in 1986.

The agroforestry system was clearly more productive than either of the sole crops in 1986, primarily because of its greater light interception since the higher conversion efficiency of the alley millet was insufficient to offset its greatly reduced light interception. This applied to an even greater extent in 1987 when the *Leucaena* captured 85% of the intercepted radiation in the alley crop. Because of the shift in light interception towards the photosynthetically less efficient C3 *Leucaena*, total yield from the alley system was only 3% higher than in 1986 despite a 15% increase in intercepted radiation. The yield advantage of the alley crop over sole millet was similar in both years, although the overall totals for the former (7.1 vs. 7.3 t ha^{-1}) mask an increase of 2.4 t ha^{-1} in *Leucaena* yield and an almost identical decrease of 2.2 t ha^{-1} for millet. The drastic reduction in millet yield resulted from the combination of greatly reduced light interception and conversion efficiency. This study provides an excellent example of the critical need to adopt appropriate tree–crop combinations and management practices in order to avoid severe competitive interactions and losses of crop yield.

Light interception and growth in the alley millet exhibited marked positional variation across the alleys. For instance, in 1986 the millet rows adjacent to the hedges intercepted $100-150$ MJ m^{-2} less radiation than the three central rows. As there was no systematic variation in *e*, this trend in light interception was reflected by a similar convex pattern of dry matter production; grain and fodder yields at final harvest were up to four times greater in the centre of the alley than in the rows adjacent to the hedges. In both years interception and growth were greater to the west of the hedges than to the east because of the asymmetric shape and shading effect of the hedges. A changing balance between above- and below-ground competition was also apparent since the reduction of growth in alley-cropped millet was decreased by the presence of root barriers in 1986, when the *Leucaena* was regularly pruned, but not in 1987, when it was allowed to grow unchecked, thereby increasing shading.

There was also clear evidence that the conversion coefficient for millet was much greater during vegetative growth than during the post-anthesis stage, when extensive senescence was occurring. During the initial 'high efficiency' phase, which extended until about 60–65 DAS, the e values for sole and alley millet were 1.80 and 1.72 g MJ^{-1} in 1986 and 1.46 and 1.09 g MJ^{-1} in 1987; the corresponding values for the later growth stages were respectively 0.13 and -0.04 g MJ^{-1} and 0.53 and 0.33 g MJ^{-1}. Thus the conversion coefficients for millet were initially greater, but subsequently lower in 1986 than in 1987. The shape of the relationship between dry matter accumulation and light interception therefore differed markedly between the two years (Corlett, 1989).

PAR interception

As noted earlier, the radiation reaching heavily shaded understorey crops may be severely depleted in photosynthetically active wavelengths. Estimates of e calculated using intercepted total solar radiation would therefore be expected to be lower in the alley millet than in SM because the radiation contained less PAR. Marshall and Willey (1983) suggested that the fractional interception of PAR (f_p) for a wide range of canopy types may be calculated as:

$$\ln (1 - f_p) = 1.4 \ln (1 - f) \qquad (4.15)$$

where f is the fractional interception of solar radiation. Their analysis suggests that the PAR content of the radiation reaching the alley millet in 1986 was 44.8%, as compared with 50% in the sole stand. The conversion coefficient for the alley millet was 45% higher than in sole millet when expressed on the basis of intercepted PAR (1.87 vs. 0.81 g MJ^{-1}, but only 30% higher when based on total solar radiation (1.08 vs. 0.81 g MJ^{-1}). Monteith (1993) described an alternative method for calculating PAR interception using tube solarimeters.

Periodic instantaneous measurements of light interception were also made across individual alleys in 1987 using a device known as the 'mouse' (Mathews *et al.*, 1987). This instrument consists of a quantum sensor located inside a metal tube perforated at intervals along its upper surface. The quantum sensor is drawn through the tube and, when referenced against another sensor above the canopy, the spot measurements under each aperture provide a linear transect of PAR interception within the canopy segment being sampled. This approach not only enables diurnal changes in fractional interception to be assessed, as opposed to the integrative approach provided by tube solarimeters, but also permits comparison of PAR interception by individual canopy components.

The more direct approach provided by the mouse is illustrated by Fig. 4.9, which shows transects of PAR interception above and below the alley-cropped millet in LM at 52 DAS; the shaded areas represent interception by the millet. PAR interception (f_p) by millet was greatest in the central rows and least in the heavily shaded rows to the west of the hedges (E1 and E2) throughout the day; f_p was also low in row W1 in the morning. For measurements made around midday, the transects predominantly reflect interception by millet, but in the morning and afternoon, when solar elevation is low, direct beam radiation would not have passed vertically between the above- and below-millet transects. Measurements made at these times are therefore unsuitable for row-by-row studies of interception by millet and so only transects taken at midday were used for such comparisons. The data obtained indicate that on 52 DAS f_p for millet was effectively zero for row E1, 16% for row E2 and 27–34% for the remaining three rows, again emphasizing the strong gradient in light interception across the alleys.

Case study 2 – perennial pigeonpea–groundnut systems

Groundnut (*Arachis hypogaea*) and perennial pigeonpea (*Cajanus cajan*) were grown together in different planting arrangements and also as sole crops in 1989 and 1990 on a medium-shallow alfisol. The pigeonpea was either line-planted (5.4 m alleys) or grown in a dispersed arrangement (1.8 × 1.2 m) to provide identical populations of 0.5 plants m^{-2}. The sole pigeonpea contained 9 plants m^{-2} in 1989, but this was reduced to 0.5 plants m^{-2} in 1990 because the initial population was believed to be supra-optimal for the second year; this was expected to cause extensive mortality, increasing the variability within the treatment to unacceptable levels. The reduction also established identical pigeonpea populations in all treatments. Groundnut was planted at 10-cm intervals in rows 30 cm apart to provide a population of 33 plants m^{-2}. The line-planting treatment was similar to the alley-cropping design widely practised in India for these species.

The pigeonpea was planted on 3 July 1989 and was first harvested for grain and fodder in January 1990. The pigeonpea was again cut for fodder to a height of 0.5 m in May and July 1990 and the final harvest was made in January 1991. The groundnut was grown during the rainy season; total annual rainfall was 1045 and 831 mm in 1989 and 1990, of which 937 and 528 mm was received during the rainy season. Groundnut growth and development were analysed at 10-day intervals in both years. As in the previous case study, light interception was measured throughout the season and saturation deficit, leaf and soil temperature and windspeed above the canopy were recorded as hourly means. Records of rainfall, soil moisture content, runoff and soil evaporation were maintained in order to construct

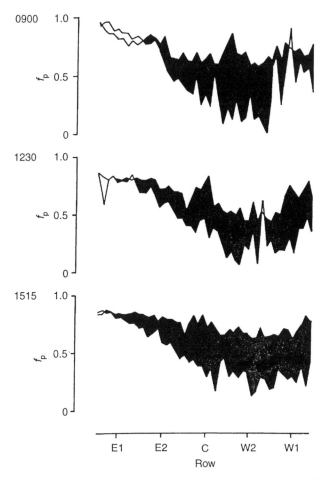

Fig. 4.9. Fractional interception of photosynthetically active radiation (f_p) above and below the millet canopy in a *Leucaena*–millet intercrop at 52 days after sowing, rainy season 1987. Shaded areas represent interception by millet (Corlett, 1989).

a soil water balance. Transpiration by pigeonpea was routinely monitored using the sap flux approach (see below).

Partitioning light interception within the line-planted system was relatively straightforward using an approach similar to that described previously. The groundnut canopy never exceeded 30 cm in height, leaving a discrete gap beneath the pigeonpea canopy, which began at a height of approximately 50 cm. Fractional interception by each component was determined using tube solarimeters positioned above and below the groundnut canopy and referenced against the incident radiation recorded by tubes located at a height of 2 m; instruments were installed on both sides of the pigeonpea rows. As the pigeonpea canopy never extended beyond tube A1 (Fig. 4.10a), interception by pigeonpea was taken as the difference between the incident radiation and that recorded by tube A1; the values

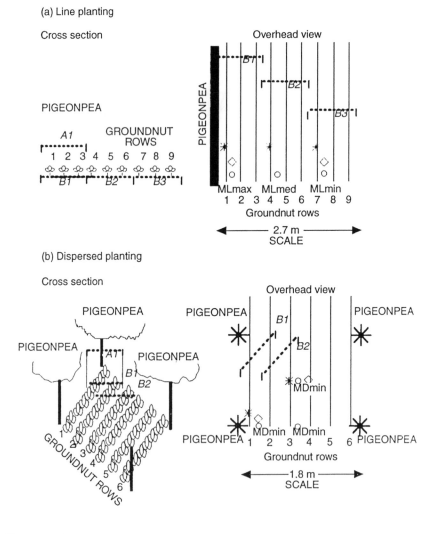

Fig. 4.10. Location of instruments in (a) the line and (b) dispersed plantings of pigeonpea showing solarimeters (-----), neutron probe access tubes (○), thermocouples (soil and leaf temperature) (∗), and psychrometers (◇) (Marshall, 1995).

obtained were divided by three to convert them to a system area basis. The tubes beneath the groundnut (B1, B2 and B3) spanned nine groundnut rows, or half of the distance between adjacent pigeonpea rows; the unmonitored half of the alley was assumed to be a mirror image of the interception pattern recorded by tubes A1 and B1–B3.

Solarimeter placement was more difficult in the dispersed treatment. Because of the limited number of instruments available, it was impossible to examine both the spatial variation in light interception with distance from the trees and the relationship between dry matter production and intercepted radiation for each component of the system. Unlike the line planting, the tube positioning required to address these issues differed. Priority was given to establishing the relationship between light interception and dry matter production as this was essential for comparisons with other systems. Sampling positions 1 and 2 (Fig. 4.10b) were chosen as they covered one-quarter of the area between pigeonpea trees and spanned the full range of distance between the trees forming the corner of each grid.

Pigeonpea interception was again taken as the difference between incident radiation and that recorded by solarimeters located below the pigeonpea canopy, as in the line-planted pigeonpea and *Leucaena* systems described previously. Ideally, solarimeters would have been placed above and below the groundnut canopy at positions 1 and 2 (Fig. 4.10b), but the lack of instruments made it necessary to omit position A2 on the assumption that, as in the line planting, the pigeonpea canopy would not extend beyond position A1 for most of the growing period. However, the pigeonpea canopy extended rapidly during the rainy season in 1990, shading the groundnut at position 2 for some time before harvest; the values recorded by tubes B1 and B2 were identical following groundnut harvest, showing that both sites were equally shaded at this time. The omission of measurements at position A2 therefore created difficulties in partitioning light interception between pigeonpea and groundnut, as is discussed below.

Values for total intercepted solar radiation, above-ground biomass and conversion coefficients are shown in Table 4.4 for all treatments. The biomass data for pigeonpea in 1989 do not include dry matter accumulated in the stems below the cutting height of 50 cm and so they and the derived conversion coefficients are underestimates of the true values. This contrasts with the simpler stem geometry of *Leucaena*, which permitted dry matter accumulation to be estimated non-destructively from measurements of stem dimensions. The estimates of light interception by groundnut shown in Table 4.4 assume that the pigeonpea covered only solarimeter position 1 in 1989, but that in 1990 its rapidly expanding canopy extended over positions 1 and 2 throughout the measurement period. The sole pigeonpea, at its initial optimum population of 9 plants m^{-2} in 1989, intercepted over seven times as much radiation (1960 MJ m^{-2}) as the line and dispersed

Table 4.4. Intercepted total solar radiation, above-ground biomass and light conversion efficiency (*e*) in perennial pigeonpea–groundnut systems, 1989 and 1990 (data from Marshall, 1995).

	Intercepted radiation (MJ m^{-2})		Biomass (t ha^{-1})		Conversion coefficient (g MJ^{-1})	
	1989*	1990[†]	1989*	1990[†]	1989*	1990[†]
Sole groundnut	943	888	3.39	2.52	0.36	0.28
Sole pigeonpea	1960	823[‡]	8.16	3.80	0.42	0.46
Line planting						
Groundnut	857	795	2.90	1.98	0.34	0.25
Pigeonpea	271	487	1.47	2.09	0.54	0.43
Total	1128	1282	4.37	4.07	0.39	0.32
Dispersed planting						
Groundnut	675	360	3.03	1.52	0.45	0.42
Pigeonpea	279	1552	1.56	4.52	0.56	0.29
Total	954	1912	4.59	6.04	0.48	0.32

* Data for 1989 calculated between groundnut sowing (2 July 1989) and first pigeonpea fodder cut (31 January 1990); [†] data for 1990 calculated between sowing and harvest for groundnut (19 July 1990 – 7 November 1990) or between the third fodder cut and final harvest for pigeonpea (8 August 1990 – 25 January 1991); [‡] calculated using *e* value for line-planted pigeonpea as insufficient solarimeters were available to quantify light interception in both treatments fully.

plantings of pigeonpea, with their much lower populations of 0.5 plants m^{-2}. Even when interception by groundnut was taken into account, light interception by sole pigeonpea was still approximately double the totals for the line and dispersed systems. However, the seasonal conversion coefficient was lower in sole pigeonpea than in the line or dispersed arrangements (0.42 vs. 0.54 and 0.56 g MJ^{-1}), with the result that the substantial light interception advantage of the sole pigeonpea was not translated into a similar biomass advantage (Table 4.4). The sole groundnut intercepted 10–40% more radiation (943 MJ m^{-2}) than the line or dispersed groundnut in 1989, but this was again not matched by a similar improvement in biomass, which was little greater than in the other treatments. *e* was higher in the dispersed treatment than in the sole and line-planted stands.

The sole pigeonpea population was reduced to 0.5 plants m^{-2} on 1 June 1990 for the reasons stated previously. As there were insufficient solarimeters to permit a rigorous study of light interception by sole pigeonpea, interception was estimated from biomass production using the conversion coefficient for the line-planted pigeonpea, on the assumption that *e* was conservative. The data for 1989 indicate that this is not necessarily true, but in the absence of direct measurements this was the only feasible approach for estimating interception. Table 4.4 shows that estimated interception by sole pigeonpea (823 MJ m^{-2}) was 70% greater than in the line-planted pigeonpea (487 MJ m^{-2}) in 1990, but almost 50% lower than in the dispersed pigeonpea (1552 MJ m^{-2}). Interception by the sole pigeonpea was also substantially lower than the totals for the line and dispersed systems (1282 and 1912 MJ m^{-2} respectively). This pattern was reflected by biomass production, although the value for dispersed pigeonpea is lower than predicted on the basis of interception because of its low e value. The combination of high interception and low conversion coefficient in the dispersed pigeonpea may reflect its bushy structure, which would have resulted in a greater proportion of the radiation being intercepted by stems and branches. The results for the sole stand suggest that, although pigeonpea exhibits great plasticity in the relationship between population and productivity (Rao and Willey, 1983), the reduced population of 0.5 plants m^{-2} in 1990 was below the optimum.

Seasonal interception and especially biomass production in groundnut were lower in 1990, probably because of the increased incidence of foliar disease. Both variables were nevertheless much greater in sole groundnut than in the dispersed system, where the greatly reduced interception was only partly redressed by the much higher conversion coefficient of the shaded groundnut. Values of e for groundnut were uniformly lower than in 1989.

The much greater total interception by the dispersed system was primarily due to the pigeonpea, which captured 81% of the intercepted radiation. However, the substantial interception advantage of pigeonpea was not reflected by increased biomass production because of its low e value. This is closely analogous to the *Leucaena*–millet system discussed earlier, in which the less efficient tree component captured most of the light in the second year, to the detriment of crop production. However, as noted earlier, the biomass and e values for pigeonpea shown in Table 4.4 are underestimates because they take no account of stem growth below the cutting height. This error would have been greatest in the dispersed pigeonpea because stem growth was greater than in the other treatments; measurements at final harvest (25 January 1991) showed that this portion of the stem had accumulated 0.62, 0.60 and 1.1 t ha^{-1} of dry matter over the experimental period in the sole, line and dispersed treatments respectively. A further potential source of error is that the estimates of interception in the dispersed system assume that the pigeonpea canopy extended over both solarimeter sites (Fig. 4.10b) throughout the measurement period. However, this assumption did not apply for some time after each fodder cut, causing pigeonpea interception to be overestimated; any overestimation of pigeonpea interception would result in underestimation of interception by groundnut, although not that of the entire system.

The uncertainty over interception by the dispersed pigeonpea applies only to periods when the groundnut canopy was present, since pigeonpea interception could be measured directly at other times using solarimeters B1 and B2; for example, interception by the dispersed pigeonpea between groundnut harvest in 1990 and the final pigeonpea harvest in January 1991 was 685 MJ m^{-2}. Figure 4.11 shows timecourses of estimated light interception by pigeonpea between 30 and 100 DAS for groundnut in 1990, calculated using various methods. Curve 1 assumes that the pigeonpea canopy extended over solarimeter sites 1 and 2 throughout the entire period, while curve 2 assumes that only site 1 was covered; these differing assumptions provide a twofold difference in estimated interception. These two curves represent the extreme scenarios since site 2 was not shaded for some time after each fodder cut but was shaded at other times following the regrowth of the side branches. The true values would therefore have been intermediate between these extremes. Alternative approaches for estimating interception by the dispersed pigeonpea are described below.

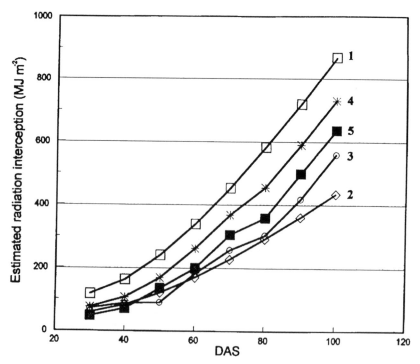

Fig. 4.11. Comparison of the estimates of radiation interception by pigeonpea in 1990 obtained using various methods. See text for details (Marshall, 1995).

Method 1. Estimating interception from crop extinction coefficient and leaf area index

As discussed earlier, fractional interception is related to the leaf area index and extinction coefficient (k) of the canopy concerned according to Equation 4.4. Method 1 made use of this relationship to estimate the mean fractional interception by the groundnut in the dispersed planting. By deducting the *measured* interception by groundnut at position 1 (Fig. 4.10b), interception by groundnut at position 2 could be estimated. Subtraction of this value from total interception measured at site 2, as determined by solarimeter B2, provided an estimate for pigeonpea interception. Equation 4.4 assumes that the canopy is homogeneous, has randomly distributed leaves and there is no effect of row structure. Under such conditions, a linear relation is obtained between L and $\ln(1 - f)$ with a gradient of $-k$. The extinction coefficient for sole groundnut was calculated from the values of L and f obtained by growth analysis between 30 DAS, when full ground cover was achieved, and 100 DAS. The k value obtained (0.44) was then used to estimate total interception by the dispersed groundnut, which was in turn used to estimate interception by pigeonpea (Fig. 4.11, curve 3). This approach assumes that groundnut interception at position 2 (Gi2) may be expressed as:

$$Gi2 = A2 - B2 \qquad\qquad (4.16)$$

where A2 and B2 denote radiation fluxes above and below the groundnut canopy at position 2. Similarly, mean interception by groundnut in the dispersed system (Gint) may be described as:

$$Gint = ((A1 - B1) + (A2 - B2))/2 \qquad\qquad (4.17)$$

where A1, A2, B1 and B2 are similarly defined. Equations 4.16 and 4.17 cannot be applied directly because of the absence of measurements at position A2, but Gi2 can be expressed in terms of parameters that were measured or can be estimated by combining the equations, i.e.:

$$Gi2 = 2(Gint) - (A1 - B1) \qquad\qquad (4.18)$$

Having estimated Gi2, the radiation received at the site of the missing solarimeters, A2, was calculated as:

$$A2 = B2 + Gi2 \qquad\qquad (4.19)$$

Interception by the pigeonpea at position 2 was then determined as the difference between A2, and the incident radiation (S_i). The values obtained allowed the overall mean interception by pigeonpea to be calculated as:

$$PPi = S_i - (A1 + A2)/2 \qquad\qquad (4.20)$$

A potential difficulty associated with this method of estimating interception is that the k value for sole groundnut (0.44) was applied to the groundnut in the dispersed system. Although previous reports suggest that the effect of shading should be small, a similar calculation of k for shaded groundnut in line planting produced a much lower value of 0.2. The origin of this difference in k between the line-planted and sole groundnut is unclear since the planting arrangement was identical and the difference in leaf area index was less than 20%, suggesting there were no major differences in ground cover and hence light penetration. A second analysis of pigeonpea interception was conducted using $k = 0.22$ to test the sensitivity of the estimates to variation in k. The values obtained (Fig. 4.11, curve 4) were much higher than with $k = 0.44$ and were closer to curve 1, which assumed that the pigeonpea canopy extended over both solarimeter positions throughout the season. The use of $k = 0.44$ produced values much closer to curve 2, which assumed only position 2 was shaded. These results support the observation that there was a period after the third fodder cut on 8 August 1990 when only position 1 was shaded, but that the pigeonpea canopy rapidly extended to cover position 2.

Method 2. Estimating radiation interception from dry matter production
As discussed earlier, when water and nutrients are non-limiting, dry matter production is linearly related to intercepted radiation according to Equation 4.7. Similarly, cumulative dry matter production (W) may be calculated as the product of cumulative intercepted radiation (ΣS_i) and the mean conversion coefficient ($\bar{e}$) for the relevant time-period, i.e.:

$$W = \Sigma S_i \bar{e} \qquad (4.21)$$

This relation was used as a basis for an alternative approach to estimating interception by pigeonpea at position A2. Mean cumulative light interception by groundnut in the dispersed system was calculated using Equation 4.21 from the above-ground biomass values obtained from the groundnut growth analysis and an estimated conversion coefficient. For this analysis, the e value for the three most shaded groundnut rows in the line planting (Fig. 4.10a, position 1) was used; the value obtained (1.0 g MJ^{-1}) was much higher than in the sole crop (0.6 g MJ^{-1}). Once the values for Gint had been derived in this way, the procedure for estimating pigeonpea interception was identical to that described in Method 1, using Equations 4.18–4.20. The results of this analysis are shown in Fig. 4.11, curve 5; the estimated interception values are at the centre of the range provided by the alternative approaches described above.

Several difficulties are involved in this approach, not least of which is choosing an appropriate value for e. For instance, the radiation reaching the groundnut in the dispersed system may have had a lower PAR content than in the line planting because the differing arrangement and canopy structure of the trees resulted in a greater proportion of the incoming radiation being filtered through the pigeonpea canopy. If this were so, the e value calculated on the basis of intercepted *total* radiation for shaded groundnut in the line planting would have been artificially high for groundnut in the dispersed system. The data available do not permit this effect to be quantified, but a sensitivity analysis showed that reductions of 10 or 20% in the e value used would produce similar reductions in the estimates of pigeonpea interception. An additional factor is that, although the seasonal mean e value was used in the analysis, e is known to be higher during the linear growth phase of crops than during the grain-filling and maturation phases, as discussed earlier. This would not alter the estimates of total seasonal interception, but would influence the shape of the timecourses obtained, since less intercepted radiation would be required to produce a specific quantity of biomass during periods when e was high.

This discussion emphasizes the importance of ensuring that the number and location of measurement sites are appropriate to the experimental objectives. The substantial uncertainties in light partitioning described above could have been avoided either by installing additional solarimeters at position A2, or by employing fish-eye photography to record the extent of the pigeonpea canopy so that corrections could have been made for the varying area of groundnut subjected to shade.

Sibbald and Griffiths (1992) described a temperate silvopastoral system involving a similar dispersed arrangement of trees with pasture grass. A number of agroforestry systems differing in the spacing and height of the trees were achieved by thinning existing Sitka spruce stands to provide inter-tree spacings of 4×4, 6×6 and 8×8 m, and the impact on microclimatic conditions and herbage production by ryegrass swards was investigated over two seasons. The ryegrass was grown in boxes filled with soil and sunk into the ground below the trees to prevent below-ground competition between the trees and grass and permit the influence of variations in radiation and microclimatic conditions to be assessed. Spatial variation within treatments was examined by sub-dividing the area between adjacent trees into nine equally sized sampling areas, in which sward boxes were installed and solar radiation, temperature and precipitation were measured using conventional approaches (tube solarimeters, platinum resistance thermometers and thermistors and raingauges). As in the tropical systems discussed earlier, instrumentation constraints made it impossible to monitor all sampling locations in every treatment, and so measurements were concentrated on the three spacing treatments at the intermediate tree height. However, in contrast to the replicated measurements previously described, Sibbald and Griffiths chose to confine their measurements to all nine sampling locations within a single replicate of each treatment in order to assess the full extent of the spatial variation.

Seasonal radiation receipts at ground level were significantly affected by tree spacing, being 22.7%, 63.2% and 68.9% of incident radiation in the 4×4 m, 6×6 m and 8×8 m spacings. Significant north–south trends in transmission were found in all spacings, but a significant east–west trend occurred only at the widest spacing. Light transmission was greatest in the central area of the 4×4 m spacing and also to the south of the trees in the 6×6 and 8×8 m treatments. Thus tree spacing had major effects on the quantity and distribution of radiation at ground level. The nature and extent of such effects will clearly vary depending on latitude and the height, spacing and canopy structure of the trees involved. Precipitation at ground-level also showed a significant north–south trend, although the positional variation was less directional than for light. Mean sward temperature was increased by 1.4°C between the closest and widest spacings, whereas the opposite trend (0.6°C) applied for mean minimum temperature. There was no significant positional variation in sward temperature within treatments, reflecting the buffering effect of the trees on air movement. The major conclusion was that the order of importance of the microclimatic changes imposed by the trees was light transmission > rainfall interception > sward temperature. Nevertheless, it was suggested that the higher minimum temperatures at the closest spacing would increase the productivity of the grass sward by extending the growing period in spring and autumn.

Measurement and Interpretation of Water Uptake and Use

Many of the philosophical and practical difficulties involved in partitioning light interception within agroforestry systems apply also to water, particularly when attempting to quantify the extent of the spatial and temporal variation. Instrumentation and technical limitations may again constrain the numbers of sites, replicates and treatments that can be examined, imposing difficult sampling decisions. These problems are compounded by the fact that the rooting environment is much more heterogeneous than the atmospheric environment, necessitating greater replication to obtain reliable estimates of water use.

Three broad approaches may be adopted to determine water use by the components of mixed communities:

1. transpiration by each component may be *estimated* using transpiration models based on light interception by each component;
2. total community water use and transpiration by one component may be *measured*, leaving transpiration by the other to be calculated as the difference;
3. transpiration by each component may be *measured* separately.

Approach (1) assumes that the Penman–Monteith equation may be modified to calculate transpiration within multispecies stands provided that radiation interception by each component can be estimated. Once this has been achieved, transpiration by each component may be calculated using a single or dual source transpiration model which requires as input variables only above-canopy net radiation, temperature and vapour pressure deficit. A detailed description of the model and its application is given by Wallace (Chapter 6, this volume and Wallace, 1995).

In approach (2), total community water use is normally determined by the soil water balance approach, which involves solving Equation 4.2 for $T_t + T_c$, and transpiration by either the tree or crop component is also measured using one of several possible techniques. Approach (3) is preferable since water use by each component is determined separately, with the result that the values obtained are subject only to the errors inherent in the techniques specifically used to obtain those values. When approach (2) is used, the estimates of transpiration obtained for each component are not statistically independent, while the values for the component derived by difference are subject to two sets of errors, those contained in the estimates of total community water use, and those for the component for which transpiration was actually measured. However, until recently, option (2) represented the only realistic approach because the methods available were technically too demanding, labour-intensive or costly to permit reliable and direct measurements of transpiration by both trees and crops throughout the season. Several studies have attempted with partial success to partition water use in agroforestry systems by determining total water use using the water balance approach, despite its potentially serious practical limitations, and measuring transpiration by one of its components (e.g. the *Leucaena*–millet and pigeonpea–groundnut systems described above). Techniques which may be applied to determine transpiration by trees or crops include diffusion porometry, small chamber gas analysis systems, deuterium labelling and sap flow techniques. Porometry and small chamber systems allow daily or seasonal timecourses of transpiration to be constructed, but are extremely labour-intensive if both components are to be monitored routinely. They are also unusable when the foliage is wet, a major problem during periods of frequent rainfall or dew. Deuterium labelling avoids these problems and permits estimation of transpiration over periods of several days, but is relatively expensive and requires sophisticated monitoring equipment. However, the advent of relatively simple and reliable sap flux techniques now permits continuous, non-destructive measurements of transpiration by trees and relatively large crop plants over extended periods. The principles and applications of the various techniques are outlined below.

Water balance

This approach is essentially an accounting procedure whereby all the terms contributing to the soil water balance are measured or otherwise accounted for, with the exception of transpiration. Total transpiration is therefore determined indirectly from the balance of all other components, namely precipitation and/or irrigation, interception losses, runoff, soil surface evaporation, deep drainage and soil moisture content within the rooting zone. Several of these terms are difficult or laborious to quantify reliably, particularly as they may show extensive spatial variation depending on windspeed and direction, rainfall intensity, solar angle and irradiance. This introduces unknown uncertainties into the estimates obtained for $T_t + T_c$ (Equation 4.2). Techniques for quantifying the components of the water balance are discussed in greater detail in Chapter 6.

Soil moisture content, a key variable in the water balance, is usually measured by gravimetric sampling in the surface horizons and by the neutron moderation approach in deeper horizons, although a new, commercially available technique, Time Domain Reflectometry (TDR), now permits rapid automated measurements at all depths (see Chapter 6). In the neutron probe approach, fast neutrons emitted from a radiation source lowered down aluminium tubes installed vertically in the soil are back-scattered as slow neutrons by hydrogen nuclei in the soil; these are predominantly associated with water molecules in most soils, but the varying quantities of other hydrogen-containing components make careful calibration essential for specific soils. With appropriate calibration against gravimetric measurements of soil moisture content, the counts for back-scattered slow neutrons may be used to provide periodic non-destructive measurements of water content throughout the soil profile; these can in turn be used as part of the water balance to estimate water uptake from specific horizons and, by summation, total water extraction from the soil profile colonized by roots.

A major difficulty with this approach is that it is often impossible with deep-rooting species or in stony soils to install neutron probe access tubes to the full depth of the rooting system without causing unacceptable damage to the surface horizons; compaction, smearing of the soil surface in contact with the tube or loose contact between the tube and the soil must be avoided since these factors alter infiltration and drainage characteristics. For instance, in the *Leucaena*–millet system described earlier, it proved impossible to install access tubes to depths greater than 85–100 cm, with the result that extraction from greater depths by *Leucaena* could not be quantified. Eastham *et al.* (1988) also concluded that unknown, though in their case probably limited, quantities of water were withdrawn by *Eucalyptus* from below their maximum measurement depth of 5.6 m. In such cases, or when the roots come in contact with the water table, community water use is underestimated to an extent related to the quantity of water extracted from horizons below the maximum measurement depth. A direct consequence is that transpiration by the tree may be grossly underestimated if calculated as the difference between apparent community water use and measurements of transpiration by the crop, no matter how accurately the latter has been determined.

An excellent example of the successful application of the water balance approach was described by Eastham *et al.* (1988), who examined the effects of tree density on water use by silvopastoral systems in Queensland, Australia. *Eucalyptus grandis* trees were planted into an established pasture dominated by *Setaria sphacelata* using a Nelder design (Fig. 4.12), which enables tree spacing to be altered systematically within a relatively small land area. Nine concentric rings of equally spaced trees provided populations of 42–3580 trees ha^{-1}, from which densities of 82 (low), 304 (medium) and 2150 trees ha^{-1} (high) were selected for detailed study.

Neutron probe access tubes installed to a depth of 5.6 m at various distances from the trees were used to establish the vertical and lateral patterns of water extraction. In conjunction with information on rainfall and deep drainage, these measurements enabled community water use to be determined. Relationships between transpiration from the pasture, open pan evaporation and soil water content were established for each tree density using small lysimeters, and these were then used routinely to estimate transpiration by the pasture from the neutron probe measurements of soil water content in the top 30 cm of the profile. These values were subtracted from total community water use to determine transpiration by the trees. The data obtained allowed both short term transpiration rates and seasonal totals to be calculated for each tree density.

Figure 4.13 shows the combined transpiration of the trees and pasture over a 2-year period commencing 14 months after the trees were planted. Water use in both years was greatest in summer and decreased in winter, although the values were generally lower between June and September 1986 because of severe drought earlier in the year. Water use was consistently greatest at high tree density during 1985 and the first half of 1986, but then became similar to the medium density due to declining soil water availability. Water use was generally least at the low tree density. Transpiration at the high tree density was close to, and occasionally exceeded, open pan evaporation during rainy periods in 1985, but was consistently below open pan evaporation in 1986, when rainfall was 361 mm below the annual average of 1100 mm. Separation of water use by the trees and pasture showed that transpiration was dominated by the pasture component when the trees were widely spaced, but the converse applied at higher densities where the trees formed a closed canopy.

Total water use increased and losses due to deep drainage decreased with increasing tree density in both years, reflecting their ability to extract water at depth. The higher tree densities encouraged deeper and more extensive rooting at depth because the increased competition between neighbouring trees inhibited the extensive lateral spread of roots in the surface horizons seen at lower densities (Eastham and Rose, 1990a). Thus,

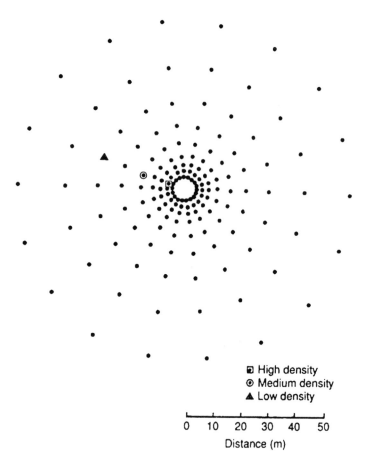

Fig. 4.12. Distribution of trees in the Nelder design; trees were sampled at high, medium and low density (Eastham and Rose, 1990a).

by January 1986, water was being extracted to the maximum measurement depth of 5.6 m at the high tree density, and by the end of 1986 soil water content was close to the wilting point at all depths between 2.0 and 5.6 m. In contrast, the trees planted at wider spacings experienced less intra-specific competition and were able to exploit water stored in the surface horizons at greater distances from the trunk, with the result that rooting density and water extraction at depth were less. Eastham and Rose (1990b)

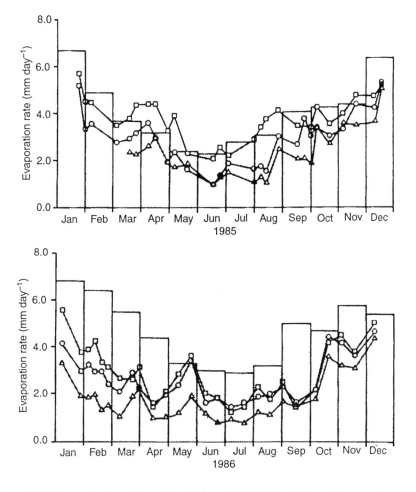

Fig. 4.13. Transpiration during 1985 (a) and 1986 (b) at high (□), medium (○) and low (△) tree densities. The histograms show monthly mean open pan evaporation (Eastham *et al.*, 1988).

showed that more of the water transpired at the highest tree density was drawn from deep in the profile due to the combined effects of the drier surface horizons, the deeper rooting system and the greater uptake per unit root length at depth; the latter factor became increasingly important as the drying cycle progressed and the surface horizons continued to dry. Proximity to the trees and increasing tree density also reduced the rooting depth and density of the pasture grasses.

Cumulative transpiration exceeded rainfall by over 300 mm at the high tree density in both 1985 and 1986, and also at the intermediate density in

1986, suggesting that these densities may be unsustainable. However, the results obtained also demonstrate that manipulation of tree density in agroforestry systems may be used to modify biomass production by the component species and control interspecific and intraspecific competition for available resources.

Diffusion porometry

The two types of diffusion porometry most commonly used for field studies are the dynamic diffusion, transit or van Bavel type, and the steady-state or null-point type. Both measure the rate of transfer of water vapour from plant tissue into a sealed chamber and can be used to determine diffusive resistance or conductance. Porometers of both types are commercially available (e.g. Delta-T, Cambridge, UK and Licor, Lincoln, Nebraska, USA). The transit type is relatively simple but requires frequent, relatively straightforward calibration. The main criticism has been that until recently the description of the underlying physical theory has been incomplete, although this has now been resolved (Monteith *et al.*, 1988; Potter, 1992). In contrast, the theoretical basis for steady-state instruments is relatively simple but calibration, though less frequently required, may be more difficult. The accuracy of steady-state porometers depends on the precision of relative humidity measurements since a 1% error at 50% r.h. introduces an error of *c.* 4% in the resistance values obtained. Another criticism is that the vigorous stirring required in the cuvettes of steady-state porometers may increase water losses and induce stomatal closure during the measurement. A similar criticism may be levelled at the opaque sensor head of Delta-T instruments which shades the enclosed leaf segment and alters its ventilation properties. The resolution and speed of measurement of the two types of instrument are comparable.

In transit porometers, a small area of leaf is temporarily enclosed in a small cup containing a humidity sensor. Dry air drawn from a silica gel reservoir is passed through the cup until relative humidity falls to a preset level and the time required for water vapour emitted from the leaf to raise humidity by a small fixed increment is then determined. This cycle is automatically repeated until a stable value is obtained. The counts obtained are converted to diffusive resistances or conductances from calibrations constructed using a plate containing perforations of known diffusive resistance. The range over which humidity cycles is adjusted to match ambient humidity as closely as possible to minimize disturbance to the stomata. Modern dynamic diffusion porometers (e.g. Delta-T Devices AP4) are microprocessor-driven, which greatly facilitates their operation and enables large quantities of data to be recorded in a choice of resistance or conductance units.

Steady-state porometers determine stomatal resistance from measurements of the equilibrium relative humidity produced by transpiration from the enclosed leaf and an inflow of dry air (e.g. Bell and Squire, 1981). Steady-state instruments require a vigorously stirred chamber which is usually relatively large, whereas the cup of transit porometers is unstirred and relatively small. Thus, transit porometers sample small areas and can measure both leaf surfaces separately, whereas steady-state instruments sample larger leaf areas and measure both leaf surfaces simultaneously. When combined with concurrent measurements of boundary layer resistance (r_a), atmospheric vapour pressure (e_a) and leaf temperature (T_l), measurements of leaf diffusive resistance (r_s) can be used to calculate the transpiration rates of individual leaves or the whole canopy if the vertical distribution of leaf area index is known. The major advantage of this approach is that it provides almost instantaneous values of transpiration, in contrast with the much poorer time resolution of neutron probe or deuterium labelling. It also permits examination of the variation in transpiration associated with differences in leaf age and position or distance from the nearest tree, which would not be possible using longer-term averaging approaches.

The canopy is usually sub-divided into two or three layers depending on canopy height when whole-canopy transpiration is to be estimated using porometry, and separate measurements of the required input variables are made for each layer. This stratification is made necessary by the steep gradients of r_s and the environmental variables influencing transpiration which may occur within the canopy (Black et al., 1985). The mean transpiration rate for each layer (E_i) is then calculated using a modified version of Equation 4.9 i.e.:

$$E_i = \frac{(v_i - v_a)L_i}{r_s + r_a} \qquad\qquad (4.22)$$

where L_i is the leaf area index of that layer and r_s and r_a are the mean stomatal and boundary layer resistances. Total transpiration is obtained by summing the values for each layer. Several workers have adopted this approach for tropical crops (e.g. Azam-Ali, 1983; Black et al., 1985; Wallace et al., 1990), while Roberts et al. (1993) applied a similar approach in conjunction with a multi-layer combination model to calculate transpiration from an Amazonian rainforest; the results obtained agreed well with measurements made directly using the eddy correlation approach (see Chapter 5).

The procedures employed may vary in detail depending on the crop involved. Azam-Ali (1983) arbitrarily divided the canopy of millet into three layers (0–20, 20–50 and > 50 cm) and measured r_s on both leaf surfaces of five leaves in all layers present on specific sampling dates. The mean r_s values for each layer were then used in the analysis. In contrast, Black et al. (1985; Fig. 4.14) derived mean values from more detailed r_s profiles in groundnut, in which many more leaves were present, biasing sampling towards the top of the canopy where r_s was much smaller and transpiration rates were consequently greater. Although Azam-Ali ignored transpiration by the leaf sheaths in his analysis, Squire and Black (1982) demonstrated that the sheaths of wheat should be included because they form a major component of total green area index and contribute significantly to canopy conductance.

Figure 4.14 shows the diurnal timecourses for canopy conductance (g_c) and transpiration (E_t) calculated for groundnut stands which had been unirrigated or subjected to progressive drought (Black *et al.*, 1985). Canopy conductance was greatly reduced in the unirrigated stand due to a combination of increased stomatal resistance and reduced leaf area index. The reduction in E_t was much smaller than that in g_c because the conservative effect of increased stomatal resistance was offset by concomitant increases in leaf temperature and hence $v_i - v_a$ in Equation 4.22. The results clearly demonstrate the danger of attributing conservative benefits to increases in stomatal resistance without a full knowledge of the other variables influencing transpiration. The reduction in E_t in the unirrigated stand was entirely attributable to its lower L. Azam-Ali (1983) reached a similar conclusion for millet. Total water use by the irrigated and unirrigated stands on 71 DAS, calculated from the areas under the curves in Fig. 4.14b, was 7.4 and 6.2 mm day^{-1}, similar to reported values for other groundnut crops grown in the field under conditions of high evaporative demand.

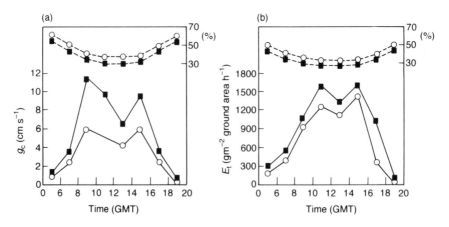

Fig. 4.14. Diurnal variation in (a) canopy conductance (g_c) and (b) transpiration (E_t) in irrigated (closed symbols) and unirrigated (open symbols) stands of groundnut at 71 days after sowing. Dashed lines show the fraction of g_c and E_t contributed by the five youngest mature leaves (Black *et al.*, 1985).

Sampling procedures must be sufficiently rigorous to cope with the extensive spatial variation in r_s that occurs between leaves of similar age and position within the canopy depending on their exposure to sunlight. The value of r_s may also show substantial variation within individual leaves depending on the proximity of water supplies and local differences in insolation; several measurements within each sampled leaf may therefore be necessary in species with large leaves such as C4 cereals. Even in small leaves such as those of groundnut, r_s may vary significantly depending on whether the porometer cup is oriented parallel or transverse to the midrib (Black et al., 1985). The frequency of sampling required to obtain satisfactory estimates of transpiration over daily or longer periods is strongly weather-dependent. Where weather conditions remain relatively uniform within and between days, measurements at 4-h intervals on 10 days spread over a 31-day period may be sufficient to provide reasonably accurate estimates of transpiration, as compared with parallel neutron probe measurements (Azam-Ali, 1983). However, much more frequent sampling is required in areas where varying weather conditions cause major changes in r_s on a daily or hourly basis (Black et al., 1985). When assessing the errors involved in this approach, Azam-Ali concluded that each variable in Equation 4.22 can be measured to within ±5–10% and that the combined error in daily estimates of transpiration would be c. 10–20%. Porometric and neutron probe estimates of seasonal water use agreed to within 4–28%, which is within the likely scale of errors.

The major shortcoming of the porometry approach arises from its highly intensive, yet discontinuous nature and the great spatial and temporal variability of the input variables. Repeated measurements are therefore necessary to establish reliable diurnal and seasonal patterns. However, despite these problems, porometry (or portable infrared gas analysis (IRGA)) using commercially available instruments from ADC (Hoddesdon, Herts, UK), PP Systems (Hitchin, Herts, UK) or Licor (Lincoln, Nebraska, USA) represents one of the few available approaches for partitioning water use within agroforestry systems. Although porometry cannot provide the continuous record of transpiration offered by sap flux techniques, it may still be the most practical alternative in situations where sap flux, deuterium labelling or water balance approaches are inappropriate. It is also well suited for studies of the impact of positional variations in shading intensity and competition for water on transpiration by crops growing in agroforestry systems.

Sap flux measurements

Methods that provide direct, continuous and non-destructive measurements of transpiration by intact plants under field conditions have been a major but elusive objective for many years. Most work directed at attaining this goal has concentrated on sap flux methods, including the magneto-hydrodynamic and heat pulse or heat balance approaches.

Magnetohydrodynamic method

This approach permits sap flow to be measured non-destructively over extended periods in species with woody stems. The system described by Sheriff (1972) was suitable for stems up to 9 mm in diameter, but is capable of modification to suit larger stems. The technique involves applying a uniform magnetic field perpendicular to the direction of flow through the stem. As the fluid passes through the line of magnetic flux, a voltage is induced whose magnitude and sign may be used to determine the volume and direction of sap flow. The measurements are independent of physical variables such as temperature and fluid viscosity. The technique involves inserting platinum electrodes into opposite sides of the stem so that their tips are close to the xylem. An electromagnet is positioned with its poles perpendicular to both the stem and the electrodes, so that a line linking the poles bisects the centre of the stem. When the electromagnet is energized, the voltage output is linearly related to sap flux, though the slope of the relation varies with species and stem diameter; individual calibration over a range of stem diameters is therefore required. The technique has been less thoroughly tested than the heat pulse method and has not been validated under field conditions.

Heat pulse

The heat pulse technique was first devised by Huber (1932) and has since been adopted and modified by numerous workers for a range of woody species. The technique has most frequently been applied to relatively large trees, but modified versions have been used with apparent success in small cotton and poplar plants (Heine and Farr, 1973; Stone and Shirazi, 1975). The technique involves injecting brief pulses of heat using small resistance probes or plates embedded in the sapwood and then monitoring their axial progress using thermocouples or thermistors installed downstream in the flowpath. The time required for the pulses to travel from their source to the point of detection is used to estimate sap velocity. When sap flow is rapid, the pulse moves predominantly by convective transfer in the xylem sap, while conductive transfer through the cell walls and the sap itself is relatively small. However, the conductive component becomes increasingly important as transpiration slows, making it essential to apply corrections for conductive transfer if positive heat pulse velocities are to be avoided at times of zero sap flow.

The volume of sap flowing cannot simply be determined as the product of velocity and xylem cross-sectional area since vessels or tracheids of differing age, size or position within the stem may conduct water at very different rates, if at all. At its simplest, the heat pulse technique determines the velocity of sap movement in the most rapidly conducting elements. Differences in the velocity and volume of sap flow between different sections of the xylem and variation in the cross-sectional area of functional xylem with transpiration, season and the physiological health of the conducting tissue may pose problems. For example, in thick stems containing several functional xylem rings, sap velocity and hence sap flux decline rapidly towards the centre of the stem (Cohen *et al.*, 1981). Similarly, sap flux may vary around the stem circumference (Heimann and Stickan, 1993) depending on local variation in the thickness of the annual ring involved. This may make it necessary to install heat sources and sensors at various depths and locations around the stem to obtain reliable estimates of sap flux, thereby increasing the complexity of the measurements. Such problems are greatest with thick stems.

The heat pulse technique has frequently been used for relative studies rather than absolute measurements of sap fluxes in trees. Several reports suggest that the velocities obtained may systematically underestimate the true values by a factor of two to seven (Sabatti *et al.*, 1993), a discrepancy which has variously been attributed to lack of thermal homogeneity within the flow pathway, embolisms, or inadequacies in the model used to derive sap flow from the heat pulse values. However, there is evidence that the discrepancy is sufficiently predictable for satisfactory corrections to be made using calibrations established against other directly quantitative methods for determining transpiration. For example, Cohen *et al.* (1981) concluded from comparisons of lysimetric and heat pulse measurements of transpiration in *Citrus* made under contrasting environmental conditions that a fixed correction factor could be applied. This approach has subsequently been used in several field studies (e.g. Cohen *et al.*, 1983; Moreshet *et al.*, 1983), and provides good agreement with estimates of transpiration obtained using a meteorological model based on porometric measurements of canopy conductance and the relevant climatic variables (Cohen *et al.*, 1993); the latter approach was described in the preceding section. Similarly, Sabatti *et al.* (1993) calibrated the heat pulse system against measurements of water uptake obtained by sealing a water-filled tank around the base of the trunk of *Quercus* trees and girdling the stem to a depth of 3 cm to permit the entry of water. Close linear correlations were obtained between sap velocity and water uptake for both standard trees and coppice sprouts, although the slope of the relationship differed greatly because much more water was transported at a given sap velocity in standard trees because of their much larger conducting area. For example, the maximum observed velocity of 30 m h^{-1} corresponded to sap fluxes of *c.* 1.5 and 5.0 l h^{-1} in coppice shoots and standard trees. A close correlation between sap velocity and sap flux has also been shown for *Eucalyptus* (Doley and Grieve, 1966).

Heat balance

A fundamentally different approach to determining stem sap flow is based on the measurement of a heat balance, rather than heat pulse velocity. The method was initially devised by Vieweg and Ziegler (1960) but numerous variants have been described by other workers. In this approach the sap flux is determined by balancing the fluxes of heat into and out of a segment of stem. In most heat balance systems the heater input is automatically and continuously adjusted to maintain a fixed temperature differential between the heated stem segment and the unheated section below. As sap flux changes, so too must the heat input required to compensate for the altered convective flux of heat away from the heated area and to maintain the heat balance constant. The changes in heat input provide the basis for the measurement of sap flow. Since it is a null method, the technique is sensitive to small changes in sap flux and has a fast response time. However, Sakura-tani (1981, 1984, 1987, 1990) and Baker and van Bavel (1987) adopted a different approach in which the heat input was maintained constant, rather than varying as sap flux changes, and the heat fluxes out of the heated area were determined from the measured temperature gradients. The heat fluxes can again be used for direct calculation of sap flux through the stem.

The principles and design of the two major variants of the heat balance approach are described by Ishida *et al.* (1991) and Baker and van Bavel (1987). The system described by Ishida *et al.* is technically simpler than those adopted by Sakuratani and Baker and van Bavel for use with small plants, is relatively inexpensive and can measure sap fluxes in the range *c.* 20–700 g h^{-1} in woody species and large crop plants (e.g. sorghum, cotton). A similar system has been used routinely for several years at ICRISAT and ICRAF in a range of agroforestry systems (e.g. Marshall *et al.*, 1994; Howard *et al.*, 1995). In this approach, an annular heater wound around the stem is used to provide a known quantity of heat, and thermocouples are employed to measure heater temperature, ambient temperature and stem temperature 2 cm above and below the heater. The stem temperature thermocouples may be surface-mounted in small stems or inserted into fine holes drilled into the sapwood of larger stems. Heater temperature is maintained at a fixed increment, usually 2–5°C above ambient, by varying the voltage applied. The entire installation is enclosed in polystyrene and reflective aluminium foil to minimize radial heat exchange, and is sealed on to the stem at the top using quick-setting silicone rubber to prevent penetration by water resulting from stem flow during rain events. Campbell 21X microloggers are used to control the heat balance system and record the data obtained; each logger is capable of controlling three heat balance installations and the entire system may be powered by car batteries or solar panels in areas where mains electricity is not available.

Under steady-state conditions, the heat input into the heated stem segment (Q_h) is balanced by the heat fluxes out of the system in the form of upwards (Q_u) and downwards (Q_d) conductive fluxes, radial heat losses (Q_r) and the component of principal interest in sap flow studies, the convective flux (Q_c). The energy balance of the heater may be described as:

$$c\, dT_h/dt = Q_h - Q_r - Q_u - Q_d - Q_c \tag{4.23}$$

where c is the heat capacity of the stem and T_h is the heater temperature. The heat input, Q_h, is determined by the resistance of the heater coil and the voltage applied, and Q_r, Q_u and Q_d can all be estimated (Ishida *et al.*, 1991), or possibly ignored when transpiration is rapid and these terms make a negligible contribution to the heat balance relative to the convective component (Q_c). The convective heat flux can be calculated as:

$$Q_c = C_w \mathcal{J}_w (T_u - T_d) \tag{4.24}$$

where C_w is the specific heat of water (4.18 J g^{-1} K^{-1}), $\mathcal{J}_w$ is the flux of water through the stem (g s^{-1}) and T_u and T_d are the averaged cross-sectional temperatures recorded by thermocouples inserted into the stem above and below the heater. When steady-state conditions prevail (i.e. T_h remains constant), the term on the left of Equation 4.23 drops out and sap flux may be calculated as:

$$\mathcal{J}_w = (Q_h - Q_r - Q_u - Q_d)/[C_w(T_u - T_d)] \tag{4.25}$$

The heat balance approach permits daily transpiration by the components of agroforestry systems to be followed continuously and more accurately than is possible using diffusion porometry or soil water balance approaches. For example, Fig. 4.15 shows typical diurnal trends for individual trees in the perennial pigeonpea–groundnut system described earlier. During the dry season (Fig. 4.15a), shortly after the plants had been cut for fodder, leaf area was small in both treatments and there was no difference in transpiration. However, in the rainy season (Fig. 4.15b), transpiration was five times greater at midday in the dispersed planting than in the line arrangement. Because of their wider spacing, the dispersed plants developed a fuller canopy and exploited the soil profile more thoroughly for water. Transpiration followed the diurnal changes in vapour pressure deficit closely, though irradiance, soil moisture availability and leaf area were also important in determining water use. Such detailed studies of transpiration in relation to microclimatic conditions contribute significantly to our understanding of the success or failure of specific agroforestry systems, and may provide valuable information for crop models.

The daily values obtained in this way can be used to calculate cumulative transpiration (Fig. 4.15c), which again emphasizes the consistently greater water use of the dispersed pigeonpea. These data indicate that transpiration by the pigeonpea amounted to 60% of the annual rainfall in the dispersed system and 30% in the line-planted treatment, to

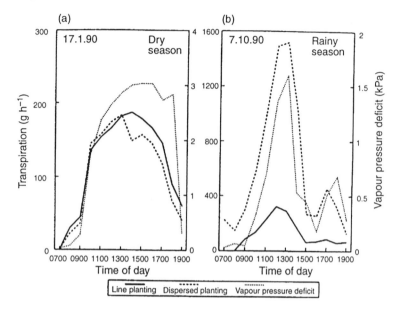

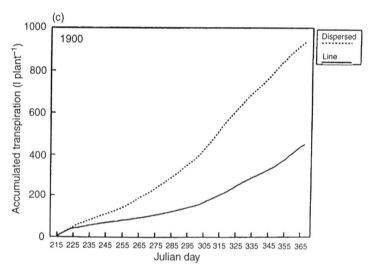

Fig. 4.15. Typical diurnal timecourses of transpiration and leaf to air vapour pressure difference in perennial pigeonpea during (a) the dry and (b) the rainy seasons, and (c) the seasonal timecourses for accumulated transpiration; $n=3$ (Marshall *et al.*, 1994).

which must be added the water used by the groundnut component. However, although water use and dry matter production by pigeonpea differed greatly between treatments, the water use ratio (WUR) differed by less than 10% (Table 4.5). This study clearly demonstrates the considerable potential of agroforestry for maximizing resource capture and overall system productivity.

Most heat balance systems have been designed for use with woody shrubs and trees or relatively large herbaceous plants with massive stems and long internodes (e.g. sunflower, sorghum and maize). Commercial heat balance systems (Dynamax Dynagage, Houston, Texas, USA) are available to suit stems ranging between 5 and 125 mm in diameter. These relatively large systems can measure sap fluxes between about 15 g h^{-1} and 15 kg h^{-1} with a typical accuracy of $\pm 10\%$, although some workers claim better than $\pm 5\%$ (van Bavel, 1988; Steinberg et al., 1989). However, estimated sap fluxes may not always agree with the true transpiration rate, at least in the short term, because of the water storage capacity of plant tissues. As a result, transpiration exceeds absorption, and hence sap flux measured at the stem base, during the early part of the day as plant water status declines, but the opposite situation prevails later in the day. The quantity of water involved is small in herbaceous crops such as groundnut, amounting to perhaps 1 g plant^{-1} or one-third of the hourly transpiration rate during the rainy season, but may amount to several kilograms in large woody species, with the dehydration or rehydration phases persisting for days or weeks during prolonged drying or wetting cycles. Under such conditions, instantaneous sap flux measurements may not provide an accurate measure of the true transpiration, even though the daily or weekly totals agree closely.

Although the heat balance approach may provide reliable measurements of transpiration in relatively small stems, great care is required to avoid serious errors in larger stems. For example, Khan and Ong (1995) reported that sap flux and gravimetric measurements of transpiration by 9-month-old potted *Grevillea robusta* seedlings with 20-mm diameter stems

Table 4.5. Dry matter production, transpiration and water use ratio (WUR) in line-planted and dispersed arrangements of perennial pigeonpea (Marshall, 1995). See text for details.

	Dry matter (t ha^{-1})	Transpiration (mm)	WUR (kg ha^{-1} mm^{-1})
Dispersed	4.51	517	8.70
Line-planted	2.09	262	8.00

agreed to within 7%. However, sap flux measurements consistently overestimated transpiration in 50–90-mm diameter stems, typically by 10–47% but by up to threefold, unless appropriate corrections were made for several sources of error. For example, the assumption of uniform mixing of heat introduced by the heater coil across the functional xylem increasingly breaks down as stem diameter increases. The large quantities of solar radiation incident upon the stems in tropical regions may also introduce serious errors, particularly when the tree canopy is sparse and the trunks large, as in dispersed planted agroforestry systems. Khan and Ong (1995) concluded that these errors may be overcome by applying various corrections and precautions. These include establishing the temperature bias within the trunk with the heater switched off, covering the trunk with 20-mm thick polystyrene sheeting covered in aluminium foil to minimize radiative heating, inserting thermocouples into the stem to quantify the radial temperature gradient across the sapwood, and accounting for the radial and conductive heat losses, which were estimated to comprise up to 32% of the heater input in large trunks. Application of these precautions to 50–90 cm trunks of *Grevillea* improved the accuracy of sap flux measurements of transpiration to ±7%. Thus, cross-calibration of the heat balance approach against other absolute quantitative techniques such as lysimetry is essential for each new species examined, although few institutes currently possess the facilities necessary to cope with large established trees.

Attempts have been made with varying success to modify the technique for use with small plants, including cotton and sunflower (Baker and van Bavel, 1987), soybean, sunflower and rice (Sakuratani, 1981, 1990). These again suggest that, with sufficient care, the heat balance technique is capable of measuring low sap fluxes with an accuracy of ±10%. Sakuratani's most recent work with rice, under laboratory conditions, appears particularly significant since he was able to resolve flow rates between 1 and 14 g h^{-1} with similar accuracy. Sap fluxes through single stems of small plants grown in agroforestry or intercropping systems (e.g. groundnut or millet) are of a similar order. However, there are two major problems in applying the heat balance approach to very small plants. First, the conductive and radial fluxes become increasingly important components of the heat balance as transpiration, and hence the convective heat flux (Q_c in Equation 4.23), decreases. Thus the radial and conductive fluxes must be quantified accurately in small plants if transpiration is to be estimated reliably. If this is not achieved, the heat balance approach may seriously overestimate true sap fluxes in very small stems and provide spurious negative values at night when transpiration approaches zero. Second, the physical size of the heat balance installation may preclude its use for plants with very short internodes since its overall length cannot be reduced much below 5–6 cm without creating instability. An additional consideration is that the sap flux measurements for species which tiller or branch near the stem base pertain only to that stem; however, total water use per plant or per unit land area may still be estimated with appropriate information on shoot numbers and plant populations.

Separating Above- and Below-Ground Interactions

Unlike intercropping, below-ground interactions have been shown to be highly significant in agroforestry systems (Ong *et al.*, 1991). The most common method for demonstrating below-ground interactions involves the installation of vertical barriers to separate the root systems of the tree and crop components. For example, Singh *et al.* (1989) installed polythene root barriers to a depth of 0.5 m in a *Leucaena* alley-cropping system in India before the cropping season. They observed that the presence of barriers prevented yield reductions in both sorghum and cowpea, but not in the longer duration castor, probably because its roots competed with those of *Leucaena* below the 0.5 m maximum depth of the barrier.

Root barriers have been adopted in other agroforestry studies but the results obtained have not always been clear-cut. One explanation may be the limited effectiveness of the relatively shallow root barriers used in most studies (0.5–1.0 m). For instance, excavation of the site in the *Leucaena* study in India revealed that tree roots had penetrated under the barriers and grown upwards into the cropping area after a single season (Corlett, 1989). Similar excavation at ICRAF's Machakos Station, Kenya showed that barriers installed to a depth of 1 m were ineffective after 12 months. It is therefore extremely important in any study using root barriers to determine by excavation whether these are indeed effective in preventing the lateral spread of tree roots. It is preferable to cut the roots by regular trenching, although it is practically impossible to prevent the lateral spread of all tree roots.

As tree roots may extend laterally by at least 2 m each year, interference between neighbouring plots is a formidable problem in agroforestry experiments involving fast-growing tree species, particularly when considering the capture of below-ground resources. For example, in the humid forest zone of southeastern Nigeria, Hauser (1993) estimated that the roots of *Senna siamea* exploited an area 6.1 times greater than the plot size of 18 × 20 m, while a slower growing species, *Dactylodenia barteri*, exploited an area 2.3 times greater than the plot size. Roots of *S. siamea* invaded the neighbouring sole cassava control plots, in which tree roots greatly outnumbered cassava roots. Thus, root interference may lead both to overestimation of the performance of the agroforestry treatments and underestimation of the performance of the sole crop control, providing an unrealistic assessment of the advantage of the agroforestry treatment. A similar problem was revealed in the semi-arid tropics by Rao *et al.* (1992), who also used shallow root barriers and found that root interference in the sole crop control caused a 32% overestimation of the alley crop yield.

Future Directions

Current attempts at ICRAF and elsewhere to use resource capture principles to understand the complex mechanisms involved in tree–crop interactions should provide a scientific basis for designing more productive and sustainable agricultural systems. It has been argued that such efforts, which are both difficult and expensive, must be capable of extrapolation to other sites and climates in order to be truly cost-effective (Anderson *et al.*, 1993). Experience gained over the last two decades has shown that resource capture principles have been useful in the analysis of crop performance under various climatic and management conditions (cf. Monteith *et al.*, 1994). As in intercropping, resource capture research in agroforestry should focus on a few representative systems in order to generate general principles.

Research on alley cropping continues to dominate the attention of agroforesters, although this system is not the most representative of the various agroforestry systems practised by farmers. Alley cropping does, however, represent the most potentially intensive agroforestry system because the trees or shrubs are pruned regularly to provide mulch or fodder and reduce competition with crops. In contrast, most farmers prefer to plant trees around farm boundaries where the tree–crop interaction is less and the limited tree management required is largely confined to the dry season when demand for labour is low. It is reasonable to assume that resource capture by unpruned trees will differ from that of frequently pruned trees because of their fuller canopy and deeper rooting profile (Hairiah *et al.*, 1992). Recent measurements at Machakos, Kenya show that a system containing pruned *Leucaena* trees produced only a small increase in the total seasonal water use when compared with sole maize (from 30% to 38% of rainfall), whereas unpruned *Leucaena* used 60% (Howard *et al.*, 1995b). Greater resource use may therefore be expected with unpruned trees, although competition with adjacent crops would also increase. Thus, a major priority in future studies should concern the resource capture of upper storey trees which experience few prunings. Furthermore, tree density should be lower than in alley-cropping systems so that the competitive effects of shading and root interactions are minimized. The farmed parklands of West Africa provide an excellent example of a dispersed tree system which deserves further research attention (Kessler, 1992).

There is increasing evidence that agroforestry systems have a greater potential than intercropping for improving soil properties, both chemical and physical, but the major obstacle is to translate such improvements into increased crop yields (Ong, 1995). The main problem is competition between trees and crops, which appears to be a common feature of fast-growing species such as *Leucaena leucocephala*. Experience from intercropping studies indicates that competition can be reduced by combining species with contrasting phenologies to provide temporal complementarity, so that major demands for growth resources occur at different times. This approach has been used in the exploitation of perennial pigeonpea as a multi-purpose tree species (Daniel and Ong, 1989), although pigeonpea suffers the practical disadvantage that it seldom survives beyond 2 years. Nevertheless, this is probably the most feasible approach for reducing competition since the labour requirement for pruning is a major deterrent to the adoption of agroforestry systems.

Spatial sharing of below-ground resources has a high priority in agroforestry research but is potentially one of the most intractable problems, in common with other aspects of root performance in mixed communities. Agroforestry again appears to offer greater scope for spatial complementarity than intercropping because the trees have deeper root systems than the associated crops. The growing body of information on the rooting profiles of multipurpose trees (e.g. Jonsson *et al.*, 1988; Dhyani *et al.*, 1990) suggests, however, that most fast-growing trees are likely to compete with field crops for water and nutrients (Szott, 1996). Exciting exceptions to this rule appear to be *Grevillea robusta* and *Paulownia fortunei*, both of which are non-leguminous, which may explain why farmers have planted these species so extensively. Research on spatial sharing of below-ground resources has only recently begun (e.g. Rao *et al.*, 1993; Huxley *et al.*, 1994), largely because of the difficulties of distinguishing between the roots of trees and crops and determining their relative capacities to absorb water and nutrients. Recent advances in sap flow technology have made it possible to attach miniature gauges to single tree roots to quantify their contribution to the overall absorption of water by individual trees (Ong and Khan, 1993), and analysis of the transpiration of young *Grevillea* trees indicates that at least 80% of the transpired water may be drawn from below the crop rooting zone (Howard *et al.*, 1995).

Finally, researchers should not ignore the extensive knowledge and wisdom of farmers and pastoralists, who may not fully understand the principles of resource capture, but are well aware of which tree species are compatible with crops and which are not (Hitimana *et al.*, 1994). For example, the remarkable compatibility of *Grevillea robusta* with crops was already well known to farmers in the eastern African highlands long before formal agroforestry research began into its rooting behaviour.

Acknowledgements

F.M.M. and J.E.C. gratefully acknowledge the postgraduate scholarships provided by the UK Natural Environmental Research Council and the Isle of Man Board of Education. Sections of this review were written during a visit to ICRAF by C.R.B. with partial support from the UK Overseas Development Administration Research Grant R5810 'Resource utilisation of trees and crops in agroforestry systems'. Thanks also to Jackie Johnson for her excellent secretarial support in producing the manuscript.

References

Anderson, L.S., Muetzelfeldt, R.I. and Sinclair, F.L. (1993) An integrated research strategy for modelling and experimentation in agroforestry. *Commonwealth Forestry Review* 72, 166–174.

Angus, J.F., Hasegawa, S., Hsiao, T.C., Liboon, S.P. and Zandstra, H.G. (1983) The water balance of post-monsoonal dryland crops. *Journal of Agricultural Science, Cambridge* 101, 699–710.

Aresta, R.B., and Fukai, S. (1984) Effects of solar radiation on growth of cassava (*Manihot esculenta* Crantz). 2. Fibrous root length. *Field Crops Research* 9, 361–371.

Azam-Ali, S.N. (1983) Seasonal estimates of transpiration from a millet crop using a porometer. *Agricultural and Forest Meteorology* 30, 13–24.

Azam-Ali, S.N. (1995) Assessing the efficiency of radiation use by intercrops. In: Sinoquet, H. and Cruz, P. (eds) *The Ecophysiology of Tropical Intercropping.* INRA Editions, Paris, pp. 305–318.

Azam-Ali, S.N., Gregory, P.J. and Monteith, J.L. (1984) Effects of planting density on water use and productivity of pearl millet (*Pennisetum typhoides*) grown on stored water. 2. Water use, light interception and dry matter production. *Experimental Agriculture* 20, 215–224.

Azam-Ali, S.N., Simmonds, L.P., Nageswara Rao, R.C. and Williams, J.H. (1989) Population, growth and water use of groundnut maintained on stored water. 3. Dry matter, water use and light interception. *Experimental Agriculture* 25, 77–86.

Azam-Ali, S.N., Mathews, R.B., Williams, J.B. and Peacock, J.M. (1990) Light use, water uptake and performance of individual components of a sorghum/groundnut intercrop. *Experimental Agriculture* 26, 413–427.

Azam-Ali, S.N., Crout, N.M.J. and Bradley, R.G. (1993) Perspectives in modelling resource capture by crops. In: Monteith, J.L., Scott, R.K. and Unsworth, M.H. (eds) *Resource Capture by Crops.* Proceedings of 52nd University of Nottingham Easter School. Nottingham University Press, Loughborough, pp. 125–148.

Baker, J.M. and van Bavel, C.H.M. (1987) Measurement of mass flow in the stems of herbaceous plants. *Plant, Cell and Environment* 10, 777–782.

Baker, N.R. (1991) A possible role for photosystem II in environmental perturbations of photosynthesis. *Physiologia Plantarum* 81, 563–570.

Batchelor, C.H. and Roberts, J.M. (1983) Evaporation from the irrigation water, foliage and panicles of paddy rice in north-east Sri Lanka. *Agricultural Meteorology* 29, 11–26.

Bell, C.J. and Squire, G.R. (1981) Comparative measurements with two water vapour diffusion porometers (dynamic and steady state). *Journal of Experimental Botany* 32, 1143–1156.

Belsky, A.J., Mwonga, S.M. and Duxbury, J.M. (1993) Effects of widely spaced trees and livestock grazing on understorey environments in tropical savannas. *Agroforestry Systems* 24, 1–20.

Biscoe, P.V., Scott, R.K. and Monteith, J.L. (1975) Barley and its environment. III. Carbon budget of the stand. *Journal of Applied Ecology* 12, 269–291.

Black, C.R., Tang, D.-Y., Ong, C.K., Solon, A. and Simmonds, L.P. (1985) Effects of soil moisture stress on the water relations and water use of groundnut stands. *New Phytologist* 100, 312–328.

Brown, S.C., Keatinge, J.D.H., Gregory, P.J. and Cooper, P.J.M. (1987) Effects of fertiliser, variety and location on barley production under rainfed conditions in Northern Syria. 1. Root and shoot growth. *Field Crops Research* 16, 53–66.

Cohen, Y., Fuchs, M. and Green, G.C. (1981) Improvement of the heat pulse method for determining sap flow in trees. *Plant, Cell and Environment* 4, 391–397.

Cohen, Y., Fuchs, M. and Cohen, S. (1983) Resistance to water uptake in mature citrus trees. *Journal of Experimental Botany* 34, 451–460.

Cohen, Y., Fuchs, M. and Moreshet, S. (1993) Estimating citrus orchard resistance from measurement of actual and potential transpiration. In: Borghetti, M., Grace, J. and Raschi, A. (eds) *Water Transport in Plants Under Climatic Stress.* Cambridge University Press, Cambridge, pp. 228–237.

Connor, D.J., Jones, T.J. and Palta, J.A. (1985) Response of sunflower to strategies of irrigation. II. Growth, yield and efficiency of water use. *Field Crops Research* 10, 15–26.

Cooper, P.J.M., Keatinge, J.D.H. and Hughes, G. (1983) Crop evapotransportation – a technique for calculation of its components by field measurements. *Field Crops Research* 7, 299–312.

Cooper, P.J.M., Gregory, P.J., Tully, D. and Harris, H.C. (1987) Improving water use efficiency of annual crops in rainfed farming systems of West Asia and North Africa. *Experimental Agriculture* 23, 113–158.

Corlett, J.E. (1989) *Leucaena*/millet alley cropping in India: microclimate and productivity. PhD thesis, University of Nottingham, UK.

Corlett, J.E., Ong, C.K., Black, C.R. and Monteith, J.L. (1992a) Above and below-ground interactions in a *Leucaena*/millet alley cropping system. I. Experimental design, instrumentation and diurnal trends. *Agricultural and Forest Meteorology* 60, 53–72.

Corlett, J.E., Black, C.R., Ong, C.K. and Monteith, J.L. (1992b) Above and below-ground interactions in a *Leucaena*/millet alley cropping system. II. Light interception and dry matter production. *Agricultural and Forest Meteorology* 60, 73–91.

Daniel, J.N. and Ong, C.K. (1989) Perennial pigeonpea: a multipurpose species for agroforestry systems. *Agroforestry Systems* 10, 113–119.

Dhyani, S.K., Narain, P. and Singh, R.K. (1990) Studies on root distribution of five multipurpose tree species in Doon Valley, India. *Agroforestry Systems* 12, 149–161.

Doley, D. and Grieve, B.J. (1966) Measurement of sap flow in a eucalypt by thermoelectric methods. *Australian Forest Research* 2, 3–27.

Eastham, J. and Rose, C.W. (1990a) Tree/pasture interactions at a range of tree densities in an agroforestry experiment. I. Rooting patterns. *Australian Journal of Agricultural Research* 41, 683–695.

Eastham, J. and Rose, C.W. (1990b) Tree/pasture interactions at a range of tree densities in an agroforestry experiment. II. Water uptake in relation to rooting patterns. *Australian Journal of Agricultural Research* 41, 697–707.

Eastham, J., Rose, C.W., Cameron, D.M. and Rance, S.J. (1988) The effect of tree spacing on evaporation from an agroforestry experiment. *Agricultural and Forest Meteorology* 42, 355–368.

El-Sharkawy, M.A. and Cock, J.H. (1984) Water use efficiency of cassava. I. Effects of air humidity and water stress on stomatal conductance and gas exchange. *Crop Science* 24, 497–501.

Evans, J.R., von Caemmerer, S. and Adams III, W.W. (1988) *The Ecology of Invasions by Plants and Animals.* Methuen, London, 181 pp.

Field Crops Research (1993) Intercropping – bases of productivity. 34 (Special issue), 239–470.

Fukai, S. and Trenbath, B.R. (1993) Processes determining intercrop productivity and yields of component crops. *Field Crops Research* 34, 247–271.

Gardner, W.R. (1965) Dynamic aspects of soil-water availability to plants. *Annual Review of Plant Physiology* 16, 323–342.

Green, C.F. and Deuchar, C.N. (1985) On improved solarimeter construction. *Journal of Experimental Botany* 36, 690–693.

Gregory, P.J. and Reddy, M.S. (1982) Root growth in an intercrop of pearl millet/groundnut. *Field Crops Research* 5, 241–252.

Hairiah, K., van Noordwijk, M., Santoso, B. and Syekhfani, M.S. (1992) Biomass production and root distribution of eight trees and their potential hedgerow intercropping on an ultisol in southern Sumatra. *Agrivita* 15, 54–68.

Harris, D., Hamdi, Q.A. and Terry, A.C. (1987) Germination and emergence of *Sorghum bicolor*: genotypic and environmentally induced variation in the response to temperature and depth of sowing. *Plant, Cell and Environment* 10, 501–508.

Hauser, S. (1993) Root distribution of *Dactyladenia (Acioa) barteri* and *Senna (Cassia) siamea* in alley cropping on Ultisol. 1 Implication for field experimentation. *Agroforestry Systems* 24, 111–122.

Hegde, D.M. and Saraf, C.S. (1979) Effect of intercropping and phosphorus fertilisation of pigeon-pea on soil temperature and soil moisture extraction. *Indian Journal of Agronomy* 24, 217–220.

Heimann, J. and Stickan, W. (1993) Heat pulse measurements on beech (*Fagus sylvatica* L.) in relation to weather conditions. In: Borghetti, M., Grace, J. and Raschi, A. (eds) *Water Transport in Plants Under Climatic Stress.* Cambridge University Press, Cambridge, pp. 174–180.

Heine, R.W. and Farr, D.J. (1973) Comparison of heat-pulse and radioisotope tracer methods for determining sap-flow velocity in stem segments of poplar. *Journal of Experimental Botany* 24, 649–654.

Hiebsch, C.K. and McCollum, R.E. (1987) Area and time equivalency ratio: A method of evaluating the productivity of intercrops. *Agronomy Journal* 79, 15–22.

Hitimana, L., Franzel, S. and Akyeanpong, E. (1994) On and off station with farmers. *Agroforestry Today* 61, 11–12.

Howard, S.B., Ong, C.K., Rao, M.R., Mathura, M. and Black, C.R. (1995) Partitioning of light and water in *Leucaena*/maize agroforestry systems. In: Sinoquet, H. and Cruz, P. (eds) *The Ecophysiology of Tropical Intercropping.* INRA Editions, Paris, pp. 123–135.

Huber, B. (1932) Beobachtung und Messung pflanzicher Saftstrome Berichte. *Deutsche Botanische Gesellschaft* 50, 89–109.

Hulugalle, N.R. and Lal, R. (1986) Soil water balance of intercropped maize and cowpea grown in a tropical hydromorphic soil in western Nigeria. *Agronomy Journal* 77, 86–90.

Huxley, P.A., Pinney, A., Akunda, E. and Muraya, P. (1994) A tree/crop interface orientation experiment with a *Grevillea robusta* hedgerow and maize. *Agroforestry Systems* 26, 23–45.

ICRAF (1994) *International Centre for Research in Agroforestry Annual Report.* ICRAF, Nairobi.

Ishida, T., Campbell, G.S. and Calissendorff, C. (1991) Improved heat balance method for determining sap flow rate. *Agricultural and Forest Meteorology* 56, 36–48.

Jena, D. and Misra, C. (1988) Effect of crop geometry (row proportions) on the water balance of the root zone of a pigeonpea and rice intercropping system. *Experimental Agriculture* 24, 385–391.

Jonsson, K., Fidjeland, L., Maghembe, J.A. and Hogberg, P. (1988) The vertical distribution of fine roots of five tree species and maize in Morogoro, Tanzania. *Agroforestry Systems* 6, 63–69.

Keating, B.A. and Carberry, P.S. (1993) Resource capture and use in intercropping; solar radiation. *Field Crops Research* 34, 273–301.

Kessler, J.J. (1992) The influence of karite (*Vitellaria paradox*) and nere (*Parkia biglobosa*) trees on sorghum production in Burkina Faso. *Agroforestry Systems* 17, 97–134.

Khan, A.A.H. and Ong, C.K. (1995) Correction of systematic errors in estimates of transpiration obtained using a constant temperature heat balance technique. *Experimental Agriculture* 31, 461–472.

Kushwaha, B.L. and De, R. (1987) Studies of resource use and yield of mustard and chickpea grown in intercropping systems. *Journal of Agricultural Science, Cambridge* 108, 31, 461–472.

Le Houéron, H.N. (1972) Africa – the mediterranean region. In: McKell, C.M., Blaisdell, J.P. and Goodin, J.R. (eds) *Wildland Shrubs – their Biology and Utilization.* USDA Forest Service, Ogden, Utah, pp. 26–36.

Loomis, R.S. (1985) Systems approaches for crop and pasture research. In: Yates, J.J. (ed.) *Proceedings of Third Australian Agronomy Conference.* Australian Society of Agronomy, Melbourne, Australia, pp. 1–8.

Loomis, R.S. and Connor, D.J. (1992) *Crop Ecology: Productivity and Management in Agricultural Systems.* Cambridge University Press, 600 pp.

Mandal, B.K., Ray, P.K. and Das Gupta, S. (1986) Water use by wheat, chickpea and mustard grown as sole crops and intercrops. *Indian Journal of Agricultural Science* 56, 187–193.

Marshall, B. and Willey, R.W. (1983) Radiation interception and growth in an intercrop of pearl millet/groundnut. *Field Crops Research* 7, 141–160.

Marshall, F.M. (1995) Resource partitioning and productivity of perennial pigeonpea/groundnut agroforestry systems in India. PhD thesis, The University of Nottingham, UK.

Marshall, F.M., Black, C.R. and Ong, C.K. (1994) Heat balance measurements of transpiration in perennial pigeonpea–groundnut agroforestry systems. In: Monteith, J.L., Scott, R.K. and Unsworth, M.H. (eds) *Resource Capture by Crops*. Proceedings of 52nd University of Nottingham Easter School. Nottingham University Press, Loughborough, pp. 426–429.

Mathews, R.B., Saffell, R.A. and Campbell, G.S. (1987) An instrument to measure light distribution in row crops. *Agricultural and Forest Meteorology* 39, 177–184.

Mathews, R.B., Harris, D., Nageswara Rao, R.C., Williams, J.H. and Wadia, K.D.R. (1988) The physiological basis for yield differences between four genotypes of groundnut (*Arachis hypogaea*) in response to drought. I. Dry matter production and water use. *Experimental Agriculture* 24, 191–202.

Monteith, J.L. (1981a) Does light limit crop production? In: Johnson, C.B. (ed.) *Physiological Processes Limiting Plant Productivity*. Butterworths, London, pp. 23–39.

Monteith, J.L. (1981b) Evaporation and surface temperature. *Quarterly Journal of the Royal Meteorological Society* 107, 1–27.

Monteith, J.L. (1986) Significance of coupling between saturation vapour pressure deficit and rainfall in monsoon climates. *Experimental Agriculture* 17, 113–126.

Monteith, J.L. (1993) Using tube solarimeters to measure radiation intercepted by crop canopies and to analyse stand growth. *Applications Note* TSL-AN-4-1. Delta T Devices Ltd, Cambridge, UK.

Monteith, J.L., Campbell, G.S. and Potter, E.A. (1988) Theory and performance of a dynamic diffusion porometer. *Agricultural and Forest Meteorology* 44, 27–38.

Monteith, J.L., Ong, C.K. and Corlett, J.E. (1991) Microclimatic interactions in agroforestry systems. *Forest Ecology and Management* 45, 31–44.

Monteith, J.L., Scott, R.K. and Unsworth, M.H. (eds) (1994) *Resource Capture By Crops*. Proceedings of 52nd University of Nottingham Easter School. Nottingham University Press, Loughborough, 469 pp.

Moreshet, S., Cohen, Y. and Fuchs, M. (1983) Response of mature 'shamouti' orange trees to irrigation of different soil volumes at similar levels of available water. *Irrigation Science* 3, 223–236.

Morris, R.A. and Garrity, D.R. (1993) Resource capture and utilization in intercropping: water. *Field Crops Research* 34, 303–317.

Morris, R.A., Villegas, A.N., Polthanee, A. and Centeno, H.S. (1990) Water use by mono-cropped and intercropped cowpea and sorghum grown after rice. *Agronomy Journal* 82, 664–668.

Natarajan, M. and Willey, R.W. (1980a) Sorghum–pigeonpea intercropping and the effects of plant population density. I. Growth and yield. *Journal of Agricultural Science, Cambridge* 95, 51–58.

Natarajan, M. and Willey, R.W. (1980b) Sorghum–pigeonpea intercropping and the effects of plant population density. II. Resource use. *Journal of Agricultural Science, Cambridge* 95, 59–65.

Natarajan, M. and Willey, R.W. (1985) Effect of row arrangement on light interception and yield in sorghum–pigeonpea intercropping. *Journal of Agricultural Science, Cambridge* 104, 263–270.

Natarajan, M. and Willey, R.W. (1986) The effects of water stress on yield advantages of intercropping systems. *Field Crops Research* 13, 117–131.

Nilson, T. (1971) A theoretical analysis of the frequency of gaps in plant stands. *Agricultural Meteorology* 8, 25–38.

ODA (1987) *Microclimatology in Tropical Agriculture*, Vol. 1 Final Report, Research Schemes R3208 and R3819. Overseas Development Administration, London, 202 pp.

Odongo, J.C.W., Ong, C.K., Khan, A.A.H. and Sharma, M.M. (1996) Productivity and resource use in a perennial pigeonpea/groundnut agroforestry system. *Agroforestry Systems* (in press).

Ofori, F. and Stern, W.R. (1987) Cereal–legume intercropping systems. *Advances in Agronomy* 41, 41–90.

Ong, C.K. (1995) The 'dark side' of intercropping: manipulation of soil resources. In: Sinoquet, H. and Cruz, P. (eds) *The Ecophysiology of Tropical Intercropping*. INRA Editions, Paris, pp. 45–66.

Ong, C.K. and Black, C.R. (1994) Complementarity of resource use in intercropping and agroforestry systems. In: Monteith, J.L., Scott, R.K. and Unsworth, M.H. (eds) *Resource Capture by Crops*. Proceedings of 52nd University of Nottingham Easter School. Nottingham University Press, Loughborough, pp. 255–278.

Ong, C.K. and Khan, A.A.H. (1993) A direct method for measuring water uptake by individual tree roots. *Agroforestry Today* 5, 2–5.

Ong, C.K., Simmonds, L.P. and Mathews, R.B. (1987) Responses to saturation deficit in a stand of groundnut (*Arachis hypogaea* L.). 2. Growth and development. *Annals of Botany* 59, 121–128.

Ong, C.K., Corlett, J.E., Singh, R.P. and Black, C.R. (1991) Above and below-ground interactions in agroforestry systems. *Forest Ecology and Management* 45, 45–57.

Ong, C.K., Odongo, J.C.W., Marshall, F. and Black, C.R. (1992) Water use of agroforestry systems in semi-arid India. In: Calder, I.R., Hall, R.L. and Adlard, P.G. (eds) *Growth and Water Use of Forest Plantations*. Wiley, Chichester, pp. 347–358.

Osmond, C.B., Winter, K. and Ziegler, H. (1982) Functional significance of different pathways of CO_2 fixation in photosynthesis. In: Lange, O.L., Nobel, P.S., Osmond, C.B. and Zeigler, H. (eds) *Physiological Plant Ecology II. Water Relations and Carbon Assimilation* (Encyclopedia of Plant Physiology, new series, vol. 12B). Springer-Verlag, Heidelberg, 747 pp.

Passioura, J.B. (1983) Roots and drought resistance. *Agricultural Water Management* 7, 265–280.

Phillips, W.S. (1963) Depths of roots in soil. *Ecology* 44, 424.

Potter, E.A. (1992) Improvements in the theory and calibration of Delta-T porometers. *Applications Note*. Delta-T Devices Limited, Cambridge, UK, 18 pp.

Rao, M.R. and Willey, R.W. (1983) Effects of pigeonpea plant population and row arrangement in sorghum/pigeonpea intercropping. *Field Crops Research* 7, 203–212.

Rao, M.R., Sharma, M.M. and Ong, C.K. (1990) A study of the potential of hedgerow intercropping in semi-arid India using a two-way systematic design. *Agroforestry Systems* 11, 243–258.

Rao, M.R., Ong, C.K., Pathak, P. and Sharma, M.M. (1992) Productivity of annual cropping and agroforestry systems on a shallow alfisol in semi-arid India. *Agroforestry Systems* 15, 51–64.

Rao, M.R., Muraya, P. and Huxley, P.A. (1993) Observations of some tree root systems in agroforestry intercrop situations, and their graphical representation. *Experimental Agriculture* 29, 183–194.

Reddy, M.S. and Willey, R.W. (1981) Growth and resource use studies in an intercrop of pearl millet/groundnut. *Field Crops Research* 4, 13–24.

Reddy, M.S., Floyd, C.N. and Willey, R.W. (1980) Groundnut in intercropping systems. In: *Proceedings of International Workshop on Groundnuts.* ICRISAT, Patancheru, India, pp. 133–142.

Rees, D.J. (1986a) Crop growth, development and yield in semi-arid conditions in Botswana. II. The effects of intercropping *Sorghum bicolor* with *Vigna unguiculata. Experimental Agriculture* 22, 169–177.

Rees, D.J. (1986b) The effects of population density and intercropping with cowpea on the water use and growth of sorghum in semi-arid conditions in Botswana. *Agricultural and Forest Meteorology* 37, 293–308.

Roberts, J., Cabral, O.M.R., Fisch, G., Molion, L.C.B., Moore, C.J. and Shuttleworth, W.J. (1993) Transpiration from an Amazonian rain forest calculated from stomatal conductance measurements. *Agricultural and Forest Meteorology* 65, 175–196.

Sabatti, M., Scarascia Mugnozza, G.E., Valentini, R. and Del Lungo, A. (1993) Water relations and water transport in coppice vs. single stem *Quercus cerris* L. trees. In: Borghetti, M., Grace, J. and Raschi, A. (eds) *Water Transport in Plants Under Climatic Stress.* Cambridge University Press, Cambridge, pp. 191–204.

Sakuratani, T. (1981) A heat balance for measuring water flux in the stem of intact plants. *Journal of Agricultural Meteorology* 37, 9–17.

Sakuratani, T. (1984) Improvement of the probe for measuring water flow in intact plants with the stem heat balance method. *Journal of Agricultural Meteorology* 40, 273–277.

Sakuratani, T. (1987) Studies on evapotranspiration from crops. (2) Separate estimation of transpiration and evaporation from a soybean field without water shortage. *Journal of Agricultural Meteorology* 42, 309–317.

Sakuratani, T. (1990) Measurement of the sap flow rate in stem of rice plant. *Journal of Agricultural Meteorology* 45, 277–280.

Sheriff, D.W. (1972) A new apparatus for the measurement of sap flux in small shoots with the magnetohydrodynamic technique. *Journal of Experimental Botany* 23, 1086–1095.

Sibbald, A.R. and Griffiths, J.H. (1992) The effects of conifer canopies at wide spacing on some under-storey meteorological parameters. *Agroforestry Forum* 3, 8–14.

Sinclair, T.R., Shiraiwa, T. and Hammer, G.L. (1992) Variation in crop radiation-use efficiency with increased diffuse radiation. *Crop Science* 32, 1281–1284.

Singh, B.P. (1985) Effect of intercropping with pearl millet on productivity, profitability and water use on aridisols. *Indian Journal of Agronomy* 30, 408–413.

Singh, S. and Russell, M.B. (1981) Water use by a maize/pigeonpea intercrop on a deep vertisol. In: *Proceedings of the International Workshop in Intercropping*, 10–13 January 1979. ICRISAT, Hyderabad, India, pp. 271–282.

Singh, S., Narwal, S.S. and Chander, J. (1988) Effect of irrigation and cropping systems on consumptive water use, water use efficiency and moisture extraction patterns of summer fodders. *Indian Journal of Agriculture* 6, 76–82.

Singh, R.P., Ong, C.K. and Saharan, N. (1989) Above and below-ground interactions in alley cropping in semi-arid India. *Agroforestry Systems* 9, 259–274.

Sinoquet, H. and Andrieu, B. (1993) The geometrical structure of plant canopies: characterisation and direct measurement methods. In: *Crop Structure and Light Microclimate: Characterisation and Applications*. INRA Editions, Paris, pp. 131–158.

Sivakumar, M.V.K. and Virmani, S.M. (1980) Growth and resource use of maize, pigeonpea and maize/pigeonpea intercrop in an operational watershed. *Experimental Agriculture* 16, 377–386.

Sivakumar, M.V.K. and Virmani, S.M. (1984) Growth and resource use of maize, pigeonpea and maize/pigeonpea intercrop in an operational research watershed. *Experimental Agriculture* 16, 377–386.

Squire, G.R. (1990) *The Physiology of Tropical Crop Production*. CAB International, Wallingford, Oxford, UK, 236 pp.

Squire, G.R. and Black, C.R. (1982) Stomatal behaviour in the field. In: Jarvis, P.J. and Mansfield, T.A. (eds) *Stomatal Physiology*. Cambridge University Press, Cambridge, pp. 223–245.

Squire, G.R., Gregory, P.J., Monteith, J.L., Russell, M.B. and Piara Singh. (1984a) Control of water use by pearl millet (*Pennisetum typhoides* S & H). *Experimental Agriculture* 20, 135–139.

Squire, G.R., Marshall, B., Terry, A. and Monteith, J.L. (1984b) Response to temperature in a stand of pearl millet. 6. Light interception and dry matter production. *Journal of Experimental Botany* 35, 599–610.

Steinberg, S., van Bavel, C.H.M. and McFarland, M.J. (1989) A gauge to measure mass flow rate of sap in stems and trunks of woody plants. *Journal of the American Society for Horticultural Science* 114, 466–472.

Stewart, J.I. (1983) Crop yields and returns under different soil moisture regimes. In: Holmes, J.C. and Talini, W.M. (eds) *More Food from Better Technology*. FAO, Rome, pp. 427–438.

Stone, J.F. and Shirazi, G.A. (1975) On the heat-pulse method for the measurement of apparent sap velocity in stems. *Planta* 122, 169–177.

Suwanarit, A., Verasan, V., Suwannarat, C. and Chaochong, S. (1984) Comparative uses of maize–mungbean intercrop, sole maize and sole mungbean. *Kasetsart Journal* 18, 117–121.

Szott, L.T. (1996) Spatial and temporal patterns of nutrient uptake by trees in agroforestry systems – the nutrient pumping hypothesis. *Agroforestry Systems* (in press).

Tanner, C.B. (1981) Transpiration efficiency of potato. *Agronomy Journal* 73, 59–64.

Trenbath, B.R. (1974) Application of a growth model to problems of the productivity and stability of mixed stands. In: *Proceedings of 12th International Grassland Congress*, Vol. 1. Moscow, pp. 546–558.

Trenbath, B.R. (1976) Light-use efficiency of crops and the potential for improvement through intercropping. In: Willey, R.W. (ed.) *Proceedings of International Workshop on Intercropping*. ICRISAT, Hyderabad, India, pp. 141–154.

Trenbath, B.R. (1986) Resource use by intercrops. In: Francis, C.A. (ed.) *Multiple Cropping Systems*. Macmillan, New York, pp. 57–81.

Tsay, J.S., Fukai, S. and Wilson, G.L. (1987) The response of cassava (*Manihot esculenta*) to spatial arrangement and to soybean intercrop. *Field Crops Research* 16, 19–31.

van Bavel, C.H.M. (1988) Absolute measurement of the sap flow rate in stems of intact plants. In: *Biophysical Measurements and Instrumentation – A Symposium Honoring Champ B. Tanner.*

Vandermeer, J. (1989) *The Ecology of Intercropping*. Cambridge University Press, Cambridge, UK, 237 pp.

Vieweg, G.I.I. and Ziegler, I.I. (1960) Thermoelektrische Registrierung der Geschwindigkeit des Transpirationsstromes. *Berichte Deutsche Botanische Gesellschaft* 73, 221–226.

Wallace, J.S. (1991) The measurement and modelling of evaporation from semi-arid land. In: Sivakumar, M.V.K., Wallace, J.S., Renard, C. and Giroux, C. (eds) *Soil Water Balance in the Sudano-Sahelian Zone*. Proceedings of Niamey Workshop, 1991. IAHS Publication No. 199, pp. 131–148.

Wallace, J.S. (1995) Towards a coupled light partitioning and transpiration model for use in intercrops and agroforestry. In: Sinoquet, H. and Cruz, P. (eds) *The Ecophysiology of Tropical Intercropping*. INRA Editions, Paris, pp. 153–162.

Wallace, J.S., Roberts, J.M. and Sivakumar, M.V.K. (1990) The estimation of transpiration from sparse dryland millet using stomatal conductance and vegetation area index. *Agricultural and Forest Meteorology* 51, 35–49.

Willey, R.W. (1979) Intercropping – its importance and research needs. I. Competition and yield advantages. *Field Crop Abstracts* 32, 1–10.

Willey, R.W., Natarajan, M., Reddy, M.S. and Rao, M.R. (1986) Cropping systems with groundnut: Resource use and productivity. In: *Agrometeorology of Groundnut*. Proceedings of International Symposium, ICRISAT Sahelian Centre, Niamey, Niger, 1985, pp. 193–205.

Yoshida, S. and Hasegawa, S. (1982) *Drought Resistance in Crops with Emphasis on Rice*. International Rice Research Institute, Los Banos, Philippines.

Young, A. (1989) *Agroforestry for Soil Conservation*. CAB International, Wallingford, and ICRAF, Nairobi, 276 pp.

5

Microclimatic Modifications in Agroforestry

A.J. BRENNER

School of Natural Resources and Environment, University of Michigan, Dana Building, 430 E. University, Ann Arbor, Michigan 48109-1115, USA

Introduction

Some of the most important effects of combining woody and non-woody plants in mixtures result from changes in the microclimate which in turn influence the growth of all components of the system. If we are to understand and predict the results of combining tree and crops under different circumstances, it is important to know the underlying principles that control these changes and to appreciate their potential effects on plant growth and development. To explain why particular agroforestry systems work in one environment and not another, and how to manage them, requires a 'process-based' approach in order to cover the large number of possible plant combinations and wide range of ecozones. Only by taking this approach can the unjustified extrapolation of agroforestry systems from one area to another be avoided, and the development of appropriate management practices be achieved.

A large number of changes occur when a tree is introduced into a farmer's field; those of particular importance to this chapter are changes in the radiation balance under trees and the surface wind pattern (Fig. 5.1). The combined effect of these changes controls the energy balance of both the overstorey and the understorey, thus influencing plant water use and productivity.

To understand the main microclimate modifications that occur in agroforestry the first part of the chapter explains the relationships between radiation, wind, air humidity and temperature and their effect on evaporation of water and plant growth in the context of agroforestry

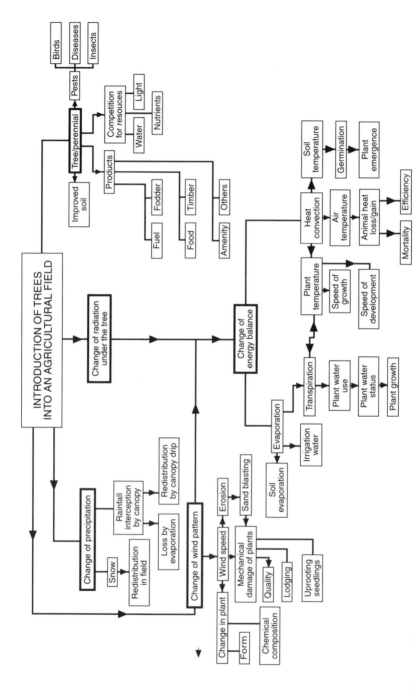

Fig. 5.1. The changes in a farming system on introduction of trees. The flow diagram shows causal relationships by lines with arrows and subdivisions by lines without arrows.

practices. These principles are then applied specifically, as an example, to windbreaks and some comments are made on other functions of windbreaks, i.e. protection against wind damage and soil erosion.

Other consequences that result from growing trees on crop land are shown in Fig. 5.1, many of which are covered in this book. All these factors need to be considered when planning an agroforestry system, but they fall outside the scope of this chapter.

Specific Microclimatic Changes

Radiation

In all agroforestry systems the planting of perennial woody components changes the average radiation incident on understorey plants. These changes are often considered a disadvantage although some authors have found that they are advantageous (Vandenbeldt and Williams, 1992). In order to determine why shading affects different systems in different ways we need to look first at the different radiation wavelengths involved.

Solar radiation
Light, or photosynthetic quantum flux density (Q) (0.4–0.7 μm), although essential for the growth of plants, only covers a small range of wavelengths of the electromagnetic spectrum (Fig. 5.2). Solar radiation (S) (0.1–4 μm) is emitted by the sun and arrives at the surface of the earth's atmosphere at a relatively constant 1355 W m^{-2}. However, values of S incident upon a flat plane at the earth's surface are substantially lower varying with time of day, time of year, latitude and optical density of the atmosphere. S is reduced while moving through the atmosphere, by the scattering and absorption of energy from the solar beam by components such as ozone, water vapour and carbon dioxide, according to Beer's law (Equation 5.1):

$$S = S_0 e^{-\tau m} \tag{5.1}$$

where S_0 is solar radiation at the surface of the atmosphere, τ is the turbidity coefficient of the atmosphere and m is the optical air mass that radiation has to travel through to reach the earth's surface. Values of τ range between 0.07 in very clear air to 0.60 in dusty atmospheres (Monteith and Unsworth, 1990).

Time of day and year and latitude influence the angle of the sun which changes the optical mass of air by ($1/\sin(a)$) where a is solar elevation. Equations for calculating solar elevation at any time or place can be found in most meteorological texts.

In most situations the solar radiation incident upon a system is not

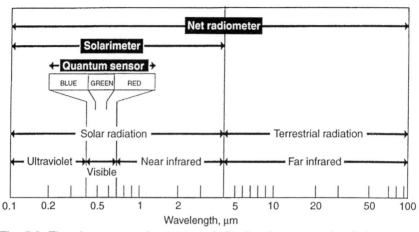

Fig. 5.2. The electromagnetic spectrum indicating the range of radiation types and the spectral sensitivity of radiation sensors (adapted from Reifsnyder and Darnhofer, 1989).

calculated using Equation 5.1 because of the need to know τ and m, so S constitutes one of the basic meteorological variables required for any type of 'process based' analysis. The solar radiation absorbed by a non-transmitting surface equals $S(1 - \alpha)$ where α is the reflectance of the surface or albedo. Very little information is available on the albedo of agroforestry systems, but information available for surfaces analogous to agroforestry systems is given in Table 5.1.

Longwave radiation
Terrestrial or longwave radiation (Fig. 5.2), although not important in terms of photosynthesis, plays a major part in plant energy balances. Every object above absolute zero emits longwave radiation (R_l), and the energy emitted from a body is dependent on its temperature (T) in Kelvins accord-

Table 5.1. Albedos for vegetated surfaces (from Monteith and Unsworth, 1990).

Vegetation type	Agroforestry equivalent	Latitude ($°$)	Daily mean
Grass	Silvopastoral parkland	52	0.24
Wheat	Low tree density silvoarable	52	0.26
Maize	Low tree density silvoarable	7	0.18
Evergreen scrub	Scattered multipurpose trees	32	0.21
Guinea savanna	Scattered multipurpose trees	9	0.19
Coniferous woodland		51	0.16
Tropical rainforest	Homegarden	7	0.13

ing to Equation 5.2:

$$R_1 = \varepsilon\sigma T^4 \tag{5.2}$$

where ε is emissivity of the surface and σ is the Stefan–Boltzmann constant $(5.67 \times 10^{-8} \text{ W m}^{-2} \text{ K}^{-4})$; ε equals 1 for a black body, 0.1 for polished copper and varies between 0.94 and 0.99 for vegetation.

Net radiation
Energy absorbed by a leaf or agroforest (net radiation, R_n) can be calculated as the balance of long- and shortwave radiation (Equation 5.3):

$$R_n = S(1 - \alpha) + R_{1,d} - R_{1,u} \tag{5.3}$$

where $R_{1,d}$ is longwave radiation absorbed by the surface and $R_{1,u}$ is longwave radiation emitted by the surface. Longwave radiation fluxes, although large (usually between 300 and 400 W m^{-2}), are relatively constant over the day, thus daytime R_n is mainly influenced by incoming S. This means that during the day there is usually a reduction of R_n under a tree canopy (this may not be true when there is no direct beam radiation). Under clear sky conditions $(R_{1,d} - R_{1,u})$ is around -100 W m^{-2} (Monteith and Unsworth, 1990), causing nighttime cooling of the atmosphere. This is because the sky is cooler than the soil or vegetation, however under a tree canopy downward longwave radiation fluxes would be similar to upward longwave fluxes from crops, thus rates of cooling of understoreys are considerably slower. This explains why less frost is observed under trees or in forests than in open fields, and may be an important function of 'shade trees' in coffee and tea plantations where these are susceptible to frost or chilling damage.

The calculation of net radiation absorbed by a canopy often uses Beer's law (Equation 5.1) substituting K, the extinction coefficient of the canopy, for τ and L, the leaf area index, for m (see Chapter 6 for the development of these ideas in an agroforestry context).

Instrumentation

Wavelength considerations There are a variety of sensors available for measuring radiation and it is important to choose the sensor that is sensitive to the wavelengths of interest (Fig. 5.2). In general, the standard instruments in most meteorological configurations are solarimeters or pyranometers which measure the whole spectrum of solar radiation, and usually have glass domes or filters. Values of Q and R_n can, if necessary, be derived from values of S. Quantum sensors are preferable for comparison with measurements of growth or rates of photosynthesis. Most quantum sensors have filters allowing measurement of radiation between 0.4 and 0.7 μm; some sensors even reproduce the wavelength sensitivity of plant leaves. Another common sensor is the red:far-red ratio sensor, which measures radiation of wavelengths 0.66 and 0.70 μm, which are important for plant development and morphogenesis. Net radiometers have plastic domes, which transmit longer wavelengths than glass, and measure the difference between upward and downward fluxes of energy over the 0.1–100 μm spectrum. More details of instruments and background to these methods can be found in Sheehy (1985), Pearcy (1989) and Bingham and Long (1990).

Spatial considerations Solarimeters and net radiometers are generally hemispherical or tubular sensors. Hemispherical sensors have the advantage of possessing no directional bias and are used for absolute values and reference measurements. Linear sensors have the advantage of spatial averaging and are suitable for measurements under crops, giving relative rather than absolute measurements. All radiation sensors need to be levelled and kept clean, otherwise errors are introduced.

The positioning of sensors to get meaningful results in agroforestry systems presents a problem that is not experienced in systems with uniform canopies, that of extreme spatial variation. Various ways of characterizing the variability of the transmission of radiation through overstoreys have been attempted. Sinclair and Jarvis (1994) measured spatial variability by regularly randomizing nine quantum sensors using a grid system that covered the area between four trees in a respaced Sitka spruce agroforest. Hemispherical photographs have also given measurements of the spatial variation of shading by tree canopies (Bonhomme and Chartier, 1972). Overstorey absorption of radiation in single trees has been measured using a rotating hemisphere of net radiometers called the 'Whirligig' (Green, 1993). Although these approaches have been shown to work they require a large amount of instrumentation. An alternative approach is to measure the incoming direct beam and diffuse solar radiation and model the radiation fluxes within the system on a computer in a complex (Wang and Jarvis, 1990) or simple way (see Chapter 6).

Wind

Trees change the wind pattern in a field both by altering the horizontal wind speed and turbulence; thus they absorb momentum and force the air to flow around them. The velocity of the air flow increases with distance from an object that absorbs momentum, whether leaf or agroforest, and if the extent of the surface is sufficient, an air-flow profile develops that is characteristic of that surface. This characteristic profile defines the boundary layer, and affects the fluxes of energy and mass to and from the surface. A relatively simple level from which to start to scale up boundary layers for agroforests is a leaf.

Boundary layer conductance at a leaf and individual canopy scale
Values of leaf boundary layer conductance can often be estimated from engineering equations based on leaf dimensions and wind speeds (Kreith, 1973). These predict transfer from real leaves reasonably well under laminar flow conditions in wind tunnels (Dixon and Grace, 1984). Forced convection for heat transfer from a single sided flat plate in laminar flow is given by:

$$g_a^h = (0.664D_h^{0.67}u^{0.5})/(d^{0.5}v^{0.17}) \qquad (5.4)$$

where g_a^h is boundary layer conductance for heat transfer, D_h is diffusivity of heat (21.5 mm^2 s^{-1} at 20°C), u is wind speed of air outside the influence of the plate, d is the characteristic dimension of the leaf, and v is kinematic viscosity (15.5 mm^2 s^{-1} at 20°C). Forced convection for heat transfer from a single sided flat plate in turbulent flow is given by:

$$g_a^h = (0.036D_h^{0.67}u^{0.8})/(d^{0.2}v^{0.47}) \qquad (5.5)$$

Except at very low wind speeds over large leaves, free convection (air movements caused by temperature gradients) is substantially smaller than forced convection (Monteith and Unsworth, 1990).

Pathways for water vapour and heat from leaf to air are usually assumed to be identical. If leaves have stomata on only one side, the conductance for heat transfer is twice that for vapour transfer. At 20°C the ratio of the conductances g_a^h/g_a^v is 0.93, where g_a^v is the boundary layer conductance for water vapour, but is often taken as 1 (Grace, 1977). Although in many cases equations can be used to calculate g_a, field measurements with heated leaf replicas (Brenner and Jarvis, 1995) indicated that for cereal leaves g_a varied between Equations 5.4 and 5.5 because of changes in turbulence of the airstream. In agroforestry where turbulence would be expected to increase, relative to a monocrop, the dependability of Equations 5.4 and 5.5 is reduced, thus the best way of obtaining g_a is directly. Direct measurement of g_a for whole Sitka spruce canopies showed that g_a per tree increased with spacing from 2 to 8 m (Teklehaimanot *et al.*, 1991). However, there are not enough data to decide whether relationships derived by this work are universally applicable.

Boundary layer conductance at the agroforest scale
Boundary layer conductance depends on surface roughness (widely spaced trees are aerodynamically rougher than pastures), extent of surface and speed and turbulence of incident air flow. Measured over extensive areas of forest or crop wind speed shows a logarithmic decrease with height characterized by a roughness length (z_0), often taken as 0.1 of the height of the vegetation (h), and displacement height (d), often taken as 0.67–0.70h. Wind, humidity, temperature and CO$_2$ profiles are used for the calculation of the fluxes of water, heat and carbon dioxide across the boundary layer,

using Fick's law of diffusion:

$$F_i = g_i \partial i / \partial x \qquad (5.6)$$

where F_i is the flux of quantity i, g_i is the conductance to quantity i and $\partial i / \partial x$ is the concentration gradient of quantity i. In terms of transfer across boundary layers the term g_i is often known as the boundary layer conductance (g_a). A simple calculation of g_a from wind profiles can be made from the knowledge of wind speed u at height z above the surface:

$$g_a = (k^2 u / \ln((z - d)/z_0)^2) \qquad (5.7)$$

where k is Von Karman's constant (0.41). This approach can be used to predict water, heat, and carbon transfer assuming that the conductances for all three quantities are equivalent. However these calculations require specific conditions that will rarely be found in agroforests. Most agroforestry systems have non-uniform and complex canopy structures, and wind profiles are not logarithmic, for example, in a field with widely spaced trees (Fig. 5.3a) or behind a windbreak (Fig. 5.3b). New equations developed for sparse canopies have more relevance to agroforests although these still need validation in the field (Shuttleworth and Gurney, 1990; and see Chapter 6). A characteristic boundary layer might develop above an extensive and uniform agroforestry system at around 1 m of characteristic boundary layer for each 200 m of system (Monteith and Unsworth, 1990), but many agroforestry systems are small in extent, thus its boundary layer would be constantly in transition between the agroforest and the surrounding vegetation.

Instrumentation

Instrumental and spatial considerations Cup anemometers are generally used for measuring wind speed and vary in shape, size and specification. Anemometers with light polystyrene cups allow very sensitive measurement of changes in wind speed and consequently give better measurements of turbulence than more robust metal and plastic cups. Better measurements of turbulence and low wind speeds are achieved with hot wire or hot film anemometers, but these are delicate and easily broken in the field. Sonic anemometers that use the Doppler effect to measure wind speed are being used increasingly; although more expensive than other methods they have an extremely fast response time (50 Hz) and can be used for studies of turbulence and calculation of fluxes by eddy correlation (Grace, 1985, 1989).

Placement of sensors is crucial to the result gained. Often wind speed is measured at a height of 2 m above the height of the vegetation. This is appropriate for uniform vegetation and should probably be maintained as a reference for heterogeneous vegetation. It does not however give sufficient information to derive the conductance of an understorey or overstorey canopy. Although the equations of Shuttleworth and Gurney (1990)

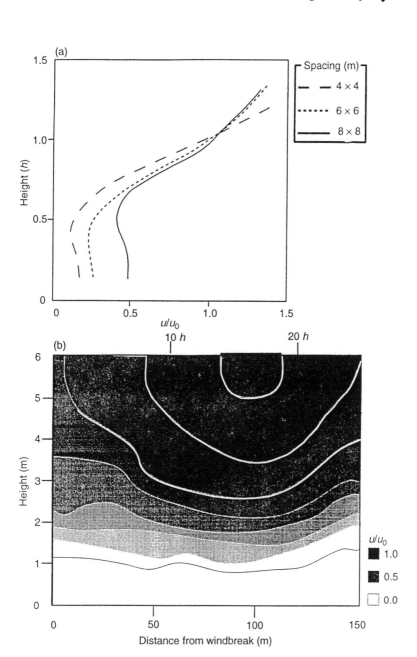

Fig. 5.3. (a) Change of normalized wind speed (u/u_0) with height within a re-spaced Sitka spruce stand. Normalized to wind speed at the top of the canopy (Green, 1991). (b) The pattern of normalized wind speed (u/u_0) in the lee of a 6-m high windbreak, normalized to wind speed $6h$ windward of the windbreak and 5 m height (Brenner *et al.*, 1995a).

provide an estimate of boundary layer conductance in a two element model, accuracy of these estimates for agroforestry systems is not known. It is thus advisable to measure wind speed at the height of the understorey since it gives the actual conditions the understorey is experiencing (Brenner *et al.*, 1995a). An alternative approach, if the instrumentation is available, is to measure wind profiles, model the system in a wind tunnel or use a fluid dynamics model (Green, 1991). The data are lacking to derive the relationship between wind speed, turbulence and boundary layer conductance for agroforestry. Consequently until these relationships are known it is more reliable to measure boundary layer conductance directly.

Direct measurement of boundary layer conductance   Boundary layer conductance can be measured directly for leaves by weighing a blotting paper replica saturated with water before and after being exposed to an air stream, then applying Equation 5.14 assuming that $r_s = 0$, and measuring all the other variables (Grace and Wilson, 1976). The boundary layer conductance of the leaf can also be measured by heating a metal replica of a leaf, and measuring its cooling curve (Dixon and Grace, 1984). Alternatively the difference in temperature between a heated and unheated replica allows calculation of g_a from Equation 5.13 (Brenner and Jarvis, 1995). Boundary layer conductance of whole tree canopies has been measured by suspending trees from a balance and measuring the rate of drying after fully wetting the canopy with a spray system (Teklehaimanot *et al.*, 1991). Direct measurements of g_a are generally preferable although require more work, but combined with development of new equations for predicting g_a from heterogeneous surfaces, e.g. scrubland, may lead to more useful solutions to this problem for agroforestry.

Humidity and temperature

Definition of terms
In order to look at how changes in the boundary layer and radiation affect the energy balance of plants it is necessary to first define some terms that will be used later in this chapter (for further information see Monteith and Unsworth, 1990; Jones, 1992). There are several ways of describing the water vapour content of the air: mass concentration of water vapour in air or absolute humidity (c_w) (g m^{-3}), water vapour pressure of air, e (kPa), vapour pressure deficit, D (kPa), and relative humidity, h_r (%). The saturation of air with water vapour is highly temperature dependent (Fig. 5.4), and an equation that describes this relationship

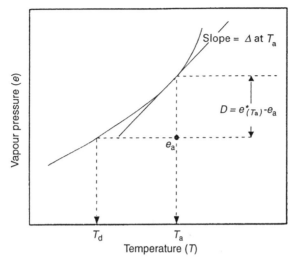

Fig. 5.4. The relationship between saturated water vapour pressure and temperature. The vapour pressure deficit (D_a), dew-point (T_d) and slope of the curve (s) are marked on the graph for air at temperature T_a and vapour pressure e_a.

(Murray, 1967) is:

$$e_{(T)}^* = 0.611 \ \exp^{(17.27(T-273)/(T-36))} \tag{5.8}$$

where T is air temperature in K, and $e_{(T)}^*$ is the saturated water vapour pressure at temperature T in kPa. Another important quantity in micro-meteorology is the slope of this relationship or Δ (for use see Chapter 6; for calculation see Jones, 1992).

Humidity is often expressed as the difference between the potential water holding capacity of the air and the actual vapour pressure of the air and is called the water vapour pressure saturation deficit of the air (D):

$$D = e_{(T_a)}^* - e_a \tag{5.9}$$

The relative humidity (h_r) is:

$$h_r = 100(e_a/e_{(T_a)}^*) \tag{5.10}$$

Vapour pressure deficit is debatably a more useful term for plant physiologists than relative humidity since it describes the gradient for vapour transfer which relative humidity does not. Another term often used to define the humidity of the air is the dew-point (T_d), which is the temperature of the air at which water will condense (Fig. 5.4).

Instrumentation

Humidity sensors Psychrometers have been used extensively for measuring humidity, from air temperature (T_a) and temperature of a wet surface evaporating at its potential rate (T_w)

$$e_a = e^*_{(T_w)} - \gamma(T_a - T_w) \tag{5.11}$$

where γ is the psychrometric constant. Psychrometers must be ventilated with a fan (> 4 m s^{-1}), the wet surface kept saturated and clean and both sensors protected from radiative heating. Capacitance sensors that measure relative humidity have low power requirements and need little maintenance compared with psychrometers. Their accuracy depends upon their calibration, which is rarely linear between 5 and 95% and may change with age and contamination. They often have errors associated with the heating or cooling of their radiation shield if it is not ventilated. So for low maintenance systems capacitance sensors are preferable, whereas psychrometers may give more accurate results. Other more expensive instruments such as infrared gas analysers and dew-point hygrometers have been used successfully in field situations, but in general these require a higher level of maintenance and expertise when compared with psychrometers and capacitance sensors. Further information on these techniques can be found in Day (1985) and Bingham and Long (1990).

Temperature sensors Although any material that changes its properties with temperature can be used as a sensor, temperature is most often measured by thermistors (resistors that change their resistance with temperature) and thermocouples (a combination of wires of different metals that when connected generate a voltage that is related to the temperature difference between the two junctions). In order to use thermocouples for absolute measurement one junction must be at a known temperature. Thermocouples are cheap and easy to make but have a low voltage output of about 40 μV °C^{-1} and consequently require a sensitive meter. Thermistors require an excitation current and a linearizing thermistor but the output voltage can be as large as required by the sensing device, thereby making it less prone to error by electrical noise than thermocouples. These two systems are generally the cheapest methods for measuring temperature. Platinum resistance thermometers are considered more accurate, and infrared thermometers are being used increasingly for surface temperature measurements. These methods are discussed fully in Bell and Rose (1985) and Ehleringer (1989).

Spatial considerations Positioning of sensors influences the result of measurements since temperature and humidity change with height and horizontal position within an agroforest. Standard height of 2 m above the plant surface is useful as a reference value but may not be easily related to air temperature at plant height or plant temperature (Brenner et al., 1995b). If these quantities are needed it is better to measure them directly, or at the mean flow height (Chapter 6; Fig. 5.8). The spatial variation of air humidity and temperature at mean flow height will depend upon the degree of clumping of the vegetation and the mixing within the system. Evidence from sparse canopies in dryland Spain suggest that, even in well-coupled canopies, air temperatures vary between open areas between bushes and within canopies by 0.6°C, and vapour pressure deficits can vary by 0.2 kPa (A.J. Brenner, unpublished data). In agroforestry systems this is probably less of a problem than when the understorey is dry bare soil. Leaf temperature will vary within the canopy so measurements should be stratified to take into account the variation that exists. The variation will be less in open well-coupled than dense self-shaded canopies.

Energy balance

Energy fluxes
Changes in wind speed and radiation caused by introducing trees have very important effects on the energy balance of the plant. Plants must lose the same amount of energy they absorb if they are to remain at a constant temperature. Although a certain amount of energy is stored as chemical bond energy, photosynthesis ($8–32$ W m^{-2}) and physical storage of heat (for a 5°C change of temperature around 17.5 W m^{-2}), energy is lost mainly by evaporation and convection (Jones, 1992). Thus the energy balance of a leaf can be calculated as:

$$R_n = \lambda E + H \tag{5.12}$$

where λE is evaporation from the leaf and H is convective heat loss from the leaf (Fig. 5.5). Convective heat loss and evaporation can be calculated from Fick's law (Equation 5.6) and converted to energy units by inserting the appropriate gradients and heat capacities:

$$H = (\rho c_p g_a (T_1 - T_a)) \tag{5.13}$$
$$\lambda E = (\rho c_p / \gamma)(e_1 - e_a) g_1 \tag{5.14}$$

where g_1 is the leaf conductance which is $(g_a g_s)/(g_a + g_s)$, $(T_1 - T_a)$ is leaf to air temperature difference and $(e_1 - e_a)$ is leaf to air vapour pressure difference, ρ is density of air (1.204 kg m^{-3} at 20°C), c_p is specific heat capacity of air at constant pressure (1010 J kg^{-1} K^{-1}), γ is psychrometric constant ($Pc_p/0.622\lambda$), P is atmospheric pressure (variable, usually around 100 kPa), λ is the latent heat of vaporization of water (2454 J g^{-1} K^{-1} at 20°C) and g_a and g_s are the boundary layer and stomatal conductances,

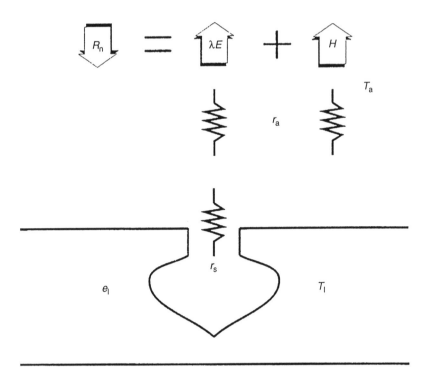

Fig. 5.5. A simplified view of the energy balance of a leaf. The heat flux (H) from the leaf is related to the temperature difference ($T_1 - T_a$) and the resistance to heat transfer (r_a). The evaporation (λE) from the leaf is related to the vapour pressure difference ($e_1 - e_a$) and the resistance to vapour transfer ($r_a + r_s$). When the leaf is in thermal equilibrium $R_n = \lambda E + H$.

respectively. Values of e_1 are usually taken as $e^*_{(T_1)}$ which is a reasonable assumption unless the plants are severely water stressed.

These equations allow the calculation of the changes of evaporation and leaf temperature with changes in net radiation and wind speed. The overall effect of a change in the boundary layer conductance of a leaf on its energy balance may be quite complex. For example in Figs. 5.6a and b, at low stomatal conductance, evaporation is controlled by g_s rather than by boundary layer conductance, g_a, but heat loss is only controlled by g_a. Therefore a decrease in g_a causes a larger decrease in the conductance for heat transfer than water transfer. Consequently the leaf heats up until a new equilibrium is established to dissipate the absorbed radiation. This increases the gradients of temperature and vapour pressure; however, because the relationship between saturated vapour pressure and tempera-

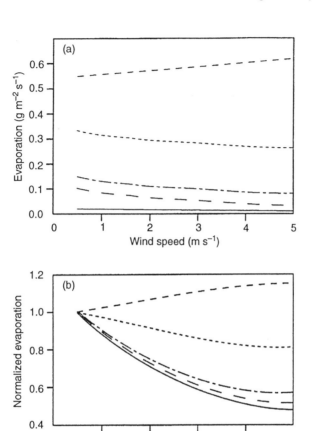

Fig. 5.6. (a) Change of leaf transpiration with wind speed (u) for a range of stomatal conductances. (b) Change of leaf transpiration relative to transpiration at $u = 0.5$ m s^{-1}. $g_s = \infty$ (– – –), $g_s = 19.5$ mm s^{-1} (- - - - -), $g_s = 3.9$ mm s^{-1} (– - – - –), $g_s = 2.0$ mm s^{-1} (— —), $g_s = 0.4$ mm s^{-1} (———). $R_n = 450$ W m^{-2}, $D = 4$ kPa and $T_a = 35°C$.

ture is non-linear (Fig. 5.4), the increase in the gradient for water vapour transfer is larger than that for heat transfer. Thus for water transfer the gradient increases and the conductance (dominated by g_s) stays the same, leading to more energy being dissipated by evaporation than previously. This is contrary to our intuitive expectation of a decrease in water loss with a decrease of wind speed.

Stomatal conductance
Nearly all land plants have stomata, some species have stomata on both sides (amphistomatous) and others have stomata on the lower side only

(hypostomatous). The main environmental variables to which stomata respond are to photosynthetic quantum flux density (Q), vapour pressure deficit (D), leaf water status (Ψ), leaf temperature (T_l) and internal CO_2 concentration (C_i) (Fig. 5.7). Shading by overstoreys reduce Q, D and T_l causing changes in g_s. Competition for water between overstorey and understorey changes leaf water status and shelter changes microclimate (see later in this chapter). So plants growing under trees may have different conductances from those grown in monoculture, changing their evaporation and photosynthetic rates. Conductance of a canopy is generally taken as average stomatal conductance multiplied by plant leaf area index. Table 5.2 presents normal leaf stomatal conductances, canopy conductances and boundary layer conductances for three types of vegetated surfaces. More advanced calculations divide the canopy into layers and calculate water loss from each layer (see Chapter 6).

In agroforestry systems water use may be divided into use by overstorey (^{t}) and understorey inside (^{c1}) and outside (^{c2}) the influence of the overstorey. The resistance network and areas of understorey (^{c1} and ^{c2}). A schematic view of the resistance network and respective areas of under storey (^{c1} and ^{c2}) are shown in Fig. 5.8. If spatial aggregation of overstorey vegetation does not make a significant difference to overall evaporation and productivity the system would tend toward the formulation described in detail in Chapter 6. Instrumentation for stomatal conductance is also described in Chapter 6.

Effect of the energy balance on plant growth

Air, leaf and soil temperatures Many developmental processes are temperature controlled with their rate increasing linearly above a base temperature (Jones, 1992). The rate of germination of millet seed, for example, increases linearly with soil temperature from 10–12°C to an optimum temperature of 32–33°C, then decreases linearly to a lethal temperature at around 48°C (ODA, 1987). It has been suggested that one of the major causes of improved crop growth under a canopy of *Faidherbia albida* is reduction of soil temperatures at the beginning of the

Table 5.2. Stomatal (g_s), canopy (g_c) and boundary layer (g_a) conductances for a variety of vegetated surface (From Jarvis, 1981).

Vegetation type	g_s (mm s^{-1}) on a leaf area basis	g_c (mm s^{-1}) on a ground area basis	g_a (mm s^{-1}) on a ground area basis
Grassland/heathland	10	20	5–20
Agricultural crops	20	50	20–50
Plantation forest	6	20	100–330

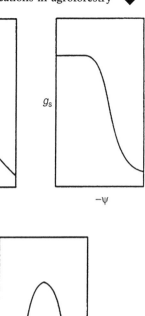

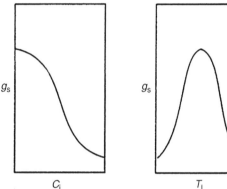

Fig. 5.7. Hypothetical responses of stomatal conductance (g_s) to photosynthetic quantum flux density (Q), vapour pressure deficit (D_a), leaf water potential (Ψ) intercellular CO_2 concentration (C_i) and leaf temperature (T_l) (adapted from Jarvis, 1976).

season, as a result of shading of the soil by the tree canopy since in the semi-arid tropics soil temperatures can exceed 50°C (Vandenbeldt and Williams, 1992). Soil temperature particularly affects germination and early growth of cereals since the meristem remains below ground level for the first 3 weeks of plant development (Ong, 1983a; Corlett *et al.*, 1992).

Optimum temperatures for growth processes depend upon the species and process. For example, leaf extension in millet was found to correspond well to meristem temperatures, with the rate of expansion decreasing above 32°C (Ong, 1983b; Terry *et al.*, 1983). However, optimum temperatures for grain yields and tillering were lower, between 20°C and 27°C (Fussell *et al.*, 1980). Temperature also affects the duration of the growth stages, so that advantages of faster rates of increase may be offset by shorter duration of that growth stage (Ong and Monteith, 1985).

Vapour pressure deficits and stomatal conductances An increase in humidity leads to a lower loss of water per unit carbon assimilated by the plant. However, because of the response of stomata to D (Fig. 5.7), an increase in humidity may increase the stomatal conductance and therefore increase total water loss from the plant. This may be a problem if available water is used up too early in the season, so preventing the crop from reaching maturity. Under irrigated conditions there is practically always a reduction of evaporation in shelter. However, as explained earlier (Fig. 5.6) this does not occur when plants are water stressed.

In an agroforestry system vapour pressure deficits could increase (if there is dry soil between the trees) or decrease (if there is a transpiring crop between the trees) depending on the relative increases of temperature and vapour pressure (Equation 5.9). The relative increases of these variables depend upon evaporation which is highly dependent on the leaf area, i.e. dry bare soil will cause temperatures to increase faster than vapour pressure, but rapidly transpiring vegetation will cause vapour pressure to increase faster than temperature. These effects have been demonstrated by Wallace *et al.* (1990) and are discussed further in Chapter 6.

Effects of Wind Speed on Quality of Products and Plant Growth

In many cases quality as well as the total yield is of economic interest. Shelter within agroforestry systems may limit mechanical damage or improve quality in other ways. Such improvements with shelter have been noted in various crops, e.g. more palatable pasture, less fibrous oats with higher protein content, higher sugar content in sugar beet, larger and finer tobacco leaves, non-spoiled asparagus, higher sugar levels in citrus, improved flower set in avocados, and higher exportable crop in kiwi fruit (Baldwin, 1988). Jaffe (1973) showed mechanical rubbing of leaves inhibited wheat growth by 11%, considerably less than more sensitive crops such as maize (28%) and beans (45%).

Case Study – Windbreaks

The processes outlined above operate for all agroforestry systems but comprehensive micrometeorological measurements are rare, notable exceptions being the work on alley cropping by Corlett *et al.* (1992) and on widely spaced trees by Green (1991) and Teklehaimanot *et al.* (1991) referred to in

Chapter 4. However, the extensive work on windbreaks provides a good deal of relevant information so that this practice has been selected as a case study. For windbreaks there is a wealth of field data available for a large variety of crops, systems and environments which serve to illustrate the principles already outlined. Windbreaks also allow microclimate changes and their influence on yield to be studied with only a minimum area of competition at the tree–crop interface.

Large differences in crop responses to shelter have been reported, the following section outlines some of these responses, more comprehensive reviews of the subject are given by Caborn (1957), Stoeckler (1962), Van Eimern *et al.* (1964), Grace (1977) and Kort (1988). Terms used are h for height of the windbreak, x for distance in the lee of the windbreak and z for height above the ground.

Micrometeorological changes in the lee of windbreaks

Radiation, temperature and humidity

Shading of crops by adjacent trees depends upon the extent of the tree canopy, the amount of leaf area, leaf angle and transmission and reflectance characteristics of the tree canopy. Extensive reviews of the light interception by canopies are given by Russell *et al.* (1989) and in agroforestry systems by Jackson (1989) as well as being discussed in Chapters 3 and 4 in this book.

General agreement exists that shelter increases air temperature during the day and decreases it at night; the day-time temperature increases usually exceeding night-time decreases because positive net radiation during the day usually exceeds negative net radiation during the night. These effects result from a reduction of heat flux away from the surface in sheltered areas (Equation 5.13), thus the amount of cooling during the day and warming during the night is likely to be less than in unsheltered areas. Some reported temperature increases measured behind windbreaks are presented in Table 5.3.

Table 5.3. Temperature increases behind windbreaks.

Source	Change of temp. (°C)	Distance from windbreak	Comments
Aslyng (1958)	1.6 and 0.8	$4h$ and $8h$	midday temp.
Bates (1911)	2.2	mean	daytime, sunny days
Brown and Rosenberg (1972)	1.8	mean	daytime, sunny days
Ujah and Adeoye (1984)	1.8	$2h$	maximum temp.
Long and Persaud (1988)	−1.2	mean	nighttime
Van Eimern *et al.* (1964)	>1	mean	daytime, soil at 0.05 m

Humidity

Decreases in vapour pressure deficit, D, in shelter have been reported for sugar beet (Brown and Rosenberg, 1972) and millet (Long and Persaud, 1988) because of high rates of evaporation from the crop, but on cloudy days changes were negligible, whereas Carr (1985), for tea and Aslyng (1958) for pasture, measured increases in D because of increases in temperature. An increase in vapour pressure of the air, e_a, is expected in shelter because of the slower transfer of water vapour away from its source, although the magnitude varies with the transpiration rate of the underlying vegetation. The deficit D may thus increase or decrease depending on the relative increases of T_a and e_a (Equation 5.9).

Stomatal conductance, evaporation and transpiration

Both higher and lower values of stomatal conductance have been reported in the literature for sheltered crops. Further analysis shows that, where D decreases in shelter, g_s increases, i.e. for bean (Rosenberg, 1966), turnips and sugar beet (Marshall, 1974) and wheat (Skidmore *et al.*, 1974), or under irrigation (Frank and Willis, 1972; Frank *et al.*, 1974), whereas under conditions where D increased in shelter, g_s was lower (Radke and Hastrom, 1973; Carr, 1985).

Evaporation measured by pans and piche evaporimeters in the lee of windbreaks indicate a reduction of evaporation of between 10 and 40%, as indicated in Fig. 5.6 for $g_s = \infty$ (Bates, 1911; Long and Persaud, 1988). Under dryland conditions the saving of water by shelter may be small but under irrigated conditions with regular wetting of the soil surface reductions are substantially larger.

Dixon and Grace (1984) showed that in three tree species transpiration rates decreased with increasing wind speed resulting from a reduction of the leaf to air vapour pressure difference as a consequence of reduced leaf temperature. Of the few direct transpiration measurements made behind windbreaks Miller *et al.* (1973) calculated that shelter reduced transpiration by 20% for a 1-m high irrigated soybean crop because of a decrease in $e_l - e_a$ (Equation 5.14). Radke and Hagstrom (1973) also found that, while the sheltered soybeans were well watered, stomatal conductance was higher but transpiration lower than in the unsheltered plants; as water availability decreased, stomatal conductance decreased but transpiration increased in shelter when compared with the unsheltered control (Fig. 5.6).

Shelter may bring about an increase in leaf area as well as an increase in transpiration per unit leaf area, so a sheltered crop may deplete water reserves earlier than an unsheltered crop. In a water-limited environment, this may cause the sheltered crop to become stressed while unsheltered plants remain unstressed (Jensen, 1954). Alternatively the rapid early growth of sheltered plants may stimulate increased rooting, enabling plants to access a greater soil volume, thereby increasing the availability of water

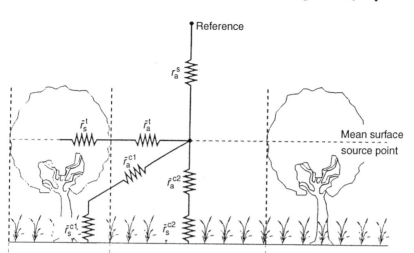

Fig. 5.8. A schematic view of the resistances to heat and vapour transfer in an agroforestry system. Fluxes from the tree (^{t}) and crop (^{c}) are controlled by mean resistance (r) of the canopy ($_s$) and the boundary layer ($_a$). The crop may be divided into the area under the influence of the tree (^{c1}) and the area outside the influence of the tree (^{c2}), which occupy fractional area A and $(1 - A)$, respectively. The resistance to fluxes moving from the mean surface source point to the reference point above the system is r_a^s.

and nutrients (Stoeckler, 1962). Thus, the possible consequences of shelter for water use are several and make generalized statements unreliable.

Interpretation of yield responses to shelter

Crop type
The importance of shelter for plant growth varies with average day temperature; winter cereals respond more than spring cereals (Stoeckler, 1962; Kort, 1988), increases in growth of tea (Carr, 1985) and cotton (Barker *et al.*, 1989) in shelter were larger when average temperatures were lower. This may be because the low temperatures were further away from the optimum temperature for growth. Thus an increase in temperature caused by shelter has a larger effect on a crop when ambient temperatures are lower.

Sheltered millet had a consistently lower harvest index than unsheltered millet since (in this crop) shelter benefits vegetative growth more than reproductive growth, i.e. smaller grains, lighter panicles, etc. (Brenner *et al.*, 1995b). This can be partly explained because the optimum temperature for millet grain yield is lower than that for vegetative growth, affecting duration rather than rate of grain filling. This may also be the reason why the 'most responsive' crops to shelter are often leafy horticultural crops such as tobacco, tea and fodder crops, e.g. alfalfa, clover (Stoeckler, 1962; Sturrock, 1984; Baldwin, 1988), whereas cereals are described as 'moderately responsive', e.g. barley and millet, or 'less responsive', e.g. spring wheat and maize (Kort, 1988). Advantages caused by an increase in temperature bringing about faster canopy expansion in shelter may be offset by a shorter canopy duration, and the increase in final harvest yield resulting from shelter may be moderated. In times of end-of-season drought, earlier maturation may increase final yields, but in wet years grain yields may be reduced because early maturation increases bird and insect damage, and provides no advantage when compared with the slower development and longer duration of the canopy on the unsheltered field (Brenner *et al.*, 1995b). Similar decreases of harvest index in shelter have been observed with maize (Bates, 1911), millet (Long and Persaud, 1988) and wheat (Vora *et al.*, 1982).

However, harvest index does not always decrease in shelter. In some crops, e.g. soybeans (Ogbuehi and Brandle, 1982) and mustard (Vora *et al.*, 1982) it increased. In soybean, pod formation started earlier in shelter but maturity was reached at the same time as in the unsheltered crop, increasing the duration of pod filling.

Overall climate and irrigation
It is often suggested that there are larger benefits of shelter on crop production in drier years (Brown and Rosenberg, 1972; Sturrock, 1984; Kort, 1988), based on crop data from the USSR (Van Eimern *et al.*, 1964) and reduction of evaporation behind windbreaks. However, Stoeckler (1962) has pointed out that maximum yields in dry years were found at distances from the windbreak where maximum snow had accumulated, i.e. water availability was increased rather than transpiration being decreased. These conclusions, drawn from areas where snow is important, may not be relevant to tropical regions. Frank *et al.* (1974, 1977) in the USA, Carr (1985) in Kenya and Reddi *et al.* (1981) in India reported yield increases behind windbreaks when moisture was available, whereas when water was scarce shelter reduced crop growth.

The combination of irrigation and shelter can increase crop yields to a larger extent than the sum of their individual influences (Sturrock, 1984; Barker *et al.*, 1989). This may results from reduced evaporative loss of irrigated water and increased stomatal conductance (Rosenberg, 1966; Frank *et al.*, 1974). Stomatal conductance, g_s, is higher in shelter in wet conditions, but as drought ensues g_s decreases in shelter (Radke and Hagstrom, 1973; Frank *et al.*, 1974, 1977). This would suggest that non-irrigated crops in semi-arid environments may not benefit from shelter as much as those under irrigation. Shelter has also been known to decrease yield in wet years because of improved microclimate for fungal growth or insect infestation (Shah, 1961; Sturrock, 1984).

Wind damage and soil erosion

Protection from storm events may, in many cases, be more important than cumulative microclimate changes. Strong winds in arid and semi-arid regions cause sandblasting of seedlings, reducing growth, slowing down establishment and uprooting young seedlings. Lodging and seed shatter in tall cereals are well known for reducing yield at the end of the growing season (Sturrock, 1984).

Reduction of soil erosion generally reduces plant damage at the beginning of the growing season. Soil erosion results from three processes: suspension (movement of fine soil particles, 2–100 μm diameter, of high plant nutrient value), saltation (the hopping movement of particles, 100–500 μm, cause abrasion of the soil surface) and surface creep (the rolling of sand-sized particles, > 500 μm) (Lyles, 1988). Wind erosion is mainly caused by saltation so the main aim of erosion control is to prevent this.

Wind-borne soil erosion can be calculated by the soil loss equation (Ticknor, 1988), which indicates that erosion is a function of: (i) soil erodibility, (ii) soil ridge roughness, (iii) climate, (iv) unsheltered travel distance across field and (v) vegetation cover. To reduce soil loss it is necessary to reduce the force of the wind on erodible aggregates by reducing field width and wind speed, and creating aggregates that are resistant to wind forces by establishing and maintaining vegetative cover, creating non-erodible aggregates, roughening the land surface and reducing the slope (Ticknor, 1988).

Techniques for reducing soil erosion

Isolated fields   Soil loss depends upon the length of the unsheltered run of wind and this can be reduced by dividing the field up into smaller fields that have no soil entering up-wind. For example, if two windbreak systems are established on a soil where a 50-m run under normal wind conditions causes a tolerable level of erosion, and if it is assumed that up to $10h$ behind the windbreak reduction of wind speed is sufficient to prevent soil erosion, then in System 1 (Fig. 5.9) the protected area is 150 m, whereas in System 2 the protected area is 60 m. The distance between windbreaks for System 1 is $15h$ and System 2, $60h$. If we assume that the competition between windbreak and crop may extend $1.5h$ either side of the windbreak in System 1 the fraction of land over which some yield reduction is expected would be 0.2 whereas in System 2 it would be 0.05. Consequently, if soil erosion is the primary concern

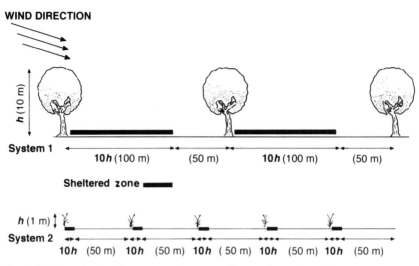

WIND DIRECTION

Fig. 5.9. Two windbreak systems for controlling soil erosion. System 1 consists of 10-m tall trees, System 2 consists of 1-m tall perennial grasses. The windbreaks shelter a zone of 10h in their lee, and larger than acceptable erosion occurs if the length of open field is greater than 50 m.

close-spaced low windbreaks are preferable to widely spaced tall ones. Additionally, if windbreaks consist of tall unpruned trees, then the below-ground distance (relative to the tree height) over which competition could be felt would substantially exceed that of low pruned windbreaks.

Threshold velocity Unlike the effects of microclimatic changes, soil erosion occurs only when wind speed exceeds a threshold value. This threshold value depends upon the size and density of the soil particles moved. Windbreak design must therefore consider the maximum likely wind speeds of the area and the threshold velocity of the soil for soil erosion rather than average values of wind speed. Furthermore it may only be necessary to reduce wind speed to just below the threshold velocity, i.e. not as low as if the microclimatic benefits of shelter were of interest (Lyles, 1988).

Windbreaks and management for erosion control Stubble mulching, or leaving surface plant residues in the field, is often used to reduce soil erosion because it not only binds the soil but increases the roughness of the soil surface. It has been shown to reduce erosion from 50.8 t ha^{-1} on a particularly erodible soil to 5.2 t ha^{-1} (Tibke, 1988). Combining windbreaks and mulching gives better results than mulching only, and adding perennial grasses improves protection further. Windbreaks of 1–3 rows are thought to be as effective as wide windbreaks for erosion control. The species selected should have a porosity and height that offer maximum protection to coincide with the maximum erosion hazard, and field sizes should be reduced whenever possible. Windbreaks should be positioned perpendicular to the problem wind and on the windward slope of a hillside to be most effective.

Some Management Considerations

In this chapter windbreaks have been used as a 'model' agroforestry system. However, the same physical principles apply to other agroforestry systems worldwide (e.g. alley cropping, scattered trees, homegardens, etc.). Consideration of the major interactions between tree and crop in the system under study and, where possible, calculation of these effects will lead to a better understanding of the behaviour and management of agroforestry systems.

For example, the effect of shading, via increased leaf area of an overstorey, reduces the energy available for photosynthesis (see Chapters 3 and 4) but also reduces the temperatures of soil and understorey leaves; this may increase or decrease productivity depending on the temperature relative to the optimum for a specific plant growth process. Shading also reduces the risk of frost because of the increase in the downward flux of longwave radiation relative to an open sky, and reduces energy available for evaporation from soil and crop. The relative importance of these processes in terms of productivity varies between environments.

Shelter, the reduction of boundary layer conductance, increases daytime temperatures and daytime vapour pressure, but as noted previously daytime vapour pressure deficits may increase or decrease. If there is an increase in D at the understorey surface then g_s generally decreases causing a decrease in plant growth. If there is a decrease in D then g_s increases and the amount of carbon fixed per unit water transpired increases. The effect of shelter on transpiration depends upon the relative magnitude of the range of variables that control it (Equation 5.14).

Pruning, species selection, planting density and arrangement allow the agroforester to reduce or increase tree transpiration, crop shading and shelter, which in turn influence temperature and humidity. It is impossible to make general recommendations on how these management options should be manipulated. However, given a specific set of conditions and objectives it is possible, with the ideas contained within this chapter, to make relatively accurate decisions as to how specific systems should be managed to achieve their particular goal.

Acknowledgements

I would like to thank Drs Janet Corlett, Peter Huxley and Jim Wallace for their helpful comments on the manuscript and Mr A. Laville for his help with the figures.

References

Aslyng, H.C. (1958) Shelter and its effect on climate and water balance. *Oikos* 9, 282–310.

Baldwin, C.S. (1988) The influence of field windbreaks on vegetable and speciality crops. In: Brandle, J.R., Hintz, D.L. and Sturrock, J.W. (eds) *Windbreak Technology.* Elsevier, Amsterdam, Netherlands, pp. 191–203.

Barker, G.L., Hatfield, J.L. and Wanjura, D.F. (1989) Cotton phenology parameters affected by wind. *Field Crops Research* 12, 33–47.

Bates, C.G. (1911) *Windbreaks: The Influence and Value.* US Department of Agriculture, Forest Service Bulletin 86, Washington DC.

Bell, C.J. and Rose, D.A. (1985) The measurement of temperature. In: Marshall, B. and Woodward, F.I. (eds) *Instrumentation for Environmental Physiology.* Cambridge University Press, Cambridge, UK, pp. 79–99.

Bingham, M.J. and Long, S.P. (1990) *Equipment for Crop and Environmental Physiology: Specifications, Sources and Costs,* 3rd edn. Department of Biology, University of Essex, UK.

Bonhomme, R. and Chartier, P. (1972) The interpretation and automatic measurement of hemispherical photographs to obtain sunlit foliage area, and gap frequency. *Israel Journal of Agricultural Research* 22, 53–61.

Brenner, A.J. and Jarvis, P.G. (1995) A heated leaf replica technique for determination of leaf boundary layer conductance in the field. *Agricultural and Forest Meteorology* 72, 261–275.

Brenner, A.J., Jarvis, P.G. and Vandenbeldt, R.J. (1995a) Tree–crop interactions in a Sahelian windbreak system. 1 Dependence of shelter on field conditions. *Agricultural and Forest Meteorology* 75, 215–234.

Brenner, A.J., Jarvis, P.G. and Vandenbeldt, R.J. (1995b) Tree–crop interactions in a Sahelian windbreak system. 2 Growth response of millet in shelter. *Agricultural and Forest Meteorology* 75, 235–262.

Brown, K.W. and Rosenberg, N.J. (1972) Shelter effects on microclimate, growth and water use by irrigated sugar beets in the great plains. *Agricultural Meteorology* 9, 241–264.

Caborn, J.M. (1957) *Shelterbelts and Microclimate.* Forestry Commission Bulletin No. 29, Edinburgh, UK.

Carr, M.K.V. (1985) Some effects of shelter on the yield and water use of tea. In: Grace, J. (ed.) *Effects of Shelter on the Physiology of Plants and Animals.* Swets and Zeitlinger B.V., Netherlands, pp. 127–144.

Corlett, J.E., Black, C.R., Ong, C.K. and Monteith, J.L. (1992) Above- and below-ground interactions in a leucaena/millet alley cropping system. I Light interception and dry matter production. *Agricultural and Forest Meteorology* 60, 73–91.

Day, W. (1985) Water vapour measurement and control. In: Marshall, B. and Woodward, F.I. (eds) *Instrumentation for Environmental Physiology.* Cambridge University Press, Cambridge, UK, pp. 59–78.

Dixon, M. and Grace, J. (1984) Effect of wind on the transpiration of young trees. *Annals of Botany* 53, 811–819.

Ehleringer, J.R. (1989) Temperature and energy budgets. In: Pearcy, R.W., Ehleringer, J., Mooney, H.A. and Rundel, P.W. (eds) *Plant Physiological Ecology.* Chapman and Hall, London, pp. 117–136.

Frank, A.B. and Willis, W.O. (1972) Influence of windbreaks on leaf water status in spring wheat. *Crop Science* 12, 668–672.

Frank, A.B., Harris, D.G. and Willis, W.O. (1974) Windbreak influence on water relations, growth, and yield of soybeans. *Crop Science* 14, 761–765.

Frank, A.B., Harris, D.G. and Willis, W.O. (1977) Growth and yield of spring wheat as influenced by shelter and soil water. *Agronomy Journal* 69, 900–903.

Fussell, L.K., Pearson, C.J. and Norman, M.J.T. (1980) Effect of temperature during various growth stages on grain development and yield of *Pennisetum americanum. Journal of Experimental Botany* 121, 621–633.

Grace, J. (1977) *Plant Response to Wind.* Academic Press, London.

Grace, J. (1985) The measurement of wind speed. In: Marshall, B. and Woodward, F.I. (eds) *Instrumentation for Environmental Physiology.* Cambridge University Press, Cambridge, UK, pp. 101–114.

Grace, J. (1989) Measurement of wind speed near vegetation. In: Pearcy, R.W., Ehleringer, J., Mooney, H.A. and Rundel, P.W. (eds) *Plant Physiological Ecology.* Chapman and Hall, London, pp. 57–73.

Grace, J. and Wilson, J. (1976) The boundary layer over a *Populus* leaf. *Journal of Experimental Botany* 27, 231–241.

Green, S.R. (1991) Air flow through and above a forest of widely spaced trees. PhD thesis, University of Edinburgh.

Green, S.R. (1993) Radiation balance, transpiration and photosynthesis of an isolated tree. *Agricultural and Forest Meteorology* 64, 201–221.

Jackson, J.E. (1989) Tree and crop selection and management to optimize overall system productivity especially light utilization in agroforestry. In: Reifsnyder, W.S. and Darnhofer, T.O. (eds) *Meteorology and Agroforestry.* ICRAF, Nairobi, Kenya, pp. 163–173.

Jaffe, M.J. (1973) Thigmomorphogenesis: The response of plant growth and development to mechanical stimulation. With special reference to *Bryonia dioica. Planta* 114, 143–157.

Jarvis, P.G. (1976) The interpretation of variation in leaf water potential and stomatal conductance found in canopies in the field. *Philosophical Transactions of the Royal Society* B273, 593–610.

Jarvis, P.G. (1981) Stomatal conductance, gaseous exchange and transpiration. In: Grace, J., Ford, E.D. and Jarvis, P.G. (eds) *Plants and their Atmospheric Environment.* Blackwell Scientific Publications, Oxford, pp. 175–204.

Jensen, M. (1954) *Shelter Effect. Investigations into the Aerodynamics of Shelter and its Effects on Climate and Crops.* The Danish Technical Press, Copenhagen.

Jones, H.G. (1992) *Plants and Microclimate,* 2nd edn. Cambridge University Press, Cambridge, 428 pp.

Kort, J. (1988) Benefits of windbreaks to field and forage crops. In: Brandle, J.R., Hintz, D.L. and Sturrock, J.W. (eds) *Windbreak Technology*. Elsevier, Amsterdam, pp. 165–190.

Kreith, F. (1973) *Principles of Heat Transfer*, 3rd edn. Intext Educational Publishers, NY, London.

Long, S.P. and Persaud, N. (1988) Influence of neem (*Azardirachta indica*) windbreaks on millet yield, microclimate, and water use in Niger, West Africa. In: Unger, P.W., Sneed, T.V., Jordan, W.R. and Jensen, R. (eds) *Challenges in Dryland Agriculture – A Global Perspective*. Texas Agricultural Experimental Station, Texas, pp. 313–314.

Lyles, L. (1988) Basic wind erosion processes. In: Brandle, J.R., Hintz, D.L. and Sturrock, J.W. (eds) *Windbreak Technology*. Elsevier, Amsterdam, pp. 91–101.

Marshall, J.K. (1974) Effects of shelter on the growth of turnips and sugar beet. *Journal of Applied Ecology* 11, 327–346.

Miller, D.R., Rosenberg, N.J. and Bagley, W.T. (1973) Soybean water use in the shelter of a slat-fence windbreak. *Agricultural Meteorology* 11, 405–418.

Monteith, J.L. and Unsworth, M.H. (1990) *Principles of Environmental Physics*. Edward Arnold, London, UK.

Murray, F.W. (1967) On the computation of saturated vapour pressure. *Journal of Applied Meteorology* 6, 203–204.

ODA (1987) *Microclimatology in Tropical Agriculture*, Vol 1. Final Report, Research schemes R3208 and R3819. Overseas Development Administration, London.

Ogbuehi, S.N. and Brandle, J.R. (1982) Influence of windbreak-shelter on soybean growth, canopy structure, and light relations. *Crop Science* 22, 269–273.

Ong, C.K. (1983a) Response to temperature in a stand of pearl millet (*Pennisetum typhoides* S. & H.). I Vegetative development. *Journal of Experimental Botany* 34, 322–366.

Ong, C.K. (1983b) Response to temperature in a stand of pearl millet (*Pennisetum typhoides* S. & H.). 4 Extension of individual leaves. *Journal of Experimental Botany* 34, 1731–1739.

Ong, C.K. and Monteith, J.L. (1985) Response of pearl millet to light and temperature. *Field Crops Research* 11, 141–160.

Pearcy, R.W. (1989) Radiation and light measurements. In: Pearcy, R.W., Ehleringer, J., Mooney, H.A. and Rundel, P.W. (eds) *Plant Physiological Ecology*. Chapman and Hall, London, pp. 97–116.

Radke, J.K. and Hagstrom, R.T. (1973) Plant-water measurements on soybeans sheltered by temporary corn windbreaks. *Crop Science* 13, 543–548.

Reddi, G.H.S., Rao, Y.Y. and Rao, M.S. (1981) The effect of shelterbelt on the productivity of annual field crops. *Indian Forester* October 1981, 624–629.

Reifsnyder, W.S. and Darnhofer, T. (1989) Meteorology and agroforestry. *Proceedings of ICRAF/WMO/UNEP workshop on Application of meteorology to Agroforestry Systems Planning and Management*, ICRAF, Nairobi, 1989.

Rosenberg, N.J. (1966) Microclimate, airmixing and physiological regulation of transpiration as influenced by wind shelter in an irrigated bean field. *Agricultural Meteorology* 3, 197–224.

Russell, G., Jarvis, P.G. and Monteith, J.L. (1989) Absorption of radiation by canopies and stand growth. In: Russell, G., Marshall, B. and Jarvis, P.G. (eds) *Plant Canopies: Their Growth Form and Function*. Cambridge University Press, Cambridge, pp. 21–39.

Shah, S.R.H. (1961) The influence of excessive rainfall on the protective value of windscreens with respect to crop yields. *Netherlands Journal of Agricultural Science* 9, 262–269.

Sheehy, J.E. (1985) Radiation. In: Marshall, B. and Woodward, F.I. (eds) *Instrumentation for Environmental Physiology*. Cambridge University Press, Cambridge, UK, pp. 5–28.

Shuttleworth, W.J. and Gurney, R.J. (1990) The theoretical relationship between foliage temperature and canopy resistance in sparse crops. *Quarterly Journal of the Royal Meteorological Society* 116, 497–519.

Sinclair, F.L. and Jarvis, P.G. (1994) The influence of the two dimensional distribution of leaf area density within individual crowns of *Picea sitchensis* on light interception and transmittance. *Agroforestry Forum* 4 (3), 35–40.

Skidmore, E.L., Hagen, L.J., Naylor, D.G. and Teare, I.D. (1974) Winter wheat response to barrier-induced microclimate. *Agronomy Journal* 66, 501–505.

Stoeckler, J.H. (1962) *Shelterbelt Influence on Great Plains Field Environment and Crops*. USDA Production Research Report No. 62, Washington DC.

Sturrock, J.W. (1984) *Shelter Research Needs in Relation to Primary Production*. The report of the national shelter working party. National Water and Soil Conservation Authority, New Zealand.

Teklehaimanot, Z., Jarvis, P.G. and Ledger, D.C. (1991) Rainfall interception and boundary layer conductance in relation to tree spacing. *Journal of Hydrology* 123, 261–278.

Terry, N., Waldron, L.J. and Taylor, S.E. (1983) Environmental influences on leaf expansion. In: Dale, J.E. and Milthorpe, F.L. (eds) *The Growth and Functioning of Leaves*. Cambridge University Press, Cambridge, pp. 179–205.

Tibke, G. (1988) Basic principles of wind erosion control. In: Brandle, J.R., Hintz, D.L. and Sturrock, J.W. (eds) *Windbreak Technology*. Elsevier, Amsterdam, pp. 103–122.

Ticknor, K.A. (1988) Design and use of field windbreaks in wind erosion control systems. In: Brandle, J.R., Hintz, D.L. and Sturrock, J.W. (eds) *Windbreak Technology*. Elsevier, Amsterdam, pp. 123–159.

Ujah, J.E. and Adeoye, K.B. (1984) Effects of shelterbelts in the sudan savanna zone of Nigeria on microclimate and yield of millet. *Agricultural and Forest Meteorology* 33, 99–107.

Vandenbeldt, R.J. and Williams, J.H. (1992) The effect of soil surface temperature on the growth of millet in relation to the effect of *Faidherbia albida* trees. *Agricultural and Forest Meteorology* 60, 93–100.

Van Eimern, J., Karschom, R., Razumova, L.A. and Robertson, G.W. (1964) *Windbreaks and Shelterbelts*. WMO Technical Note 59. World Meteorological Office, Geneva, Switzerland, 188 pp.

Vora, A.B., Parappillil, A.J. and Sharma, K.S. (1982) Effect of windbreaks and shelterbelts on wheat and mustard as well as on wind velocity. *Indian Forester* March 1982, 215–220.

Wallace, J.S., Roberts, J.M. and Sivakumar, M.V.K. (1990) The estimation of transpiration from sparse dryland millet using stomatal conductance and vegetation area indices. *Agricultural and Forest Meteorology* 51, 35–49.

Wang, Y.P. and Jarvis, P.G. (1990) Description and validation of an array model – MAESTRO. *Agricultural and Forest Meteorology* 51, 257–280.

6

◆

The Water Balance of Mixed Tree–Crop Systems

J.S. Wallace

Institute of Hydrology, Wallingford, Oxon OX10 8BB, UK

Introduction

The hydrological and biological reasons for the success, or otherwise, of certain combinations of species in agroforestry practices in particular environments are still largely unknown; only some basic guidelines have been outlined by research work so far. Many of these come from agricultural intercropping studies where research into the controlling processes, and the interactions between the components, is more advanced. Successful intercrops appear to be those which make 'better' use of resources, by using more of a resource, by using it more efficiently, or both. In terms of the water use of an agroforestry system a central question is, therefore, does intercropping woody and non-woody plants increase total harvestable produce by making more effective use of rainfall? It is, at least theoretically, possible in a system where rainfall is not completely used, that the inclusion of trees or shrubs may improve the rainfall use efficiency either directly, by more rain being used as transpiration, or indirectly by improved transpiration efficiency, i.e. more dry matter produced per unit of water transpired (see Chapter 4). The former effect could be achieved by either spatial or temporal 'complementarity' in the constituent species, and the latter by modifying the plant microclimate (see Chapter 5). These two possibilities are discussed briefly below.

Direct effects: increased water use by tree and crop combination

Current cropping systems in tropical semi-arid regions often use less than half the rainfall input since there can be substantial losses of water via soil evaporation, runoff and drainage. For example, in the semi-arid regions of the Middle East and West Africa direct soil evaporation can account for 30–60% of rainfall (Cooper *et al.*, 1983; Wallace, 1991). In the Deccan plateau of India the best cropping systems only use 40% of the annual rainfall, while the rest is lost as runoff (26%) and deep percolation (33%) (Ong *et al.*, 1991). Drainage losses from tropical crops are rarely quantified and are therefore largely unknown. However, this may be one of the terms in the water balance which can be most easily modified by the presence of trees, since they can utilize water outside the rooting zone of annual crops and also outside the crop growing season (e.g. Huda and Ong, 1989). Improved rainfall utilization can also occur via the substantial reductions in runoff which can be achieved in agroforestry systems, particularly on sloping land (Young, 1989).

Indirect effects: improved soil condition and microclimate

With their elevated canopies the trees in an agroforestry system may modify the microclimate in such a way as to increase the overall water use efficiency of the understorey crop. This can happen in several ways.

Shading of the ground may significantly reduce soil evaporation. This is likely to be particularly important in the early part of the crop season under a rainfall regime of frequent and comparatively small storms. Under these conditions the direct evaporation from the soil is dominated by the radiation, humidity and wind speed at the soil surface. Any reduction in radiation or surface wind speed or increase in humidity caused by the presence of the trees should lower soil evaporation. Reduced soil evaporation allows for an increase in the amount of water available in the soil for transpiration by the crops or the trees. This is discussed in more detail in the section on soil evaporation.

Another form of microclimatic modification which may improve transpiration efficiency is the reduction in the within-canopy vapour pressure deficit as a result of transpiration from the second canopy and/or soil evaporation. It has been shown by Monteith (1988) that transpiration efficiency is inversely proportional to the vapour pressure deficit of the air, i.e. plants use more water to fix a given amount of carbon when the air is drier (see Chapter 4). The modification of in-canopy vapour pressure deficit is discussed in more detail in the section on transpiration processes.

Shading of the crop may also reduce its transpiration without a proportional decrease in photosynthesis. This would occur if the under-storey species became 'light saturated' at relatively low radiation levels, which is typical in C3 species. In this case reduced radiation input to the crop would increase the transpiration efficiency. The potential benefits would depend on the degree of shading, levels of incident radiation and the light saturation point of the canopy (see Chapter 4).

Shading of a crop by a tree canopy may reduce surface soil and air temperatures to give beneficial effects on crop growth. Crop growth and rates of development increase with temperature up to an optimum temperature and decline thereafter. For example, both millet seed germination and leaf extension rates decrease when the temperature is above 33°C (see Chapter 5). The exposure of large areas of bare soil in semi-arid crops may produce a thermal environment that is too hot for the crop and the introduction of shade may bring the system back closer to the optimum temperature. Some of the above effects will be returned to later in this chapter and more details of others are given in Chapters 4 and 5.

Clearly, there are a number of ways in which agroforestry systems could use water more efficiently than agricultural sole crops, but a clear picture of how this actually happens in any system can only be obtained through a comprehensive water balance study. Very little is currently known about the water balance of agroforestry systems, so this chapter concentrates on descriptions of water balance processes and methods, the latter having been developed mainly in sole tree and sole crop studies. From these, techniques that are likely to be most suitable in agroforestry research are highlighted and some preliminary deductions are made about the changes to the water balance of a crop that might follow the introduction of trees.

The Water Balance

A schematic representation of the water balance of an agroforestry system on a hillslope is shown in Fig. 6.1. The gross precipitation P_g is first intercepted by the tree and crop canopies, giving rise to interception losses from the trees, I_t, and crop, I_c. The presence of the plant canopies modifies the rainfall so that the input to the ground beneath the trees, P_t, is different from that beneath the crop, P_c. This input of water to the ground may infiltrate at different rates below the trees (F_t) and crop (F_c), producing different rates of surface runoff, R_t and R_c respectively. In some circumstances F_t may be sufficiently high not only to reduce R_t to zero, but also to absorb any runoff (R_c) from the cropped area. In this way the total runoff from the entire plot would be negligible, as observed in studies of total runoff on sloping land (14%) at Machakos, Kenya where runoff was < 2% of annual rainfall (Kiepe and Rao, 1994).

Water will evaporate directly from the soil surface at rates E_t and E_c. The water contents of the soil zones beneath the trees, θ_t, and the crop, θ_c, due to the different surface inputs and transpiration rates, T_t and T_c, may lead to different drainage rates, D_t and D_c. There may also be some lateral sub-surface water movement, R_s, particularly if the soil saturates for significant amounts of time. The total transpiration of tree and crop system

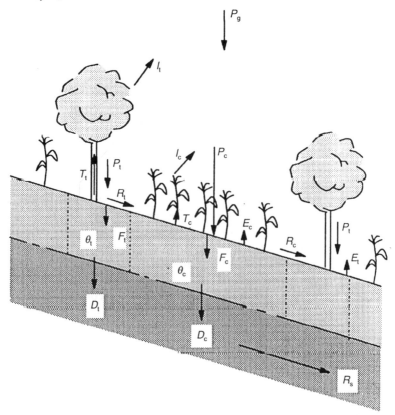

Fig. 6.1. A schematic representation of the water balance of an agroforestry system on a hillslope. Gross precipitation P_g is intercepted by the tree and crop canopies, giving rise to interception losses from the trees, I_t, and crop, I_c. Rainfall input to the ground beneath the trees, P_t, may be different from that beneath the crop, P_c. Infiltration rates below the trees, F_t, and crop, F_c, may produce different rates of surface runoff, R_t and R_c. Evaporation from the soil surface proceeds at rates E_t and E_c beneath the trees and crop respectively. The water contents of the soil zones beneath the trees, θ_t, and the crop, θ_c, due to the different surface inputs and transpiration rates, T_t and T_c, may lead to different drainage rates, D_t and D_c. There may also be some lateral sub-surface water movement, R_s.

is therefore:

$$T_t + T_c = P_g - I_t - I_c - E_t - E_c - D_t - D_c - R_t - R_c - \delta\theta_t - \delta\theta_c$$

$$(6.1)$$

In a cropped area on the same slope without trees, the equivalent total transpiration is:

$$T_c^* = P_g - I_c^* - E_c^* - D_c^* - R_c^* - \delta\theta_c^*$$

$$(6.2)$$

where the * indicates that these terms can be different in the absence of trees. The hypothesis that agroforestry systems can use rainfall more effectively can therefore be expressed mathematically as:

$$T_t + T_c > T_c^* \tag{6.3}$$

This defines the way in which the different water balance components need to be managed in order to benefit from the addition of trees to a crop. For example, soil evaporation, runoff and drainage should be minimized in the agroforestry system and this could be achieved via the utilization of as much as possible of the locally 'non-productive' components of the sole crop water balance (e.g. E_c^*, R_c^* and D_c^*). Where agroforestry is used in the upper reaches of a catchment, the effects of reducing runoff and drainage on the water supply 'downstream' may need to be taken into account.

The water balance components which have to be quantified are discussed individually in the following sections. The first part of each section presents the basic processes controlling the water balance component in question and how these may be influenced by a tree–crop mixture. The second part presents some techniques that could be used to quantify the water balance component being discussed.

Rainfall

Rainfall processes

The presence of tree crowns amongst a crop can modify the pattern, amount and energy of the rainfall input to the ground. For example, Darnhofer et al. (1989) measured rainfall at a number of distances perpendicular to a 12-m wide row of *Senna siamea* trees. Figure 6.2 shows a summary of the results from their work which illustrates several effects. First, and not surprisingly, the rainfall input beneath the trees is reduced from that in the open, by about 15% in their case. Second, there is a canopy edge effect, where the rainfall is enhanced on the windward side of the canopy and reduced on the leeward side. This is because more rain is caught on the windward side of the canopy, and this rain drips on to the ground below. Darnhofer et al. (1989) also showed that the rainfall modification effect of the trees only occurs for a distance away from the trees approximately equal to the tree height.

As well as modifying the amount and spatial distribution of rainfall input to the soil, tree canopies can also alter the kinetic energy of the rainfall by changing the size and velocity of the raindrops. This effect can be important in relation to soil erosion in situations where the amount of soil material that is eroded during a rainstorm is affected by the kinetic energy of the raindrops. Tree canopies can either increase or decrease the kinetic

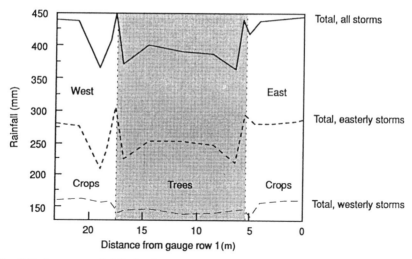

Fig. 6.2. Average rainfall distribution across a tree–crop mixture. The trees (*Senna siamea*) were planted in a 12-m wide row with two 5.5-m wide rows of maize on either side. (Reproduced from Darnhofer *et al.*, 1989).

energy of the incident rainfall, depending on the tree species and the rainfall characteristics. For example, during heavy storms with large drop sizes the tree canopy can break the drops up, thereby reducing their size and kinetic energy. On the other hand, in low intensity storms where raindrops are small, tree canopies can amalgamate drops so that the drips from the canopy are larger than the raindrops. If the tree canopy is tall enough, the kinetic energy of these drops falling from the canopy can be greater than that of the incident rainfall. Recent work by Hall and Calder (1993) has shown dramatic differences, of up to a factor of nine, in the canopy drip kinetic energies in three species, *Eucalyptus camaldulensis*, *Tectona grandis* and *Pinus caribaea*. These results are explained by the different leaf forms of the species studied. The large, flat leaves of *Tectona grandis* form larger drops than the thin *Pinus caribaea* needles. Clearly, where agroforestry trees are chosen for their role in preventing soil erosion, care must be taken to select species with leaf characteristics and canopy heights which will reduce rather than increase the risks of erosion. Management of the tree shape (e.g. by pruning) and use of mixtures of species will make the situation even more complex.

Rainfall measurement

The wide range of gauges used for measuring rainfall have been reviewed recently by Sevruk and Klemm (1989) and only the essential guiding principles are repeated here. Gross precipitation is most accurately meas-

ured using rain gauges installed with their rim at ground level and surrounded by an anti-splash baffle (see, for example, Rodda and Smith,
1986). This minimizes the effects of aerodynamic distortion and splash into
or out of the gauge. Where ground-level gauging is impractical, it should be
borne in mind that when a rain gauge is mounted above the ground surface,
there is considerable potential for systematic error in rainfall measurement.
For example, in the highly intensive convective rainfall climate of the West
African Sahel, gauges mounted 1 m above the ground recorded catches
which were up to 30% less than a ground-level gauge at the same site
(Chevallier and Lapetite, 1986). In more temperate rainfall climates underestimation is closer to 10% (World Meteorological Organization, 1982),
still a very significant fraction of the water balance.

   Until recently it has been very difficult to make accurate measurements
of the size and velocity of raindrops in the field. However, this is now
possible using a disdrometer (Illingworth and Stevens, 1987), which detects
drops as they pass through a collimated light beam. Figure 6.3 shows a
disdrometer used by Hall and Calder (1993) to study the modification of
raindrops by tree canopies in southern India. The diameter of a drop is
related to the maximum amount of light occlusion it produces in the beam,
which is detected by a photodiode. The velocity of the drop is calculated
from the time taken for the drop to cross the light beam. Hall and Calder
(1993) found that canopy drips have size spectra that are characteristic of
the tree species and are largely independent of the incident rainfall spectra.

Rainfall Interception

Interception processes

When trees are introduced into a crop their initial effect may be to reduce
the net input of rainfall to the crop by the amount of their interception loss
(I_t in Fig. 6.1). Rather than increasing rainfall use efficiency therefore, the
presence of trees will initially decrease rainfall use (in terms of the fraction
available for plant growth), and any overall gain in rainfall use efficiency will
have to be made up in other terms of the water balance. A key question is
therefore – 'How large is the interception loss in typical agroforestry canopies?' To answer this question we first need to consider the basic process
of interception in complete, closed forest canopies and then examine how
these processes may change in agroforestry canopies, which are generally
more open.

   When precipitation falls some of it is held for a time on the plant
canopy. During and after rainfall this water can evaporate from the canopy,
if there is an input of energy from the sun or from the atmosphere itself. If
rainfall persists at a rate in excess of the evaporation rate, water builds up on

Fig. 6.3. An optical disdrometer for the measurement of drop size and velocity.

the canopy to the point where it will begin to drip off the leaves and run down the plant stem. This entire process is referred to as interception, and has been recognized for some time (e.g. Horton, 1919) as an important process in the water balance of many types of vegetation, particularly trees. A landmark in the understanding and ability to predict interception came with the development of the physically based model of Rutter *et al.* (1971; 1975). In essence this model calculates a running water balance for the canopy and trunks of a forest stand using hourly rainfall and the meteorological parameters needed to calculate evaporation. It computes the rate of evaporation of intercepted water and also the amount of water reaching the ground either directly or in the form of drips from the canopy ('throughfall') and down the trunks of the trees ('stemflow'). The Rutter model has been shown to work well in a number of forests, generally of high leaf area index (e.g. Calder, 1977; Gash and Morton, 1978).

An alternative to the Rutter model is the analytical model described by Gash (1979), which requires less data than the Rutter model. The Gash model is based on a knowledge of the ability of the vegetation to store water on the canopy and the average rates of rainfall and evaporation from the wet canopy. Interception can be predicted using the Gash model and simple daily rainfall data, which are available from most standard meteorological stations (e.g. Lloyd *et al.*, 1988). Both Rutter's and Gash's original models were developed for forests with a high degree of ground cover and they perform less well in sparse tree canopies more representative of those which occur in agroforestry. Teklehaimanot and Jarvis (1991a, b) studied interception in sparse Sitka spruce stands in Scotland and found that both the Rutter and Gash models did not work well in sparse canopies without further modification. Teklehaimanot and Jarvis (1991a) reported that the main effect of greater spacing in their tree stand was an increase in the rate of wet canopy evaporation resulting from an increase in the boundary layer conductance of the trees. They also found that at the relatively wide row spacings studied the amount of water stored per tree was independent of tree spacing.

Recognizing the need to modify his original model for use in sparse forests, Gash reformulated his analytical equations to take explicit account of more open canopies (Gash et al., 1995). The new model gave much better agreement (5% underestimate) than his original model (39% overestimate) when compared with interception measurements made in a sparse pine forest in southwest France. The modified Gash model still only requires daily rainfall data and the degree of canopy cover and its application to estimate interception losses in an agroforestry system in Kenya is illustrated in Fig. 6.4. The model assumes a canopy storage per unit area of cover, which stays the same as the canopy cover changes. In the example illustrated here a value of 0.8 mm has been used. The mean rainfall rate used in the model was 2.3 mm h^{-1}; this was calculated from 642 hours of rainfall data from Machakos in 1992 and 1993. The mean evaporation rate during rainfall per unit area of cover was assumed to be 0.2 mm h^{-1}. Figure 6.4 shows that with complete canopy cover the annual interception loss is $\sim 20\%$ of rainfall and this fraction decreases slightly as rainfall increases. This interception loss is intermediate between values reported for (dense) forests in temperate climates (e.g. $\sim 40\%$, see Calder and Newson, 1979) and tropical climates (e.g. $\sim 10\%$, see Lloyd et al., 1988 and Shuttleworth, 1988). This is a consequence of the mean rainfall rate at Machakos being higher than that in temperate areas and lower than that in the humid tropics. The tendency for the annual interception fraction to decrease as rainfall increases from ~ 400 to 1000 mm has also been reported in temperate climates by Calder and Newson (1979). The analysis presented here using the revised Gash model suggests that in sparse canopies more typical of those in agroforestry systems, the annual interception loss is likely to be between 3 and 10% of rainfall.

Interception measurement

Interception is difficult to measure directly and the most common approach is to estimate it as the difference between gross rainfall and throughfall beneath the canopy plus stemflow. Throughfall has been measured using arrays of rain gauges placed beneath the tree canopy in either line transects

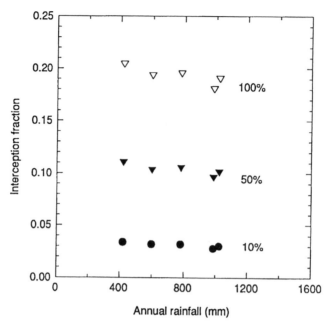

Fig. 6.4. Estimates of the annual fraction of rainfall lost as interception (I_t/P_g) made using the Gash *et al.* (1995) sparse forest model with rainfall data for 1984 to 1988 from Machakos, Kenya. Different degrees of cover are input to the model to simulate dense (100%, ▽), intermediate (50%, ▼) and sparse (10%, ●) canopies.

or random grids. Examples of the types of gauge and their spatial placement are given by Rutter *et al.* (1971, 1975), Gash and Morton (1978) and Lloyd and Marques (1988). Troughs have also been used to collect throughfall (e.g. Reigner, 1964; Leyton *et al.*, 1967), giving better line averages, but less information on point spatial variation. Lloyd and Marques (1988) examined the spatial variability of throughfall in the Amazonian rainforest in Brazil and concluded that in that situation it was better to use a random grid of rain gauges which were relocated every 2 weeks, rather than line transects. In agroforestry systems, where there is often a regular spatial arrangement of the trees, it may be better to use sampling arrangements which take this into account rather than assuming throughfall is random in space. An example of this approach is that used by Darnhofer *et al.* (1989) who arranged their rain gauges in line transects perpendicular to the tree rows in their study of a tree–crop interface. Throughfall has also been measured using troughs made from PVC drain-pipe in agroforestry systems in northern Nigeria (Sinclair *et al.*, 1994). In this study the troughs were arranged to protrude radially from beneath in-dividual tree crowns so that they collected rainfall within and outside the tree canopy.

A combined measure of throughfall and stemflow is given by the net rainfall gauge described by Calder and Rosier (1976). Here a plastic sheet is draped between a number of trees and sealed to the trunks. The sheet is arranged so that it sheds any throughfall and stemflow into a collecting trough which is connected to a large tipping bucket. Often the volume of water that is shed by such large gauges (typical area ~ 100 m^2) is so great that flow dividers are used to increase their measurement capacity. This type of gauge avoids some of the sampling problems associated with throughfall measurement with troughs and gauges, but requires substantial maintenance if used for prolonged periods of time. There is also the disadvantage that no water reaches the soil beneath the large plastic sheet and regular irrigation of the trees beneath the sheet may be necessary.

Stemflow can also be measured separately using 'collar' gauges moulded to the plant stem (see Helvey and Patric, 1965). This type of gauge has been used on trees, e.g. by Leyton et al. (1967), and on crops, e.g. by Parkin and Codling (1990). In most tree canopies examined so far stemflow is usually found to be only a small fraction of the total throughfall. Similar results have been obtained in a wheat crop by Butler and Huband (1985). However, in some crops stemflow is more significant, e.g. in maize where Parkin and Codling (1990) found stemflow was between 19% and 49% of rainfall. In canopies containing mixed species, therefore, the capture of water by individual plants in the form of stemflow may be important for survival and competition.

Comparatively few observations of interception have been made in open tree canopies more representative of agroforestry. Throughfall measurements were recorded by Monteith et al. (1991) using rain gauges in a *Leucaena leucocephala*–millet alley crop. They found that throughfall varied between 70% and 90% of rainfall during the season depending on crop cover, wind direction and distance from the *Leucaena* hedgerow. These results are similar to those reported by Darnhofer et al. (1989) in a *Senna siamea*–maize agroforestry system. Some work has also been done on solitary trees, for example by Aston (1979). He determined the storage capacity of small trees by measuring their increase in mass when they were artificially wetted by spraying with water in a laboratory. Teklehaimanot and Jarvis (1991a) extended this mass balance technique to the field where they suspended Sitka spruce trees on a load cell hanging from a tall tripod. The trees were wetted using spray nozzles and they were able to calculate the storage capacity of the trees and their rate of wet canopy evaporation.

Infiltration and Runoff

Infiltration and runoff processes

When rainfall reaches the soil surface some of it may infiltrate into the soil (see Fig. 6.1). If the rainfall rate is greater than the infiltration rate the excess water starts to collect at the surface and when the surface storage is ex-ceeded, runoff will occur. Infiltration is therefore a dynamic process that changes during the course of a rainstorm depending on the soil character-istics, slope of the land and the rainfall intensity. Where the intercropping of woody and non-woody plants alters any of these factors, then the infiltra-tion and runoff may be affected.

Hoogmoed et al. (1991) demonstrated the dynamic nature of the infiltration process using a rainfall simulator on cultivated millet fields in Niger. Figure 6.5 is reproduced from their paper and shows that initially infiltration rates exceeded 100 mm h^{-1}, but they dropped rapidly to a final infiltration rate of ~ 30 mm h^{-1}. The difference between the infiltration rate and the rainfall rate gives the runoff rate. Figure 6.5 also shows the effect of tillage on infiltration. Undisturbed soil had a much lower infiltration rate than soil which had been hand tilled using a local digging scoop ('daba'). Tilling the soil also altered the surface storage, as shown by the delay (18 min) in surface runoff in the recently tilled soil. The effect of tillage also changed rapidly with time as the rains reduced surface roughness and the soil became re-compacted. Surface treatment also played an important role in determining the runoff from the agroforestry systems studied by Lal (1989a). On slopes of 7% the presence of Leucaena or Gliricidia hedgerows reduced runoff by between 64% and 80% compared with sole annual crops (maize or cowpea) on plough-tilled land. However, the least runoff occurred from sole crops grown on untilled land, illustrating the important effect of land preparation in agroforestry systems.

Soil characteristics that affect infiltration are surface crust, surface storage, saturated hydraulic conductivity and the presence or absence of plant residues. Surface crusting has a very marked effect on infiltration as demonstrated by Casenave and Valentin (1991) for the soils of the Sahelian zone of West Africa. Even in these very sandy soils surface crusts can form and generate significant runoff in comparatively flat land. For example, Roose and Bertrand (1971) and Hoogmoed and Stroosnijder (1984) report runoff of between 16% and 25% of rainfall in sandy soils with slopes of less than 3%. Vegetation cover generally increases infiltration and reduces runoff for a number of reasons. For example, in Senegal runoff decreased from 456 mm in bare soil to 264 mm in cultivated land and further to 200 mm in fallow land containing a mixture of shrubs and herbs (Lal, 1991). Vegetation cover can affect surface infiltration in two principal ways. First the canopy modification of the rainfall kinetic energy (as discussed in the section on rainfall processes) may alter soil particle detach-ment and crust formation. The second effect is via a reduction in surface crusting and improved soil hydraulic conductivity as a result of the incorporation of plant residues into the soil (Kiepe and Rao, 1994). Mulching is widely used in the tropics for conserving soil water and reducing soil erosion (e.g. see Stigter, 1984) and the distribution of plant

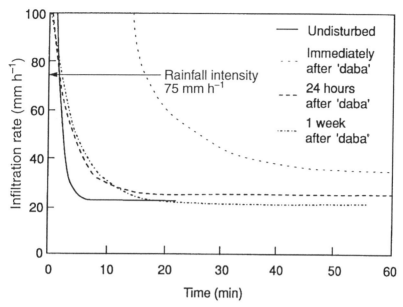

Fig. 6.5. Infiltration rate as a function of time under simulated rainfall of 75 mm h⁻¹ (fitted curves) on cultivated land at the ICRISAT Sahelian Center, Sadoré, Niger. Tilling the surface with a local digging scoop ('daba') dramatically increased infiltration. (Reproduced from Hoogmoed *et al.*, 1991.)

residues in one form or another is usually a major feature of many agroforestry systems. An additional beneficial effect of mulching is through increased activity of soil fauna, further improving the soil structure (Lavelle *et al.*, 1992).

There are a number of agroforestry practices that are designed to conserve water and reduce runoff by the direct effect trees can have on soil slope. Planting of trees or hedgerows on the contours of sloping land can have the effect of forming natural terraces as water and soil are collected on the up-slope side of the hedgerow (Fig. 6.6; Lal, 1989a; Young, 1989; Kiepe and Rao, 1994; see also Chapter 8). In time this type of agroforestry system can produce almost level terraces between the hedgerows, improving the infiltration of rainfall to crops which can be grown in the space between the hedgerows.

Infiltration and runoff measurement

There are two basic approaches to measuring infiltration, one where water is applied to the surface without any kinetic energy (i.e. by ponding) and the other where water is applied with kinetic energy (i.e. like rainfall). Ring infiltrometers are the most commonly used method of measuring the in-

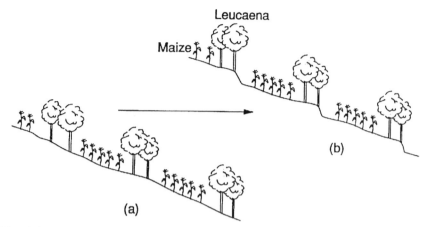

Fig. 6.6. A schematic example of the process of self-terracing which can be achieved using agroforestry systems on sloping land. In time the presence of the trees can alter the soil surface from (a) to (b). (Reproduced from Young, 1989.)

filtration of ponded water. Double ring infiltrometers, which are inserted concentrically a small distance into the soil, are used to ensure one-dimensional flow in the inner ring where the rate of fall of the water gives the infiltration rate (FAO, 1979). The method has been used in agroforestry systems (e.g. by Lal, 1989b) and has the advantage that it is simple and inexpensive, and can give quite accurate results. The disadvantage is that it does not work well in soils with fragile surface crusts, which break when the rings are pushed into the ground, or for soils with macropores where flow may not be one-dimensional. Macropores are likely to be quite common in soils under agroforestry as a result of tree roots which die (van Noordwijk *et al.*, 1991) and enhanced faunal activity (Lavelle *et al.*, 1992) due to mulching. Another device which is used to measure the infiltration of water into soils is the Guelph permeameter (Elrick and Reynolds, 1992), which applies water to the soil surface via a constant head device. This technique gives a measure of the saturated hydraulic conductivity of the soil and can be used at a range of soil depths by digging pits to the level of interest. Again this technique does not work well in soils with macropores. Other techniques for the measurement of water infiltration into soils (e.g. disc permeameters and tension infiltrometers) have been recently reviewed by White *et al.* (1992).

The rate of infiltration can be affected by changes in the soil surface condition brought about by rainfall impaction. Infiltration rates measured under rainfall are therefore more representative of the real behaviour of soils in the field. Infiltration can be measured under natural rainfall or under rainfall simulators. The latter have the advantage that a range of rainfall rates can be created on demand, whereas natural rainstorms come in uncontrolled intensities and at unpredictable times. To produce realistic rainfall, simulators have to be carefully designed and the characteristics of a range of devices has been reviewed by Peterson and Bubenzer (1986). The choice of system will depend on soil type and the precise characteristics of the rainfall that dominate the infiltration process, e.g. drop size distribution, velocity at impact and intensity. The main disadvantage of rainfall simulators is that it is difficult to reproduce very realistic rainfall in all of the above characteristics; however, despite this rainfall simulators have been widely used to study the dynamics of infiltration and runoff in a number of soils, e.g. by Casenave and Valentin (1991) and Stroosnijder and Hoogmoed (1984). Figure 6.7 shows a miniature rainfall simulator which can be used to study differences in infiltration rate beneath trees and cropped areas in agroforestry systems. In this small rainfall simulator the drops only fall a short distance so their kinetic energy is increased by using relatively large drop sizes.

Runoff can be measured by collecting the water from a given area which may range in size from small plots $< 1 \ m^2$ to large catchments. The type of information obtained changes with the scale of the runoff measurement. Runoff collected from small areas gives information about particular soils and surface treatments, whereas data from very large areas reflect the integrated effect of the complete range of soil types, surface covers and topography of the entire catchment. Small plots are useful to study the effects of rainfall on surface condition (crusting, aggregate breakdown, etc.), but larger plots are needed to study overland flow processes. Perhaps the most widely used plot size is the Wischmeier or USLE (universal soil loss equation) plot (Wischmeier and Smith, 1978). Standard plots are 1.8 m × 22 m (0.04 ha) and are usually large enough to give information on the combined effects of surface dynamics and overland flow processes. Any runoff generated during a rainstorm is intercepted at the bottom of the plot in Gerlach troughs (e.g. see Kiepe, 1995), which channel the water into collection tanks, sometimes via a large tipping bucket. To evaluate the effects of vegetation barriers or ridges on contours, runoff plots wider than the standard Wischmeier plots are required (e.g. see Lal, 1989a).

Soil Water and Drainage

Soil water and drainage processes

The amount of water that enters the soil is a consequence of the combined effects of rainfall, infiltration and runoff. Once within the soil matrix water

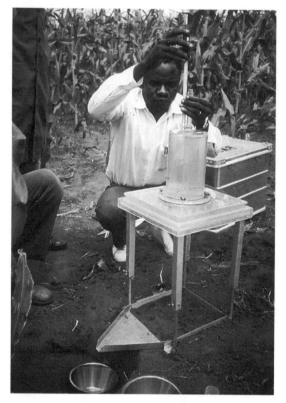

Fig. 6.7. A miniature rainfall simulator in a cropped area (maize) of an agroforestry system at Machakos, Kenya. As the drops only fall a short distance their kinetic energy is increased by using relatively large drop sizes.

can be re-evaporated from the surface, taken up by plant roots and under certain conditions excess water can flow laterally and/or leave the bottom of the soil profile as drainage (Fig. 6.1). The water content of the soil can vary with time and depth in the soil. Surface soil water content influences infiltration, runoff and soil evaporation. The water content of deeper layers affects any lateral water movement and drainage. Soil water content is also a key parameter in determining plant growth via its direct and indirect controls on transpiration and nutrient uptake. Soil water content directly affects plant water uptake by changing the hydraulic properties of the soil. Indirect effects on transpiration may also be generated via the way in which soil water content can trigger the production of plant hormones which control stomatal behaviour (Davies and Zhang, 1991). In agroforestry systems differences in soil water content beneath trees and the cropped area can arise due to differences in infiltration and runoff under the trees and crop (see Fig. 6.1) and also via the different soil water abstraction rates by tree and crop roots. Drainage patterns may also be altered due to greater abstraction by tree roots and the utilization of out of crop season soil water by trees.

Soil water and drainage measurement

Sampling strategies for measuring soil water content in agroforestry systems vary according to the spatial arrangement of the trees and crop. In two-dimensional planting arrangements where the tree component is planted in lines (Fig. 6.8a) (e.g. hedgerow and windbreak systems) soil water content measurements are often made at a series of points perpendicular to the tree line at either linear or non-linear spacing. The non-linear spacing is particularly suited to windbreak systems where the tree lines are a large distance apart (e.g. Brenner, 1991). In three-dimensional planting arrange-ments two main strategies are used. Sampling at increasing radial distances from the trees has been used by Eastham *et al.* (1988) (Fig. 6.8b) where the trees were planted in a radial 'Nelder' design and again linear spacing of sample points was used for closely spaced trees and non-linear spacing for more widely spaced trees. Another form of spatial sampling is to use two-dimensional grids. An example of this type of sampling technique which has been used in a *Grevillea robusta*–maize agroforestry system by Wallace and Jackson (1994) is shown in Fig. 6.8c. Here the sampling points are spaced on a regular grid which attempts to account for the changes in soil water content that may occur along and between the tree lines. Each sampling point is considered to be at the centre of a rectangle and because of the symmetry within the arrangement shown it is only necessary to make observations in one-quarter of the 3×4 m planting pattern. This latter grid pattern also gives samples at a range of radial distances from the trees. In practice, the precise sampling pattern and number of replications will depend on which soil parameter is required, for what purpose, and its inherent variability within the soil and the tree–crop planting arrangement. Most studies are usually forced to make some practical compromise between what is statistically desirable and what is logistically feasible.

Near-surface changes in soil water content are usually measured by gravimetric sampling, but new techniques have recently been developed which allow more rapid and automatic measurement of surface soil water content. Time Domain Reflectometry (TDR) as described by Topp and Davies (1985) is now a commercially available technique which can be used to measure soil water content. A very high frequency ($\sim$ GHz) electrical signal is sent to a wave guide placed in the soil, the reflected signal from which is a function of the soil water content. Although expensive, the main advantage of the TDR technique is its ability to monitor rapid (e.g. hourly) changes in soil water content which occur near the soil surface as rain falls, ceases and subsequent evaporation takes place. Figure 6.9 shows an example of some TDR data recorded in an agroforestry system in

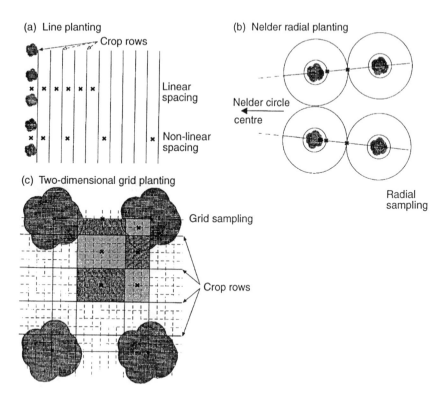

Fig. 6.8. Examples of some soil water content sampling patterns used in agroforestry systems. Measurement points (×) can be arranged (a) perpendicular to the tree line, (b) at radial distances from individual trees and (c) in a two-dimensional grid.

Kenya. The data shown are for four depths (5, 15, 25 and 35 cm) near (0.5 m) a tree and the same four depths 2.5 m away from a tree. During the rainfall event the TDR sensors show that less water infiltrated near to the tree, especially at 35 cm depth, presumably because of the sheltering effect of the canopy. The rapid decline in surface water content after rainfall ceased is due to the combined effects of drainage and evaporation, the data shown indicating a higher rate of soil water loss in the more open area between the trees. The TDR technique needs to be properly calibrated for different soils, since it is affected by bulk density and mineral and organic matter content (Roth *et al.*, 1992; Robinson *et al.*, 1994). Care therefore needs to be taken when using TDR in agroforestry systems where mulches are used which may alter the bulk density, organic matter and mineral

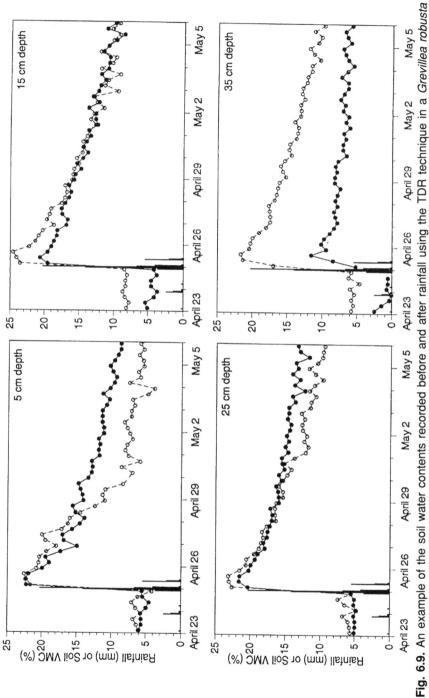

Fig. 6.9. An example of the soil water contents recorded before and after rainfall using the TDR technique in a *Grevillea robusta* agroforestry system at Machakos, Kenya. Two traces are shown for each of four depths, one near (0.3 m, ●) to a tree and the other furthest (2.5 m, ○) from the tree.

composition of the soil.

Another new technique for measuring surface soil water content is the Surface Insertion Capacitance Probe (SCIP) developed at the Institute of Hydrology in the UK (Robinson and Dean, 1993; Fig. 6.10). This manually operated device has two steel prongs which can be inserted into the soil to a depth of either 5 or 10 cm. The instrument works on the principle that the dielectric constant of the soil around the prongs varies with the soil water content. Once calibrated using the gravimetric technique the SCIP offers a very rapid means of measuring surface soil water content, which means that, for example, spatial variability in surface soil water content can be studied in great detail.

Deeper (i.e. below about 15 cm) soil water content is most frequently measured using a neutron probe (Bell, 1969, 1987). Fast neutrons from a radioactive source are slowed by water molecules in the soil and the probe counts these slow neutrons as a measure of the soil water content. The resolution of this technique limits the timescale over which it can be used as

Fig. 6.10. The Institute of Hydrology Surface Insertion Capacitance Probe (SCIP) for the measurement of soil surface water content.

water content changes cannot usually be detected for periods much shorter than about 1 week. The method has most use during periods when there is little or no rainfall, when the terms P_g, I_t, I_c, R_t and R_c in Equation 6.1 are zero and changes in soil water content can be ascribed to evaporation and drainage. In some circumstances drainage is also insignificant but, even when it is not, the simultaneous measurement of soil water potential profiles using tensiometers (Wellings and Bell, 1982) can provide sufficient information to separate evaporation and drainage.

Soil water potentials can be measured as above with tensiometers, where ceramic cups are placed in the soil and connected hydraulically to either a mercury manometer for manual reading, or to pressure transducers for automatic reading. Tensiometers have the disadvantage that they only work for soil water potentials down to ~ -80 kPa, so in many arid regions these devices may be off scale for much of the time. Soil water potential can also be measured using gypsum blocks (Wellings *et al.*, 1985), which have the advantage that they work down to much lower water potentials (~ -1500 kPa), but they can exhibit hysteresis and must be properly calibrated to obtain accurate readings. Uncalibrated gypsum blocks have been used in agroforestry systems by Huxley *et al.* (1994) to give qualitative pictures of the two-dimensional soil drying and wetting patterns that occurred during several rainy seasons in Kenya.

Drainage is the most difficult component of the water balance to measure directly. As a consequence it is often calculated as the residual in the water balance; however, this approach is subject to the problem that the accuracy of the drainage estimate is dependent on the cumulative errors in all the rest of the water balance terms. Direct measurement of drainage can be made using lysimeters. Large lysimeters contain many tens of m^3 of soil and drainage is collected directly from outlets at their base (e.g. see Forbes, 1979). Large lysimeters have also been used successfully where the subsoil forms a 'natural' impermeable layer and the outflow can be collected to measure drainage (Calder, 1976). Smaller lysimeters containing 1 m^3 or less have also been used to estimate drainage. At this size they can be weighed and/or drainage can be collected as it flows out of the bottom of the lysimeter (e.g. Agnew, 1982; Klocke *et al.*, 1985). The presence of a solid base at the bottom of a lysimeter can induced artificial saturation and, to avoid this, the most representative lysimeters use a ceramic suction plate to achieve the correct unsaturated soil water potential at their base.

Drainage can be estimated indirectly from measurements of soil water content under certain circumstances. Figure 6.11 illustrates a technique known as the 'zero flux plane' method. Simultaneous measurements of the profile of soil water content and soil water potential (Φ) are required on at least two occasions between which the drainage occurs. Under ideal conditions when there has been no rain, evaporation of water from the upper part of the soil profile produces a soil water potential profile similar to that shown in Fig. 6.11b. There is a point in the profile where $\delta\Phi/\delta z = 0$, above which water moves upwards and below which water moves downwards. At the point where $\delta\Phi/\delta z = 0$, water moves neither up nor down, hence the name 'zero flux plane' or ZFP. Once this has been identified water content changes above the ZFP can be ascribed to evaporation and water content changes below the ZFP to drainage. In practice the method can be difficult to use in the field where frequent rainfall events produce soil water potential profiles from which it is difficult to determine the zero flux plane. Under certain circumstances there may even be more than one ZFP in the profile.

Another method for estimating drainage from soil water content observations has been described by McGowan (1974), who used time series of soil water content to infer drainage. For progressively deeper layers in the soil, graphs of water content changes with time are inspected for inflection points where the rate of change in water content increases abruptly. This point is interpreted as the time when the roots of the annual crop have reached this depth in the soil. Decreases in soil water content before this time are taken as drainage and after this time as root abstraction. Again this method only works well in rather idealized conditions, i.e. in annual crops, where a progressively deepening rooting front can be assumed, and during prolonged drying spells.

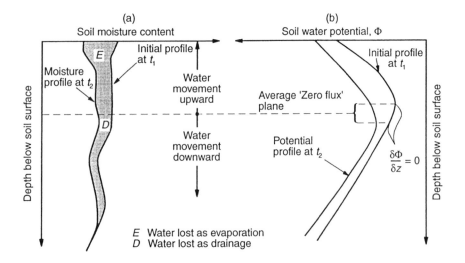

Fig. 6.11. Schematic diagram showing (b) how the zero flux plane (the point where $\delta\Phi/\delta z = 0$) is determined from profiles of soil water potential and (a) how it is used to calculate evaporation and drainage from soil water content profiles.

Soil Evaporation

Soil evaporation processes

Water that is held in the upper layers of the soil can be directly evaporated into the air if there are both sufficient water and energy available. In tropical agricultural crops, which do not cover the ground for some or all of the season, direct soil evaporation can be a significant component of the water balance. For example, in the semi-arid regions of the Middle East and West Africa, direct soil evaporation can account for 30–60% of rainfall (Cooper *et al.*, 1983; Wallace, 1991). Direct evaporation of water from bare soil occurs in two distinct phases (e.g. see Ritchie, 1972). Initially evaporation from the soil proceeds at the potential rate E_{s0} during the 'first phase' immediately following rewetting of the surface by rain. This lasts for a number of days (t_1) until the total amount of water evaporated is U, after which the 'second phase' begins where the rate of soil evaporation declines according to the square root of time. This can be expressed mathematically as:

$$\Sigma E_{s1} = \sum_{t=0}^{t_1} E_{s0} = U \qquad t < t_1 \tag{6.4}$$

$$\Sigma E_{s2} = \alpha \sqrt{(t - t_1)} \qquad t > t_1 \tag{6.5}$$

where ΣE_{s1} and ΣE_{s2} are the cumulative amounts of soil evaporation in the first and second drying phases respectively. α is assumed constant for any particular soil and is a function of the soil water diffusivity (Black *et al.*, 1969). A third phase of soil evaporation also exists where under very dry soil conditions the loss rate is determined by diffusion of water vapour through the upper soil layers (Philip, 1957). In most environments where agroforestry is practised this third phase is likely to be a small component of the total soil evaporative loss.

Shading of the ground by a tree and/or crop canopy should reduce first phase soil evaporation rates which are dependent on E_{s0} and hence the energy available at the soil surface. Second phase soil evaporation rates are determined by the soil hydraulic properties and should therefore be independent of shade. The net effect of shade on cumulative soil evaporation over periods of several weeks or more will therefore depend on the total amount of time the soil spends in first stage drying, which is a function of soil type and the frequency with which the surface is rewetted by rainfall.

The potential for conserving soil water by reducing soil evaporation is illustrated in Fig. 6.12. Here the cumulative seasonal amount of direct evaporation from bare soil at Machakos, Kenya was calculated using the simple two phase model described by Equations 6.4 and 6.5. First stage losses (ΣE_{s1}) were set (conservatively) to 3 mm with $t_1 = 1$, values appropriate to a sandy soil (Bley, 1990). For the second stage losses α was taken as 2.5, again appropriate for a sandy soil (Wallace *et al.*, 1986; Bley, 1990). Figure 6.12 shows that total bare soil evaporation is $\sim 40\%$ of rainfall. Bearing in mind that soil evaporation losses in more clayey soil are likely to be even greater than the values shown here, there is clearly substantial scope for manipulating tree–crop canopies in an attempt to reduce this unproductive loss of water. This is especially so in the rainfall climate used here because, as the simulation in Fig. 6.12 shows, most ($\sim 80\%$) of the soil evaporation occurred in the energy limited (first) phase. To illustrate the potential effect of canopy shade on soil evaporation the above model was used with a reduced first phase drying rate of 1.5 mm day^{-1} and an increased first phase drying duration (t_1) of 2 days. This reduced annual total soil evaporation by nearly 100 mm. The precise effect of canopy shade on soil evaporation will depend on soil type, the reduction in energy at the soil surface and the frequency intensity of rainfall.

The amount of water that is lost as soil evaporation may also be affected by surface mulches, which are widely used in agroforestry. Direct evaporation from the soil surface may be reduced by the mulch, if it reduces the wind speed and radiation at the soil surface. The physical structure of the mulch may also impede vapour transfer from the soil surface, further reducing evaporation. However, some mulches may also absorb water during rainfall which can subsequently be re-evaporated without entering the soil, in effect another kind of interception loss. Whether a mulch causes a net gain or loss to the soil water balance will depend on the relative effects it has on evaporation and infiltration. These are likely to vary with different types of mulch.

Soil evaporation measurement

Soil evaporation has been measured directly using small 'micro' lysimeters placed between the plants (Boast and Robertson, 1982) and the technique has been applied in semi-arid crops by Allen (1990) and Wallace *et al.* (1993). Microlysimeters are small containers filled with bare soil which are placed flush with the soil surface and weighed daily (or more frequently) to determine water loss. It is important that the soil in the lysimeters is as representative of the surrounding field soil as possible and several precautions need to be taken to ensure this. First, the soil within the microlysimeters should be as undisturbed as possible. This can be achieved by carefully pushing the hollow lysimeter casing gently into the soil, the mechanical impedance of which can be reduced by excavating a few centimetres in front of the leading edge of the lysimeter. Second, the soil in the lysimeters should be renewed daily to overcome the fact that there is no root abstraction of water within the lysimeters and that they have an artificial

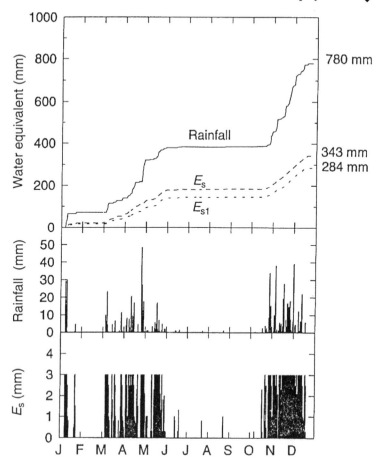

Fig. 6.12. Estimated bare soil evaporation loss at Machakos, Kenya in 1986 made using the Ritchie (1972) model. Cumulative rainfall, soil evaporation in the first phase, E_{s1}, and total soil evaporation, E_s, are also shown.

boundary to water flow at their base. A range of sizes and depths of micro-lysimeter have been used and these have been reviewed by Daamen *et al.* (1993) who also give a protocol for using microlysimeters in the field.

Other methods which have been used to measure soil evaporation include the use of a steady-state porometer adapted for use in soil (Nobel and Geller, 1987), a miniature micrometeorological (Bowen-ratio) system which measures temperature and humidity gradients very close to the soil surface (Ashktorab *et al.*, 1989; Wallace and Holwill, 1996), soil isotope profile techniques (Taupin *et al.*, 1991) and remote sensing (Chanzy and Bruckler, 1991). The micrometeorological technique is only suitable for use in flat land with comparatively large areas ($\sim$ several 100 m^2) of bare soil and would therefore only be of use in agroforestry systems with wide tree spacings (e.g. windbreaks, etc.).

Transpiration

Transpiration processes

When a vegetation canopy is dry, evaporation comes under the physiological control of the plants, since it has to pass through the stomata or cuticle of the leaves. The rate of cuticular transpiration is usually very low, since the resistance to water vapour transfer across the cuticle is at least an order of magnitude greater than when the stomata are open (Rutter, 1975). The transpiration rate therefore depends on the bulk stomatal resistance of the entire canopy, or simply the 'surface resistance'. Monteith (1965) derived an expression for the transpiration rate (λE) of vegetation which completely covers the ground as

$$\lambda E = \frac{\Delta R_n + \rho c_p D / r_a}{\Delta + \gamma (1 + r_s / r_a)} \tag{6.6}$$

where R_n is the above-canopy net radiation, γ is the psychrometric constant, Δ is the rate of change of saturated vapour pressure with temperature, ρ and c_p are the density and specific heat (at constant pressure) of air and λ is the latent heat of vaporization of water. r_a is the aerodynamic resistance to water vapour transfer from the canopy to a reference point above the crop and r_s is the bulk stomatal resistance of the entire canopy. Surface resistance is a function of the individual stomatal resistances (r_{st}) of the leaves and the quantity of leaves per unit ground area (Monteith et al., 1965; Biscoe et al., 1976), i.e.

$$r_s = \sum_i (L_i / r_{st}^i) \tag{6.7}$$

where r_{st}^i and L_i are the stomatal resistance and leaf area of the constituent canopy elements (i).

In trying to apply the Penman–Monteith equation (Equation 6.6) to calculate transpiration from mixed tree–crop canopies a major problem arises because the canopy has two distinct components rather than the one 'big leaf' envisaged by Monteith. Two key modifications are needed to the original equation so that it can be applied in mixed species canopies. First, the individual radiation interception by each of the constituent canopies needs to be calculated. Once this has been done the transpiration from each canopy can be calculated using either a single or dual source transpiration model.

Radiation interception model

An exact theoretical description of the diurnal behaviour of light interception by a plant canopy is very complex and depends on a great many
variables (e.g. canopy architecture, solar angle, ratio of diffuse to direct
radiation, etc. – see Ross, 1981). Despite this complexity much progress
has been made in homogeneous single-species canopies by assuming that
foliage is randomly distributed, in which case the fraction of light intercepted by a canopy of leaf area index L is given using Beer's law (see p. 161):

$$f = 1 - e^{-KL} \tag{6.8}$$

where K is the extinction coefficient. To allow for diurnal changes in f due
to variation in solar angle (β), K can be expressed as:

$$K = K_{min}/\sin \beta \tag{6.9}$$

where K_{min} is the minimum value of the extinction coefficient occurring
when the solar angle is 90°.

Equation 6.8 has been successfully applied in single-species canopies
uniformly distributed over the ground, so much so that several modern
instruments for indirectly estimating leaf area index rely on this equation
being applicable. Even where there is a moderate degree of inhomogeneity,
such as in a row crop where the spacing is up to one-third of the crop
height, there is reported to be little difference in light interception
compared with a closed canopy (Gijzen and Goudriaan, 1989). There are
some light interception models which explicitly deal with spatially
inhomogeneous canopies, such as those of Jackson and Palmer (1979),
Norman and Welles (1983), Gijzen and Goudriaan (1989) and Wang and
Jarvis (1990). However, the main disadvantage of these physically rigorous
models is that they demand many canopy architecture and light
transmission parameters which are not very often available in practice.

There are a number of light interception models which have been
developed for mixed-species canopies (see also Ong *et al.*, Chapter 4 this
volume). Rimmington (1984) developed a model based on the quantity and
optical properties of the foliage of the constituent species. In horizontally
well-mixed intercrops the model described by Sinoquet *et al.* (1990) is
appropriate and Sinoquet and Bonhomme (1991) developed this approach
to give a theoretical framework for dealing with spatial inhomogeneity in
intercrops. However, even the more simple of these models require basic
information on leaf inclination and light scattering properties as well as a
knowledge of the fraction of leaf area of each constituent species present in
a number of canopy layers. Again these data are rarely available.

In a pragmatic attempt to overcome this a more simple light
interception model has been developed by Wallace (1995) which only
requires the individual values of K and L for each species along with their
relative heights (h_1/h_2). The fraction of light intercepted by each

component, f_i, is given by:

$$f_i = f_i^s + \frac{h_i^2}{h_1^2 + h_2^2}(f_{i}^{d}j - f_i^s) \tag{6.10}$$

where f_i^d and f_i^s are the amounts of light species i would intercept if its foliage was either completely above (f_i^d) or completely below (f_i^s) the other canopy. The amounts of light f_i^d and f_i^s are therefore given by the expressions:

$$f_i^d = 1 - e^{-K_i L_i} \tag{6.11}$$

and

$$f_i^s = (1 - e^{-K_i L_i}) \prod_{j \neq i}^{2} e^{-K_j L_j} \tag{6.12}$$

Figure 6.13 shows an example of the output of Equations 6.10, 6.11 and 6.12 for a mixture of two species both with a leaf area index of 3 and an extinction coefficient of 0.5. When species 1 is completely subordinate ($h_1/h_2 \sim 0$), species 2 intercepts the incident light first and the remainder

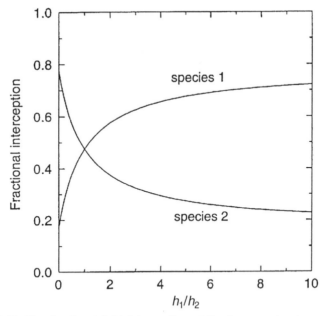

Fig. 6.13. The fractional light interception of the two species in a mixture as a function of their relative heights estimated using the Wallace (1995) model. In this example each species has the same leaf area (3.0) and extinction coefficient (0.5).

falls on species 1. As h_1/h_2 increases, species 1 becomes more light domi-nant until, ultimately, it intercepts the incident light first, the remainder falling on species 2. This simple model applies to horizontally uniform species mixtures and would need further modification for use in agro-forestry systems where the overstorey tree canopy is often spatially clumped.

Transpiration model
Once the radiation intercepted by the component canopies is known, we can use this to drive a two-source transpiration model. One assumption that is made at this stage is that the fraction of net radiation intercepted by the canopy is similar to the fraction of solar radiation intercepted. This assumption seems reasonable since Ross (1981) reported that on cloudless days net radiation profiles in a crop canopy are primarily determined by direct solar radiation. This has been confirmed by Wallace *et al.* (1991) who measured net and solar radiation interception in a sugar cane–maize mixture and found that their daily total interception differed by only 5%. Figure 6.14 shows a schematic diagram of the evaporation process from a mixture of two species. For simplicity in the following analysis soil evaporation is ignored; however, this can also be taken into account if necessary. Water diffuses from the leaves of each canopy, which have specified stomatal resistances and leaf area indices that can be combined to calculate their canopy resistance, r_s^1 and r_s^2. Transpiration from each species can then be calculated using a modified version of the Penman–Monteith Equation 6.6 i.e.:

$$\lambda E_i = \frac{\Delta f_i R_n + \rho c_p D_0 / r_a^i}{\Delta + \gamma(1 + r_s^i / r_a^i)} \qquad (6.13)$$

where f_i is the fraction of the above-canopy net radiation (R_n) which is intercepted by canopy i ($=1$ or 2). The term r_a^i is the aerodynamic resis-tance to water vapour transfer from the canopy to a point in the air between the two canopies where the vapour pressure deficit is D_0. When the 'in canopy' deficit D_0 is known, then Equation 6.13 provides an adequate description of the transpiration of each species and the total transpiration is simply the sum of the constituents. In practice, values of D_0 are rarely available and calculations have to be based on the above-canopy vapour pressure deficit, D.

A first approach is simply to use D in Equation 6.13 with a bulk mixed canopy value for the aerodynamic resistance (r_a) which is assumed to be the same for both components (i.e. $r_a = r_a^1 + r_a^a = r_a^2 + r_a^a$). However, this approach is limited since it ignores any interaction that may occur between the component fluxes of latent and sensible heat, potentially the type of microclimatic interactions that may be important in intercrops and

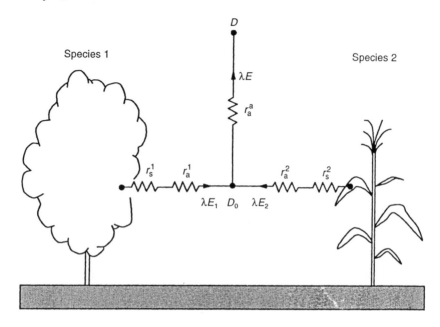

Fig. 6.14. A schematic representation of the transpiration from a two-species mixture. Each species has its canopy (r_s^1 and r_s^2) and boundary layer resistance (r_a^1 and r_a^2). Component transpiration fluxes, λE_1 and λE_2, converge within the mixed canopy where the vapour pressure deficit is D_0. The total transpiration, λE, mixes with the air at the screen height via the aerodynamic resistance r_a^a.

agroforestry. The effect of sub-canopy fluxes on canopy transpiration has been demonstrated by Wallace *et al.* (1990), who showed that in a sparse millet crop, where heat water vapour fluxes can arise from both the crop and the soil, the above modified Penman–Monteith formula can be significantly in error. It was concluded that an improved description of evaporation from a two-source system was given by the Shuttleworth and Wallace (1985) model, since it allows for interaction between the source fluxes. Shuttleworth and Wallace originally proposed a model to describe heat and water vapour transfer from a monocultural sparse crop. It can, however, be applied to any two-component system, in which case the total evaporation, λE, can be written as:

$$\lambda E = C_1 PM_1 + C_2 PM_2 \tag{6.14}$$

where PM_1 and PM_2 are terms each similar to the Penman–Monteith formula which would apply to closed canopies of the constituent species. They have the form:

$$PM_1 = \frac{\Delta A + [\rho c_p D - \Delta r_a^1 (A - A_1)]/[r_a^a + r_a^1]}{\Delta + \gamma[1 + r_s^1/(r_a^a + r_a^1)]}$$ (6.15)

and

$$PM_2 = \frac{\Delta A + [\rho c_p D - \Delta r_a^2 (A - A_2)]/[r_a^a + r_a^2]}{\Delta + \gamma[1 + r_s^2/(r_a^a + r_a^2)]}$$ (6.16)

where A is the total energy available to the mixed canopy, A_1 and A_2 are the amounts of energy available to the component species, so that:

$$A_1 = f_1 R_n$$ (6.17)

and:

$$A_2 = f_2 R_n$$ (6.18)

Since the values of f_1 and f_2 are generated by the radiation interception model, described in the previous section, the available energy and hence transpiration from each canopy are coupled to the above-ground competition for light.

The various resistance terms in Equations 6.15 and 6.16 are shown diagrammatically in Fig. 6.14. The bulk leaf boundary layer resistances of the two component species are represented by r_a^1 and r_a^2 and r_a^a is the aerodynamic resistance between the in-canopy mixing point (D_0) and the reference height (D). The coefficients C_1 and C_2 in Equation 6.14 are given by:

$$C_1 = \left[1 + \frac{1/R_2}{1/R_1 + 1/R_a}\right]^{-1}$$ (6.19)

and

$$C_2 = \left[1 + \frac{1/R_1}{1/R_2 + 1/R_a}\right]^{-1}$$ (6.20)

where

$$R_a = (\Delta + \gamma)r_a^a$$ (6.21)

$$R_1 = (\Delta + \gamma)r_a^1 + \gamma r_s^1$$ (6.22)

$$R_2 = (\Delta + \gamma)r_a^2 + \gamma r_s^2$$ (6.23)

Equations 6.14 to 6.23 provide the means to calculate evaporation from a two-species canopy when only the above-canopy net radiation, temperature and vapour pressure deficit are known.

The aerodynamic resistance terms in the above formulation are probably the least well-understood part of the Shuttleworth–Wallace model and a number of prescriptions have been proposed and discussed by Shuttleworth and Wallace (1985), Choudhury and Monteith (1988) and Shuttleworth and Gurney (1990). Under some circumstances the uncertainties in the aerodynamic resistance terms may be of limited importance where evaporation rates have been shown to be insensitive to their values (e.g. see Shuttleworth and Wallace, 1985; and Shuttleworth and Gurney, 1990). However, it would be prudent to check the sensitivity of transpiration rates from intercrop and agroforestry canopies as this would help define the amount of effort needed to specify in-canopy transfer in these systems.

The effect of between-canopy transfer of heat and water vapour is illustrated in Fig. 6.15 for the case of a dominant main canopy (species 1) with a surface resistance of 200 s m^{-1} at a leaf area index of 1 (equivalent to a closed canopy with a surface resistance of ~ 50 s m^{-1}). The diagram shows the ratio of species 1 transpiration calculated using Equation 6.13 (i.e. based on the Penman–Monteith formula) to that calculated using Equation 6.14 (based on the Shuttleworth–Wallace model) as a function of the sub-canopy (species 2) surface resistance. The calculations are performed for average weather conditions and for three different values of dominant canopy leaf area index. Since Equation 6.14 allows for flux interactions between the two canopies and Equation 6.13 does not, their ratio indicates the degree to which this is important under the conditions illustrated. When the sub-canopy resistance is very low, as it would be for a non-stressed vegetation with a high leaf area, or if the sub-canopy was wet just after rain, the Penman–Monteith equation overestimates transpiration from the dominant canopy by as much as 15%. Conversely, when the sub-canopy resistance is high, as would be the case if it were water stressed or had a low leaf area, the Penman–Monteith equation underestimates the overstorey transpiration by nearly 20%. Clearly, the net effect of flux interactions from the two canopies over a season will depend on the constituent species surface resistances and how these are affected by rainfall and soil moisture stress.

The mechanism by which the overstorey canopy transpiration is altered is via the modification of the within-canopy vapour pressure deficit (D_0) because of fluxes of heat and water vapour from the sub-canopy. When the sub-canopy resistance is low, transpiration from the sub-canopy is large enough to humidify the air around the overstorey and hence reduces its transpiration. When the sub-canopy resistance is high, the sensible heat flux from the sub-canopy is sufficient to increase the vapour pressure deficit around the overstorey, thereby increasing its transpiration rate. The magnitude of this effect is greatest when the overstorey leaf area is low, because the amount of energy reaching the sub-canopy, and hence the degree of modification of D_0, is largest.

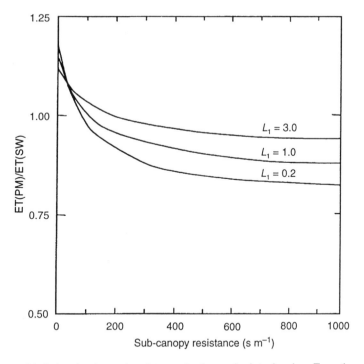

Fig. 6.15. Variation in the ratio of transpiration calculated using Equations 6.13 and 6.14, ET(PM)/ET(SW), with sub-canopy surface resistance. The three different lines are for a dominant overstorey with leaf area indices of 0.2, 1.0 and 3.0 under typical mid-day conditions.

Transpiration measurement

Plant physiological methods are commonly used to measure transpiration (see also Chapter 4). Most progress in this area has been made since the introduction of porometers. These devices are attached to intact plant leaves and measure the conductance of the epidermis to water vapour. Parkinson (1985) has provided a comprehensive review of the types of porometer available. Most of the work with porometers is directed at studying the behaviour of stomata at an individual leaf level. However, a few studies have shown how porometry can be used to measure transpiration from complete vegetation canopies in both temperate (e.g. Roberts *et al.*, 1980; Waring *et al.*, 1980) and semi-arid climates (Azam-Ali, 1983; Wallace *et al.*, 1990). Figure 6.16 shows an example of the diurnal response of the stomatal conductance (the reciprocal of stomatal resistance) of millet leaves obtained by Wallace *et al.* (1990) using a diffusion porometer. The conductances vary with leaf age and position in the canopy, but they can be combined with profiles of leaf area using Equation 6.7 to obtain the total canopy conductance. The total canopy conductance can then be combined with weather data in Equation 6.14 to calculate transpiration, as shown in Fig. 6.17. Although this technique gives very detailed information on the physiological response of the plants, being manual it is very time consuming and, in practice, cannot be used to measure transpiration continuously for long periods. Computation of transpiration over an entire season therefore requires some detailed modelling of the stomatal behaviour in response to atmospheric and soil conditions (e.g. see Jarvis, 1976).

The use of tracers in measuring transpiration is another technique which has been shown to give very good results (Waring and Roberts, 1979). For example, Waring *et al.* (1980) used a radioactive isotope injected into individual Scots pine trees as a method to compare with porometry. In the semi-arid region of southern India, Calder *et al.* (1986) have used deuterium oxide as a tracer to measure transpiration in eucalyptus trees. The isotope was injected into the tree stem and the concentration of

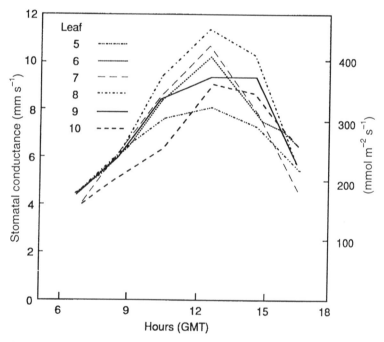

Fig. 6.16. Diurnal variation in the stomatal conductance of the leaves in a millet crop before anthesis on 9 July 1986 at the ICRISAT Sahelian Center, Sadoré in Niger. Leaves are numbered from the bottom to the top of the canopy. (Reproduced from Wallace *et al.*, 1990.)

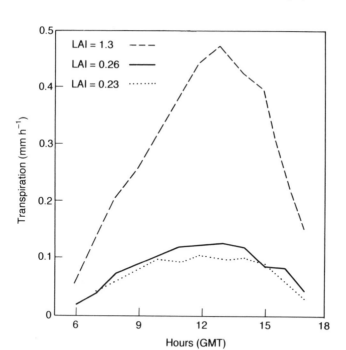

Fig. 6.17. Transpirational rates from millet at the beginning (2 July, · · ·), middle (18 July, – – –) and end (27 August, ——) of the 1986 season at the ICRISAT Sahelian Center, Sadoré in Niger. The leaf area index (LAI) on each day is also shown. (Reproduced from Wallace *et al.*, 1990.)

deuterium in the transpiration measured by putting plastic bags over samples of the leaves in the canopy. The transpiration rate over a number of days was then calculated from the area under the deuterium concentration/time curve. The method is simple, but has the disadvantage that it requires a mass spectrometer to analyse the transpirate for deuterium concentration and it only gives mean transpiration rates over a number of days.

Another form of 'tracer' used to measure transpiration is heat: a technique often referred to as the heat pulse method (see also Chapter 4). Heat is applied for a short time (a few seconds) using a thin probe inserted into the plant stem and the velocity of the transpiration stream calculated from the time required for the heat to travel a short distance from the point of application. Early work using this method encountered problems with calibration and wound corrections due to the insertion of the heat pulse probe; however, recent developments of the technique appear to have overcome many of these problems (Cohen *et al.*, 1981; Edwards and Warick, 1984). A further development of the heat tracing technique is the stem heat balance method (Sakuratani, 1981; Baker and van Bavel, 1987; Chapter 4). In this method the heat is applied using a small electric heating element wrapped around the plant stem, thereby overcoming the problems encountered with the heat pulse technique (above) caused by the insertion of the heating probe in the stem. The stem heat balance method also has the advantage that it is continuous, automatic and directly measures the mass flow rate without the need for individual calibration; however, the method is not suitable for large diameter stems greater than about 10 cm.

All of the above tracer techniques are well suited for studying transpiration in agroforestry systems. For example, the deuterium tracing method has been shown to work in *Grevillea robusta*, a popular agroforestry species in Kenya (Fig. 6.18). Six grams of deuterium were injected into five small (3 mm) holes drilled around the circumference of a 4-year-old tree. Six bags were placed on leaves at different positions in the canopy and samples of the transpirate collected every day for a period of 6 days. The concentration curves obtained are shown in Fig. 6.18, and demonstrate the need to sample the canopy carefully as the peak concentration (and hence calculated transpiration rate) of the curve depends on the position in the canopy. The mean transpiration rate for the period observed was 18 (± 3) litres day^{-1}. The heat balance method has also been used to study tree

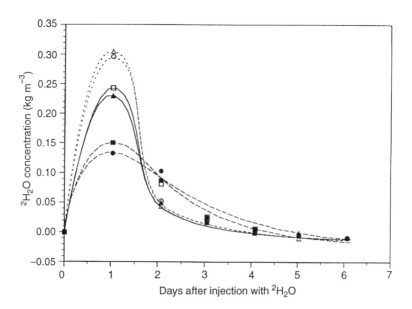

Fig. 6.18. The change in deuterium concentration in the transpiration from a 4-year-old *Grevillea robusta* tree growing at Machakos, Kenya with time after injection. The different symbols represent various positions around the canopy and the mean transpiration rate over the period shown is 18(± 3) litres day^{-1}.

transpiration in agroforestry systems, both on tree stems (e.g. Ong *et al.*, 1992) and on tree roots (Ong and Khan, 1994).

Concluding Remarks

Clearly the water balance of agroforestry systems is very complex, more so than any agricultural system studied so far. Currently very little is known about the way in which rainfall is partitioned in agroforestry systems. However, the techniques developed and the experience gained in studies of sole trees, sole crops and intercrops has made available an impressive range of techniques and models with which to take on the challenge of examining agroforestry system resource use in detail. We should also learn from our mistakes in the past. For example, previously the roles of water and nutrients have been largely studied in isolation, yet we now know that they are intimately linked and we cannot expect to understand and predict the performance of tree–crop mixtures unless we take due recognition of this. Again in the past we have tended to study above- and below-ground processes separately. Plants have to balance the above-ground demands for evaporation with the below-ground supply of water and nutrients. In mixed communities plants rarely compete for light without simultaneously competing for water and nutrients and our understanding of how mixed species systems grow and utilize resources will remain restricted unless we design experiments and models which explicitly recognize this.

What little that is known about agroforestry processes is largely about the above-ground parts of the system. As with previous studies in other agricultural systems, this is not necessarily because this is the most important part of the system, but because it is the easiest to study. In many tropical environments it is the below-ground resources that limit growth and the excuse that it is 'too difficult' to make the necessary below-ground observations is simply no longer acceptable (see Chapter 9). We must shift our priorities and apply the energy and enthusiasm of young scientists to the study of below-ground processes in combination with important above-ground phenomena.

The 'state-of-the-art' in agroforestry process modelling, although in need of further development, is still vastly in advance of the availability of the measurements to test the models. Again we can learn from the lessons of the past with monocultures and realize that the output of complex simulation models is only as good as the component process modules of which they are composed. In agroforestry systems many of these important component processes are not yet properly understood, and the emphasis of research should be on elucidating these before we confuse ourselves with the output of complex simulations of tree–crop mixtures.

Finally, we should keep an open mind about whether it is always a good idea to grow intimate mixtures of trees and crops. There may be situations where such mixtures will prove to be a net benefit, and there will also be situations where they undoubtedly will not. The challenge to agroforestry researchers is to define these situations clearly and not to be swayed by the temptation to make unsubstantiated claims in the propagation of their science.

Acknowledgements

I am grateful to a number of colleagues on whose work I have drawn on in this chapter. I would particularly like to thank John Gash who helped me with the interception analysis using his revised model and Nick Jackson for the TDR data and help with the diagrams. The water balance research project at Machakos, Kenya was carried out with the support and collaboration of ICRAF and was funded as project R4853 in the Forestry Research Programme component of the UK Overseas Development Administration's Renewable Natural Resources Strategy.

References

Agnew, C.T. (1982) Water availability and the development of rainfed agriculture in south-west Niger, West Africa. *Transactions of the Institute of British Geographers* 7, 419–457.

Allen, S.J. (1990) Measurement and estimation of evaporation from soil under sparse barley crops in northern Syria. *Agricultural and Forest Meteorology* 49, 291–309.

Ashktorab, H., Pruitt, W.O., Paw U, K.T. and George, W.V. (1989) Energy balance determination close to the soil surface using a micro-Bowen ratio system. *Agricultural and Forest Meteorology* 46, 259–274.

Aston, A.R. (1979) Rainfall interception by eight small trees. *Journal of Hydrology* 42, 383–396.

Azam-Ali, S.N. (1983) Seasonal estimates of transpiration from a millet crop using a porometer. *Agricultural Meteorology* 30, 13–24.

Baker, J.M. and van Bavel, C.H.M. (1987) Measurement of the mass flow of water in the stems of herbaceous plants. *Plant, Cell and Environment* 10, 777–782.

Bell, J.P. (1969) A new design principle for neutron soil moisture gauges: the 'Wallingford' neutron probe. *Soil Science* 108, 160–164.

Bell, J.P. (1987) *Neutron Probe Practice.* NERC Institute of Hydrology, Wallingford, Report No. 19.

Biscoe, P.V., Cohen, Y. and Wallace, J.S. (1976) Community water relations: daily and seasonal changes of water potential in cereals. *Philosophical Transactions of the Royal Society of London, B* 273, 565–580.

Black, T.A., Gardner, W.R. and Thurtell, G.W. (1969) The prediction of evaporation, drainage and soil water storage for a bare soil. *Soil Science Society of America Journal* 33, 655–660.

Bley, J. (1990) Experimentelle und modellanalytische Undersuchungen zum Wasser- und Naehrstoffhaushalt von Perlhirse (*Pennisetum americanum* L.) im Südwest-Niger. PhD thesis, University of Hohenheim, Stuttgart, Germany.

Boast, C.W. and Robertson, T.M. (1982) A 'micro-lysimeter' method for determining evaporation from bare soil: description and laboratory evaluation. *Soil Science Society of America Journal* 46, 689–696.

Brenner, A.J. (1991) Tree–crop interactions within a Sahelian windbreak system. PhD thesis, University of Edinburgh, 234 pp.

Butler, D.R. and Huband, N.D.S. (1985) Throughfall and stem-flow in wheat. *Agricultural and Forest Meteorology* 35, 329–338.

Calder, I.R. (1976) The measurement of water losses from a forested area using a 'natural' lysimeter. *Journal of Hydrology* 30, 311–325.

Calder, I.R. (1977) A model of transpiration and interception loss from a spruce forest in Plynlimon, Central Wales. *Journal of Hydrology* 33, 247–265.

Calder, I.R. and Newson, M.D. (1979) Land-use and upland water resources in Britain – a strategic look. *Water Resources Bulletin* 15, 1628–1639.

Calder, I.R. and Rosier, P.T.W. (1976) The design of large plastic-sheet net-rainfall gauges. *Journal of Hydrology* 30, 403–405.

Calder, I.R., Narayanswamy, M.N., Srinivasalu, N.V., Darling, W.G. and Lardner, A.J. (1986) Investigations into the use of deuterium as a tracer for measuring transpiration from eucalypts. *Journal of Hydrology* 84, 345–351.

Casenave, A. and Valentin, C. (1991) Influence des états de surface sur l'infiltration en zone sahélienne. In: Sivakumar, M.V.K., Wallace, J.S., Renard, C. and Giroux, C. (eds) *Soil Water Balance in the Sudano-Sahelian Zone.* Proceedings International Workshop, Niamey, Niger, February 1991. IAHS Publication No. 199. IAHS Press, Institute of Hydrology, Wallingford, UK, pp. 99–108.

Chanzy, A. and Bruckler, L. (1991) Estimation de l'évaporation à partir d'informations accessibles par télédétection micronde: approches mechaniste, semi-empirique et empirique. In: Sivakumar, M.V.K., Wallace, J.S., Renard, C. and Giroux, C. (eds) *Soil Water Balance in the Sudano-Sahelian Zone.* Proceedings International Workshop, Niamey, Niger, February 1991. IAHS Publication No. 199. IAHS Press, Institute of Hydrology, Wallingford, UK, pp. 149–158.

Chevallier, P. and Lapetite, J.-M. (1986) Note sur les écarts de mesure observés entre les pluviomètres standards et les pluviomètres au sol en Afrique de l'Ouest. *Hydrology Continental* 1 (2), 111–119.

Choudhury, B.J. and Monteith, J.L. (1988) A four-layer model for the heat budget of homogeneous surfaces. *Quarterly Journal of the Royal Meteorological Society* 114, 373–398.

Cohen, Y., Fuchs, M. and Green, G.C. (1981) Improvement of the heat pulse method for determining sap flow in trees. *Plant, Cell and Environment* 4, 391–397.

Cooper, P.J.M., Keatinge, J.D.H. and Hughes, G. (1983) Crop evapotranspiration – a technique for calculation of its components by field measurements. *Field Crop Research* 7, 299–312.

Daamen, C.C., Simmonds, L.P., Wallace, J.S., Laryea, K.K. and Sivakumar, M.V.K. (1993) Use of microlysimeters to measure evaporation from sandy soils. *Agricultural and Forest Meteorology* 65, 159–173.

Darnhofer, T., Gatama, D., Huxley, P. and Akunda, E. (1989) The rainfall distribution at a tree/crop interface. In: Reifsnyder, W.E. and Darnhofer, T. (eds) *Meteorology and Agroforestry.* Proceedings of ICRAF/WMO/UNEP Workshop on Application of Meteorology to Agroforestry Systems Planning and Management. ICRAF, Nairobi, pp. 371–382.

Davies, W.J. and Zhang, J. (1991) Root signals and the regulation of growth and development of plants in drying soil. *Annual Review of Plant Physiology* 42, 55–76.

Eastham, J., Rose, C.W., Cameron, D.M., Rance, S.J. and Talsma, T. (1988) The effect of tree spacing on evaporation from an agroforestry experiment. *Agricultural and Forest Meteorology* 42, 355–368.

Edwards, W.R.N. and Warick, N.W.M. (1984) Transpiration from a kiwifruit vine as estimated by the heat pulse technique and the Penman–Monteith equation. *New Zealand Journal of Agricultural Research* 27, 537–543.

Elrick, D.E. and Reynolds, W.D. (1992) Infiltration from constant-head well permeameters and infiltrometers. In: Topp, G.C., Reynolds, W.D. and Green, R.E. (eds) *Advances in Measurement of Soil Physical Properties: Bringing Theory into Practice.* Soil Science Society of America Special Publication Number 30, Madison, Wisconsin, USA, pp. 1–24.

FAO (1979) *Soil Survey Investigations for Irrigation.* FAO, Rome.

Forbes, C.L. (1979) The Fleam Dyke lysimeters. In: Kitching, R. and Day, J.B.W. (eds) *Report of the Institute of Geological Sciences No. 79/6.* HMSO, London, pp. 31–33.

Gash, J.H.C. (1979) An analytical model of rainfall interception by forests. *Quarterly Journal of the Royal Meteorological Society* 105, 43–55.

Gash, J.H.C. and Morton, A.J. (1978) An application of the Rutter model to the estimation of interception loss from Thetford Forest. *Journal of Hydrology* 38, 49–58.

Gash, J.H.C., Lloyd, C.R. and Lachaud, G. (1995) Estimating sparse forest rainfall interception with an analytical model. *Journal of Hydrology* 170, 79–86.

Gijzen, H. and Goudriaan, J. (1989) A flexible and explanatory model of light distribution and photosynthesis in row crops. *Agricultural and Forest Meteorology* 48, 1–20.

Hall, R.L. and Calder, I.R. (1993) Drop size modification by forest canopies – measurements using a disdrometer. *Journal of Geophysical Research* 90, 465–470.

Helvey, J.D. and Patric, J.M. (1965) Canopy and litter interception of rainfall by hardwoods of eastern United States. *Water Resources Research* 1, 193–206.

Hoogmoed, W.B. and Stroosnijder, L. (1984) Crust formation on sandy soils in the Sahel. I. Rainfall infiltration. *Soil Tillage Research* 44, 5–23.

Hoogmoed, W.B., Klaij, M.C. and Brouwer, J. (1991) Infiltration, runoff and drainage in the Sudano-Sahelian zone. In: Sivakumar, M.V.K., Wallace, J.S., Renard, C. and Giroux, C. (eds) *Soil Water Balance in the Sudano-Sahelian Zone.* Proceedings International Workshop, Niamey, Niger, February 1991. IAHS Publ. No. 199, 85–98. IAHS Press. Institute of Hydrology, Wallingford, UK.

Horton, R.E. (1919) Rainfall interception. *Monthly Weather Review* 47(9), 603–623.

Huda, A.K.S. and Ong, C.K. (1989) Crop simulation models and some implications for agroforestry systems. In: Reifsnyder, W.E. and Darnhofer, T. (eds) *Meteorology and Agroforestry.* Proceedings of ICRAF/WMO/UNEP Workshop on Application of Meteorology to Agroforestry Systems Planning and Management. ICRAF, Nairobi, pp. 115–124.

Huxley, P.A., Pinney, A., Akunda, E. and Muraya, P. (1994) A tree/crop interface orientation experiment with a *Grevillea robusta* hedgerow and maize. *Agroforestry Systems* 25, 1–23.

Illingworth, A.J. and Stevens, C.J. (1987) An optical disdrometer for the measurement of raindrop size spectra in windy conditions. *Journal of Atmospheric and Oceanic Technology* 4, 411–421.

Jackson, J.E. and Palmer, J.W. (1979) A simple model of light transmission and interception by discontinuous canopies. *Annals of Botany* 44, 381–383.

Jarvis, P.G. (1976) The interpretation of the variations in leaf water potential and stomatal conductance found in canopies in the field. *Philosophical Transactions of the Royal Society of London B* 273, 593–610.

Kiepe, P. (1995) No runoff, no soil loss: soil and water conversion in hedgerow barrier systems. PhD thesis, Wageningen Agricultural University, The Netherlands.

Kiepe, P. and Rao, M.R. (1994) Management of agroforestry for the conservation and utilization of land and water resources. *Outlook on Agriculture,* 23, 17–25.

Klocke, N.L., Heerman, D.F. and Duke, H.R. (1985) Measurement of evaporation and transpiration with lysimeters. *Transactions of the American Society of Agricultural Engineers* 28, 183–189.

Lal, R. (1989a) Agroforestry systems and soil surface management of a tropical alfisol: II: Water runoff, soil erosion, and nutrient loss. *Agroforestry Systems* 8, 97–111.

Lal, R. (1989b) Agroforestry systems and soil surface management of a tropical alfisol: V: Water infiltrability, transmissivity and soil water sorptivity. *Agroforestry Systems* 8, 217–238.

Lal, R. (1991) Current research on crop water balance and implications for the future. In: Sivakumar, M.V.K., Wallace, J.S. Renard, C. and Giroux, C. (eds) *Soil Water Balance in the Sudano-Sahelian Zone.* Proceedings International Workshop, Niamey, Niger, February 1991. IAHS Publ. No. 199. IAHS Press, Institute of Hydrology, Wallingford, UK, pp. 31–44.

Lavelle, P., Spain, A.V., Blanchart, E., Martin, A. and Martin, S. (1992) The impact of soil fauna on the properties of soils in the humid tropics. In: *Myths and Science of Soils of the Tropics.* Soil Science Society of America Special Publication 29, pp. 157–185.

Leyton, L., Reynolds, E.R.C. and Thompson, F.B. (1967) Rainfall interception in forest and moorland. In: Sopper, W.E. and Lull, H.W. (eds) *Forest Hydrology.* Pergamon, Oxford, pp. 163–178.

Lloyd, C.R. and Marques, F.A. de O. (1988) Spatial variability of throughflow and stemflow in Amazonian rainforest. *Agricultural and Forest Meteorology* 42, 63–73.

Lloyd, C.R., Gash, J.H.C., Shuttleworth, W.J. and Marques, F.A. de O. (1988) The measurement and modelling of rainfall interception by Amazonian rain forest. *Agricultural and Forest Meteorology* 43, 277–294.

McGowan, M. (1974) Depths of water extraction by roots. In: *Isotope and Radiation Techniques in Soil Physics and Irrigation Studies*. International Atomic Energy Agency, Vienna, pp. 435–445.

Monteith, J.L. (1965) Evaporation and environment. In: *State and Movement of Water in Living Organisms*. 19th Symposium Society Experimental Biology, Swansea, 1964. Cambridge University Press, Cambridge, pp. 205–234.

Monteith, J.L. (1988) Steps in crop climatology. In: Unger, P.W., Sneed, T.V., Jordan, W.R. and Jensen, R. *Challenges in Dryland Agriculture: A Global Perspective*. Proceedings of the International Conference on Dryland Farming, August 1988. Texas Agricultural Experimental Station, Texas, USA, pp. 273–282.

Monteith, J.L., Szeicz, G. and Waggoner, P.E. (1965) The measurement and control of stomatal resistance in the field. *Journal of Applied Ecology* 2, 345–355.

Monteith, J.L., Ong, C.K. and Corlett, J.E. (1991) Microclimatic interactions in agroforestry systems. *Forest Ecology and Management* 45, 31–44.

Nobel, P.S. and Geller, G.N. (1987) Temperature modelling of wet and dry desert soils. *Journal of Ecology* 75, 247–258.

Norman, J.M. and Welles, J.M. (1983) Radiative transfer in an array of canopies. *Agronomy Journal* 75, 481–488.

Ong, C.K. and Khan, A.A.H. (1994) A direct method for measuring water uptake by individual tree roots. *Agroforestry Today*, 5 (4), 2–5.

Ong, C.K., Odongo, J.C.W., Marshall, F. and Black, C.R. (1991) Water use by trees crops: five hypotheses. *Agroforestry Today* 3, 7–10.

Ong, C.K., Odongo, J.C.W., Marshall, F. and Black, C.R. (1992) Water use of agroforestry systems in semi-arid India. In: Calder, I.R., Hall, R.L. and Adlard, P.G. (eds) *Growth and Water Use of Forest Plantations*. John Wiley Sons, Chichester, UK, pp. 347–358.

Parkin, T.B. and Codling, E.E. (1990) Rainfall distribution under a corn canopy: implications for managing agrochemicals. *Agronomy Journal* 82, 1166–1169.

Parkinson, K.J. (1985) Porometry. In: Marshall, B. and Woodward, F.I. (eds) *Instrumentation for Environmental Physiology*. Cambridge University Press, Cambridge, pp. 171–191.

Peterson, A.E. and Bubenzer, G.D. (1986) Intake rate: sprinkler infiltrometer. In: Klute, A. (ed.) *Methods of Soil Analysis. Part I. Physical and Mineralogical Methods*, 2nd edn. Agronomy Monograph No. 9, American Society of Agronomy and Soil Science, Madison, Wisconsin, USA, pp. 845–870.

Philip, J.R. (1957) Evaporation, and soil moisture and heat fields in the soil. *Journal of Meteorology* 14, 354–366.

Reigner, I.C. (1964) Evaluation of the trough-type raingauge. *US Forest Service Northeast Experimental Station Research Note 20*, 4 pp.

Rimmington, G.M. (1984) A model of the effects of interspecies competition for light on dry-matter production. *Australian Journal of Plant Physiology* 11, 277–286.

Ritchie, J.T. (1972) Model for predicting evaporation from a row crop with incomplete cover. *Water Resources Research* 8, 1204–1213.

Roberts, J.M., Pymar, C.F., Wallace, J.S. and Pitman, R.M. (1980) Seasonal changes in leaf area, stomatal and canopy conductances and transpiration from bracken below a forest canopy. *Journal of Applied Ecology* 17, 490–522.

Robinson, D.A., Bell, J.P. and Batchelor, C.H. (1994) Influence of iron minerals on the determination of soil water content using dielectric techniques. *Journal of Hydrology* 161, 169–180.

Robinson, M. and Dean, T.J. (1993) Measurement of near surface soil water content using a capacitance probe. *Hydrological Processes* 7, 77–86.

Rodda, J.C. and Smith, S.W. (1986) The significance of systematic error in rainfall measurement for assessing wet deposition. *Atmospheric Environment* 20, 1059–1064.

Roose, E.J. and Bertrand, R. (1971) Contribution à l'étude de la méthode des bandes d'arrêt pour lutte contre l'érosion hydrique en Afrique de l'ouest, résultats expérimentaux et observations sur le terrain. *Agronomie Tropicale* 26, 1279–1283.

Ross, J. (1981) The radiation regime and architecture of plant stands. In: Leith, H. (ed.) *Tasks for Vegetation Sciences 3.* Dr W. Junk Publishers, The Hague, 391 pp.

Roth, C.H., Malicki, M.A. and Plagge, R. (1992) Empirical evaluation of the relationship between soil dielectric constant and volumetric water content as a basis for calibrating soil moisture measurements by TDR. *Journal of Soil Science* 43, 1–13.

Rutter, A.J. (1975) The hydrological cycle in vegetation. In: *Vegetation and the Atmosphere, Vol. 1.* Academic Press, London, pp. 111–154.

Rutter, A.J., Kershaw, K.A., Robins, P.C. and Morton, A.J. (1971) A predictive model of rainfall interception in forests I: Derivation of the model from observations in a plantation of Corsican pine. *Agricultural Meteorology* 9, 367–387.

Rutter, A.J., Morton, A.J. and Robins, P.C. (1975) A predictive model of rainfall interception in forests II: Generalization of the model and comparison with observations in some coniferous and hardwood stands. *Journal of Applied Ecology* 12, 367–380.

Sakuratani, T. (1981) A heat balance method for measuring water flux in the stem of intact plants. *Agricultural Meteorology* 37, 9–17.

Sevruk, B. and Klemm, S. (1989) Types of standard precipitation gauges. In: Sevruk, B. (ed.) *Precipitation Measurement.* Proceedings of International Workshop on Precipitation Measurement, St. Moritz. Swiss Federal Institute of Technology, Zurich, pp. 227–232.

Shuttleworth, W.J. (1988) Evaporation from Amazonian rain forest. *Proceedings of the Royal Society of London Series B* 233, 321–346.

Shuttleworth, W.J. and Gurney, R.J. (1990) The theoretical relationship between foliage temperature and canopy resistance is sparse crops. *Quarterly Journal of the Royal Meteorological Society* 116, 497–519.

Shuttleworth, W.J. and Wallace, J.S. (1985) Evaporation from sparse crops – an energy combination theory. *Quarterly Journal of the Royal Meteorological Society* 111, 839–855.

Sinclair, F.L., Jenkins, D.A., Grime, V.L., Adderley, W.P., Emmett, B., Verinumbe, I. and Adeogun, P.F. (1994) Evaluation of planting spaced trees in cultivated fields on vertisolic soils under low rainfall, with special reference to soil conditions and crop yield. *Second Annual Report on ODA Forestry and Agroforestry Research Strategy Project R4850.* School of Agriculture and Forest Sciences, University of Wales, Bangor, 108 pp.

Sinoquet, H. and Bonhomme, R. (1991) A theoretical analysis of radiation interception in a two-species plant canopy. *Mathematical Biosciences* 105, 23–45.

Sinoquet, H., Moulia, B., Gastall, F., Bonhomme, R. and Varlet-Grancher, C. (1990) Modeling the radiative balance of the components of a well-mixed canopy: application to a white clover–tall fescue mixture. *Acta Œcologica* 11, 469–486.

Stigter, C.J. (1984) Mulching as a traditional method of microclimate management. *Archives for Meteorology Geophysics Bioclimatology Series B* 35, 147–154.

Stroosnijder, L. and Hoogmoed, W.B. (1984) Crust formation on sandy soils in the Sahel. II. Tillage and its effect on the water balance. *Soil Tillage Research* 4, 321–337.

Taupin, J.D., Dever, L., Fontes, J.Ch., Guéro, Y., Ousmane, B. and Vachier, P. (1991) Evaluation de l'évaporation à travers les sols par modélisation des profils isotopiques sous climat sahélian: la vallée du Niger. In: Sivakumar, M.V.K., Wallace, J.S., Renard, C. and Giroux, C. (eds) *Soil Water Balance in the Sudano-Sahelian Zone*. Proceedings International Workshop, Niamey, Niger, February 1991. IAHS Publ. No. 199. IAHS Press, Institute of Hydrology, Wallingford, UK, pp. 159–172.

Teklehaimanot, Z. and Jarvis, P.G. (1991a) Direct measurement of evaporation of intercepted water from forest canopies. *Journal of Applied Ecology* 28, 603–618.

Teklehaimanot, Z. and Jarvis, P.G. (1991b) Modelling rainfall interception loss in agroforestry systems. *Agroforestry Systems* 14, 65–80.

Topp, G.C. and Davies, J.L. (1985) Time-domain reflectrometry (TDR) and its application to irrigation scheduling. In: Hillel, D. (ed.) *Advances in Irrigation, Vol. 3*. Academic Press, London, pp. 107–127.

van Noordwijk, M., Widianto, H.M. and Hairiah, K. (1991) Old tree root channels in acid soils in the humid tropics: important for root penetration, water infiltration and nitrogen management. *Plant and Soil* 134, 37–44.

Wallace, J.S. (1991) The measurement and modelling of evaporation from semiarid land. In: Sivakumar, M.V.K., Wallace, J.S., Renard, C. and Giroux, C. (eds) *Soil Water Balance in the Sudano-Sahelian Zone*. Proceedings International Workshop, Niamey, Niger, February 1991. IAHS Publ. No. 199. IAHS Press, Institute of Hydrology, Wallingford, UK, pp. 131–148.

Wallace, J.S. (1995) Towards a coupled light partitioning and transpiration model for use in intercrops and agroforestry. In: Sinoquet, H. and Cruz, P. (eds) *Ecophysiology of Tropical Intercropping*. INRA Editions, Paris, France, pp. 153–162.

Wallace, J.S. and Holwill, C.J. (1996) Soil evaporation from tiger-bush in Niger. *HAPEX-Sahel Special Issue of Journal of Hydrology* (in press).

Wallace, J.S. and Jackson, N.A. (1994) Interim report to the ODA Forestry Programme for the period 1 October 1992 to 30 September 1993. *Report No. OD94/2*. Institute of Hydrology, Wallingford, 59 pp.

Wallace, J.S., Gash, J.H.C., McNeil, D.D. and Sivakumar, M.V.K. (1986) Measurement and prediction of actual evaporation from sparse dryland crops. Scientific Report on Phase II of ODA Project 149. *ODA Report No. OD149/3*. Institute of Hydrology, Wallingford, 59 pp.

Wallace, J.S., Roberts, J.M. and Sivakumar, M.V.K. (1990) The estimation of transpiration from sparse dryland millet using stomatal conductance and vegetation area indices. *Agricultural and Forest Meteorology* 51, 35–49.

Wallace, J.S., Batchelor, C.H., Daabeesing, D.N., Teeluck, M. and Soopramanien,

G.C. (1991) A comparison of the light interception and water use of plant and first ratoon sugar cane intercropped with maize. *Agricultural and Forest Meteorology* 57, 85–105.

Wallace, J.S., Lloyd, C.R. and Sivakumar, M.V.K. (1993) Measurements of soil, plant and total evaporation from millet in Niger. *Agricultural and Forest Meteorology* 63, 149–169.

Wang, Y.P. and Jarvis, P.G. (1990) Description and validation of an array model – MAESTRO. *Agricultural and Forest Meteorology* 51, 257–280.

Waring, R.H. and Roberts, J.M. (1979) Estimating water flux through stems of Scots pine with tritiated water and phosphorus-32. *Journal of Experimental Botany* 30, 459–471.

Waring, R.H., Whitehead, D. and Jarvis, P.G. (1980) Comparison of an isotropic method and the Penman–Monteith equation for estimating transpiration from Scots pine. *Canadian Journal of Forest Research* 10, 555–558.

Wellings, S.R. and Bell, J.P. (1982) Physical controls of water movement in the unsaturated zone. *Quarterly Journal of Engineering Geology, London* 15, 235–241.

Wellings, S.R., Bell, J.P. and Raynor, R.J. (1985) The use of gypsum resistance blocks for measuring soil water potential in the field. *Institute of Hydrology Report No. 92*, Wallingford, UK, 32 pp.

White, I., Sully, M.J. and Perroux, K.M. (1992) Measurement of surface-soil hydraulic properties: disk permeameters, tension infiltrometers, and other techniques. In: Topp, G.C., Reynolds, W.D. and Green, R.E. (eds) *Advances in Measurement of Soil Physical Properties: Bringing Theory into Practice*. Soil Science Society of America Special Publication Number 30, Madison, Wisconsin, USA, pp. 69–103.

Wischmeier, W.H. and Smith, D.D. (1978) Predicting rainfall erosion losses – a guide to conservation planning. *USDA Agricultural Handbook no. 537*. USDA, Washington DC.

World Meteorological Organization (1982) Methods of correction for systematic error in point precipitation measurements for operational use. *Operational Hydrology Report 21*, WMO No. 589.

Young, A. (1989) *Agroforestry for Soil Conservation*. CAB International, Wallingford, and ICRAF, Nairobi, 276 pp.

7

◆

Biological Factors Affecting Form and Function in Woody–Non-woody Plant Mixtures

P. HUXLEY

Flat 4, 9 Linton Road, Oxford OX2 6UH, UK

Introduction

Many forms of agroforestry involve growing woody perennial and non-woody herbaceous plants on the same unit of land and at the same time. In such associations the woody plants are, physiologically, likely to be the dominating species, but either trees or the understorey crop or pasture may form the farmer's main economic enterprise. We already know a good deal about how to grow crops and pastures but, in order to appreciate fully the potentials and constraints of agroforestry mixtures, it is necessary to observe and understand the characteristics of the woody species, the so-called 'multipurpose trees' (Huxley, 1996). It is essential, in this respect, to relate form to function and plant behaviour, not the least because of the remarkably large number of different kinds of woody species that may be used in agroforestry, the large within-species genetic diversity found, and the many ways there are to arrange and manage them. Interactions between components in these mixtures will be highly dependent on the nature of the woody plants and, often in agroforestry, on the kind and extent of the responses that their management evokes. It will only be when we have explored these issues that the opportunities for fully exploiting environmental resources will be realized, and the constraints to resource sharing in woody–non-woody mixtures be better understood. These two issues are critical if the design and management of agroforestry systems are not to be left to trial and error and farmers' needs are to be fulfilled within well-understood and feasible biological limits.

'Woody perennial' is, itself, a term open to misinterpretation. Indeed, a number of important crop species that can be grown seasonally are, in fact, short term, woody perennials: for example, cotton, cassava, pigeonpea, castor bean, jack bean, derris, etc., although some such, e.g. cotton and pigeonpea, have been selected by breeders for short stature, compact form and early flowering so as to be grown more productively as seasonal crops in sole cropping systems. So what are some of the attributes of 'woodiness', and how best may we appreciate the constraints and opportunities that it provides? This chapter outlines some relevant aspects of plant organization that need to be taken into account when considering using and sharing environmental resource in woody–non-woody plant mixtures from this aspect, and it forms a biological base to support the specialist topics dealt with in the other chapters.

The Nature of Agroforestry: Intercropping Woody and Non-woody Plants

Tree–crop mixtures: the infinite variations

Apart from managed tree lots, and practices that rotate plant components in time (e.g. as in shifting cultivation, or improved tree fallows), there are many different ways in which woody and non-woody plants can be arranged spatially, and in different proportions, so as to incorporate both different planting densities and different levels of intimacy between the various plant components being used (Nair, 1989). Obviously, these systems have different kinds of overall geometry: for example, scattered trees on crop land, sparsely or densely aggregated, or some form of strip or zonal planting such as hedgerow intercropping (alley cropping), or boundary plantings of trees. All such systems present open, environmentally well-coupled canopy structures, unlike most densely planted sole crops. Complex plant mixtures in multi-strata 'homegardens' more closely resemble the forests of which they are sometimes a part. Where multipurpose trees are grown widely spaced in mixtures, water use will be contingent to a large extent on plant characteristics and photosynthesis will be dependent on canopy form and leaf demography (Jackson, 1980). Hence the need to understand how and when woody plant parts grow.

In agroforestry, we may intercrop plants of diverse stature and physiognomy, and potentially very disparate life cycles. Above ground, the crop species are usually relatively short statured, determinate or indeterminate in form, and they are mostly raised to be harvested in just one season although some can be 'ratooned' (e.g. as with sorghum or sugar cane). Development through a succession of phases, from germination to senescence and death, is virtually continuous; at least it is if the crop is to yield successfully. The woody species (trees, shrubs, vines, bamboos) may often be relatively taller statured and of very diverse architectural forms, single or multi-stemmed. They may be chosen because they are 'fast-' or 'slow-growing', potentially nitrogen fixing, coppice well and so on. There may often be pauses between different growth phases.

The physiological characteristics of the individual plant components used in mixtures in agroforestry systems will vary, often widely. For example, particular crop species used might have either C3 or C4 photosynthetic pathways, and C3 species may or may not benefit from, or tolerate, shading from the overstorey with regard to their growth and/or developmental processes (Chapter 4). The woody species will be C3 (Jarvis and Sandford, 1986), and fast- or slow-growing, with all the concomitant effects on water use and nutrient demands. Sensitivity to environmental stresses will vary widely.

The challenge in agroforestry is to select species, design systems and devise appropriate management regimes that will optimize environmental resource capture and resource use efficiency in a sustainable way, whilst fulfilling the farmer's objectives. But to do this effectively requires both a knowledge of the interactive responses of the components in these woody–non-woody intercrops (Fig. 7.1) and how these can be modified by phenology.[1] Vegetative growth of a whole tree will seldom be truly continuous but often seasonally periodic and, especially in seasonally arid tropical regions, with considerable variability between the start and finish of various phenophases in relation to the established crop growing seasons, even to the point where growth appears to be just intermittent. Flowering and fruiting may be continuous, periodic or intermittent. If grasses are part of an agroforestry practice, they may be either seasonal or perennial, and the latter may exhibit growth regulation. Where different patterns of activity exist among the various plant components in a mixture this will not only affect the ways in which they interact, but there will be different opportunities for resource capture and, perhaps, changes in the efficiency of environmental resource use also.

Changes due to management and time

One beneficial aspect of the woody habit is the scope it offers for the control of form through management. Trees and shrubs have 'shapeability', clearly

---

[1] Phenology refers to the timing of particular phases, e.g. in plants, growth periods, flowering, fruit maturation, leaf fall, etc., each of which is a 'phenophase'. The broader term 'plant behaviour' also covers morphological and physiological responses brought about by management and environmental changes.

Fig. 7.1. The 'tree–crop interface' between close-planted *Senna siamea* and maize. The effects of asymmetrical competition for light above ground is often readily apparent resulting in poorer crop growth next to the trees (see far sub-plot of tilled young maize). However, the roots of these *S. siamea* plants extended well beyond the cropped plot and the maize in the untilled sub-plot in the foreground is clearly stressed in the presence of the undisturbed surface roots of the trees.

important in agroforestry. A consequence of manipulating tree shape by pruning or lopping, however, can be the changes that occur in phenology. This may actually offer beneficial possibilities for changing the time of occurrence of particular phenophases (e.g. vegetative growth or flowering and fruiting) to different climatic periods and/or to periods useful to the farmer ('entrainment'). For example, the production of leafy fodder from hedgerows can be either brought forward or delayed by this means to a time when grass feed is inadequate, either through low yields or poor quality. Irrigation is another method that can be used to bring this about. However, because different woody species may not respond alike to 'perturbations' caused by management practices such as pruning we need to understand how the plant is structured to achieve the form it has. This means knowing the positions and potential numbers of active and dormant buds, and the responses of these and other meristems to climate and manipulations. Without this, management by pruning, especially, is an art rather than a science. Living tissues spread among the lignified organs enable woody plants to store large amounts of nutrient and carbohydrate resources which they access with differing degrees of success (Priestley, 1969, 1970) unlike herbaceous crop plants, such as cereals (e.g. maize; Allison and Watson, 1966), which transfer very little stored carbohydrate to grain.

Stands of broad-leaved trees may achieve leaf area indexes (LAIs) of around 6, but evergreen species will have denser canopies within which the well-shaded leaves are likely to be light adapted, unlike crop plants where lower, shaded leaves often rapidly senesce (Ong and Monteith, 1993). In conifers the LAI can reach > 15. A woody structure will intercept light, in some mature trees as much as 30–40%, much of which will not be used for photosynthesis unless some of the younger stems contain chlorophyll.

As plants mature there is a progressive and ordered change in form, but changes in branching habit occur differently depending on whether the plants are being grown in an association (e.g. as a crop, or a plantation/ woodlot or hedgerow), or as effectively free-standing plants (e.g. scattered trees in grassland or crops). Mutual shading will change dry matter distribution, i.e. between roots : shoots and structural parts (the stems/ branches) and leaf (Ledig, 1983; Cannell, 1985). Thus carbon allocation to roots will tend to be more in free-standing plants, crops or trees as compared with stands, although changes in root specific activity (e.g. through changes in soil fertility) are probably of greater significance (Cannell, 1989). In trees, flowering sites usually occur more prolifically in less shaded parts (Jackson and Palmer, 1977; Cannell, 1983), and leaves emerging and/or retained in shaded regions of the canopy will adapt photosynthetically to this condition, an added complication when considering the photosynthesis of trees with clumped foliage. Shading of the lower storey crop layer by trees will also bring these changes about in the crop, and change them morphogenetically and physiologically (e.g. bring about etiolation and/or delayed flowering at the tree–crop interface).

The effects of different environmental factors on the rate and duration of growth of a leaf canopy are relatively well understood (Squire, 1990). In woody plants there are some additional features to consider. For example, different kinds of leaves can be produced on different kinds of shoots (e.g. on long terminals or on 'spurs' (Ghosh, 1973)), and the number of leaves produced on a shoot may be predetermined by the previous season's influence on the apical meristem (e.g. in spurs and other short shoots), or by the rate of production of leaf initials on the currently active shoot meristems (on long terminals). Leaves produced sequentially on a shoot may expand at different rates.

After full expansion we can expect a decline in leaf efficiency. This is well studied with agricultural crop plants but there are fewer observations on trees, particularly in the tropics. Leverenz et al. (1982) showed that stomatal conductance of 2- and 3-year-old Sitka spruce declined to 48% and 31%, respectively, of that of 1-year-old needles, and Reich and Borchert (1988) showed a loss of stomatal sensitivity (in *Tabebuia rosea* in Costa Rica) when stomata on 2-month-old leaves closed at leaf water potentials of -1.8 MPa, but 7-month-old leaves failed to close even at -4.5 MPa. As detailed in Chapter 6, the gradients of radiation, temperature, specific humidity deficit and windspeed in a canopy of mixed species creates large variations in canopy stomatal conductances (g_s) and boundary layer conductances (g_a), with significant effects on overall transpiration. A detailed study of these factors in a Brazilian rainforest (Roberts *et al.*, 1990) records the decline of g_s with increased saturation deficit (D) as is found with temperate tree species, but they note that leaf age may have had an effect on this. A lower g_s was observed with older leaves, but they showed less response to changes in radiation and D. Reich and Borchert (1988) compared stomatal function and leaf water status of leaves of known ages on deciduous and evergreen (i.e. 'leaf-retaining') woody species on *in situ* branches and excised ones. Species showed marked differences in stomatal sensitivity and control of water loss, leaf diffusive resistance (r_l) and xylem pressure potential (P_p). The evergreen species (*Mangifera indica* and *Licania arborea*) had the most sensitive stomatal responses, and the best control of water loss, even when detached. Compared with that of young leaves, r_l increased with age, the authors suggest probably as a response to decreasing photosynthetic rate and thus higher internal CO_2. These trees continued to grow even during seasonal drought. Stomatal control was poorest in the deciduous *Tabebuia rosea*, which is leafless for 4 to 6 months of the year at this site in Costa Rica, and r_l fell with age and P_p increased.

The results of studies on the decline of photosynthesis with leaf age in crop plants is fairly well documented. For example, in maize leaf photosynthesis decreased linearly over 8 weeks to about one-third of that at full leaf expansion and was similarly reduced as the plant aged over 14 weeks (measured as P_{max} and P_{2000}, i.e. with an irradiance of 2000 $\mu E\ m^{-2}\ s^{-1}$, see Fig. 7.2a and b). The leaf P/I curves flatten to an asymptote at low irradiance levels as leaves develop beyond full expansion. Data for trees are harder to come by, but the extreme differences that can be found in leaf longevity between woody species (see 'Organ replacement in trees', p. 267) suggests that this is an area that needs to be addressed.

As Thomas (1987) points out, net carbon contribution during leaf ontogeny can be assessed to provide a 'carbon credit contour' (see Fig. 7.3). Analogously, and particularly for leaf-retaining woody species with cohorts of leaves of different ages, we need to understand the characteristics of this if even a rough prediction is to be made about the contribution to the tree as a whole by various parts of the canopy. The left hand boundary of the contour is characterized by the efficiency of the photosynthetic surface and its rate of expansion (LAI in crops) but, clearly, not only the maximum CO_2 flux attained but the time of initiation of the functional decline and its rate and duration are key parameters.

Below ground the nature of crop, grass and tree root systems can vary

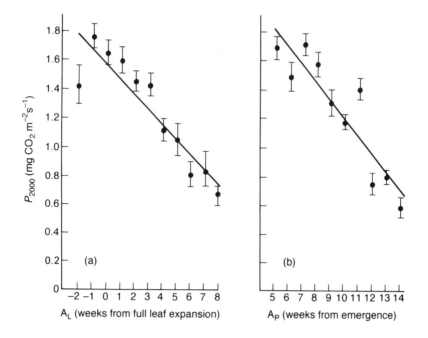

Fig. 7.2. Photosynthesis of maize at $I = 2000$ μE m^{-2} s^{-1} (P_{2000}) as a function of age (A), (a) leaf age and (b) plant age. (Source: Dwyer and Stewart, 1986.)

greatly. Crops, and especially grasses, tend to fill out the spaces that the root systems occupy, whereas at least some tree species have root systems that become decidedly clumped (see 'Below ground', p. 263). Because of the perennial nature of the root system in woody plants, and the expectancy of some degree of fine root turnover, the exploitation of the occupied volume may be much more irregular in both time and space. The extension with time of woody plant root systems so as to occupy more vertical and, perhaps, lateral space than the crop is hypothesized to be one of the major benefits of agroforestry practices (and see Chapter 9).

Choice of species

Tropical agroforestry is concerned with an enormous range of species. The crops utilized in different agroforestry practices will usually be those found in local agriculture, and the environmental and management factors that affect their growth and development will be well established, either empirically or, in some cases, through detailed investigations. Physiological responses of seasonal herbaceous crops are reasonably well understood,

242 ◆ P. Huxley

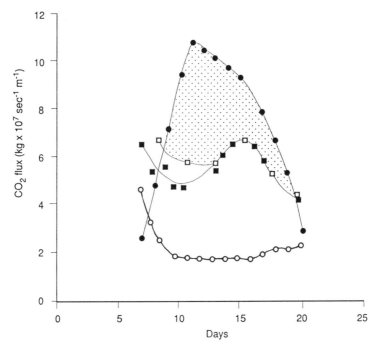

Fig. 7.3. The carbon economy of a *Phaseolus* leaf during ontogeny:
● photosynthesis; ○ dark respiration; ■ dark + photorespiration;
□ respiration + dry matter accumulated. The shaded area delimits the 'carbon credit contour'. (Source: Thomas, 1987 (data from Catsky *et al.*)).

although there are still relatively few comprehensive studies on tropical species. A key issue is the potential length of the cropping season evaluated in terms of water-, light- or temperature-limiting factors and, preferably, expressed in terms of probability. Monteith's model of cumulative crop growth designating water- and energy-limiting stages, and based on the known relationships between dry matter production and transpiration and dry matter production and intercepted light, provides one example of how physiological principles can be applied to crop prediction (Monteith, 1991). Ecozones, or specific climatic regions where particular crops are likely to do well, can be mapped (e.g. Higgins *et al.*, 1987), and the effects of altitude, for example, standardized by expressing growth in terms of thermal time. At the management level, time of sowing is the main influencing variable, with planting density, fertilization and irrigation also having some effects on the length of season, either directly through growth, or through effects on flowering and fruiting.

This level of understanding is sadly missing with regard to many of the species of trees, shrubs and other woody plants that are being used in tropical agroforestry. Seeking general climatic compatibility is clearly a priority, but this is only one of a number of considerations that can account for adaptability. Furthermore, the essence of climatic adaptability for many tropical woody perennial species is not just that the environmental resources are available at optimal or near-optimal levels throughout a 'cropping season', but that plant 'behaviour' remains sufficiently flexible so that the influence of past events can be accommodated within whatever current opportunities for resource capture and use present themselves. For trees and shrubs the climatic sequences that go to make up any particular growing season will not only impose limits to growth but also exert control over the timing of events that are necessary for growth, flowering and fruiting (Fig. 7.4) in future seasons. The interactions between history, present conditions and internal regulatory processes result, especially with tropical woody species, in apparent irregularities in the timing of growth and flowering. In any one season, if left untouched, there may be considerable differences to be observed between species on any one site and, within species, considerable disparities in phenology between seasons (Huxley and Van Eck, 1974, and see Fig. 7.18). Such shifts in the timing and duration of particular phenophases can provide the opportunities to improve environmental resource capture and resource sharing potentials in agroforestry systems.

We can view the development of any flowering plant as a transition through a series of phenophases. With a seasonal, herbaceous crop species these can be described by just a single sequence: germination, vegetative growth, maturation and the flowering processes, fruit development, and so on. Each can be identified by a process of change in the structure and/or size of particular organs, or in their potential behaviour (e.g. achieving the potential to flower). The environmental factors affecting the duration of these are studied in crop plants (and can often most usefully be expressed in terms of thermal rather than chronological time), in order to discover whether or not a particular cultivar makes the best use of the growing season, can minimize the risk of crop failure, and/or has potential for a new area. With perennials, there is a *reiteration* of sequences of growth and flowering. However, an analysis of factors that may initiate, maintain or terminate any particular phenophase can be difficult, not just because of the influence of past events, but because the timing of the activity of vegetative, and sometimes floral, organs is frequently controlled through growth regulation (dormancy mechanisms). Thus endogenous as well as exogenous factors have to be taken into account. Understanding how environment and management can affect phenology is a key issue if we are to select biologically suitable woody species for different agroforestry practices.

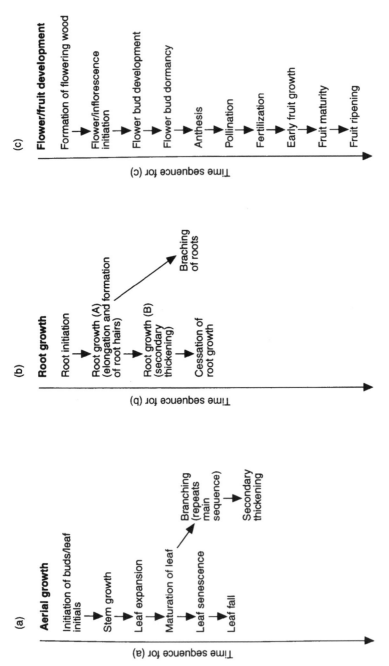

Fig. 7.4. Phenophases. Although completely interdependent aerial growth, root growth and flower and fruit development can usefully be considered as each progressing through a series of steps or 'phenophases'; these will occupy different real time periods and sequences will overlap. (Source: Huxley, 1983.)

Woody perennials are commonly categorized as 'deciduous' or 'evergreen', i.e. whether or not the canopy is retained or shed (usually seasonally, but not necessarily coinciding with the crop season). Evergreen species are sometimes called 'leaf-retainers' as this highlights a wider range of functional consequences. However, these categories are far from distinct and a species may behave in a more deciduous fashion as the aridity of the site increases (Richards, 1952; Longman and Jenik, 1974; and see Fig. 7.5). Furthermore, the longevity of leaves of different 'evergreen' species can vary greatly (see 'Organ replacement in trees', p. 267). Unfortunately for our understanding of the growth of multipurpose tree species, we know very little about the relationship between the production of new 'cohorts' of leaves, and leaf longevity and functional decline in tropical woody species.

Fig. 7.5. *Melia volkensii.* Usually deciduous in prolonged dry seasons at Machakos but, as here, only partly so if some water is available.

Clearly, these are factors that will markedly affect resource capture potentials, as well as resource use efficiency.

In the growing plant, form establishes limits within which the plant can function. Sets of such combined characteristics are often described as 'ideotypes', i.e. a biological model chosen to fit a well-defined end use that can be expected to behave in a certain way in a particular environment (Donald, 1968). The concept is also used by foresters (Dickman, 1985) to help understand the relative performance of different tree types in plantations. It provides a workable goal for those concerned with plant improvement. However, an ideotype can only be defined if it is required for a clearly specified end use: a situation that is obviously not readily compatible with the concept of 'multipurposeness' that has emerged for the woody components in agroforestry systems (Fig. 7.6)! In any case, the matter is not so simple as the study of individual species and of forest dynamics indicates. Sets of characteristics that typify a particular behaviour early on in a succession (e.g. 'pioneer' species) may be succeeded by others later. These changes in tree 'temperament', which accommodate several previous schemes, have recently been discussed by Oldeman and van Dijk (1992). In agroforestry, an understanding of the nature of such changes is clearly important, at least where unpruned trees form part of the system. What is still required for so-called 'multipurpose trees' is not just to define this concept more clearly, but to state, also, what sets of characteristics above and below ground will best describe the woody plant required for a particular site and purpose unambiguously.

The rest of this chapter is concerned with a more detailed account of some of these issues so as to place them in the context of the specialized topics covered elsewhere in this volume.

Genotype and Phenotype Responses to Environmental Change

Environmental variation and adaptability

Many tropical climates, especially in seasonally arid and semi-arid regions, exhibit considerable between- and within-season variability. Even in the 'wet' tropics relatively drier periods can occur within seasons. In arid zones, on a longer timescale, rainfall events may happen every decade or so. Spatial variation in the environment at different scales is also a feature of many tropical ecozones. This can be very marked in the high altitude tropics and, even in a rainforest, there are ecologically significant gaps ('chablis'). Thus, in any particular ecozone, surviving plant species can be expected to have evolved not only to be adapted to the limits, i.e. the range,

Fig. 7.6. *Sesbania sesban*, a 'multipurpose tree'. There are many ways of combining the plant components in different agroforestry systems and this farmer, in western Kenya, has devised a method of growing passion fruit vines using the small trees of *Sesbania sesban* he is encouraging in his plot for soil improvement and crop shelter. Before defining a multipurpose tree ideotype, in both morphological and physiological terms, it is necessary to specify the uses precisely. This can sometimes result in biologically infeasible requirements.

imposed by the main climatic factors over long periods of time but, also, to the degree of uncertainty of climatic variability.

Of course, the evolution of plant characteristics is also a consequence of the development of competitive strategies, and so it depends also on the presence or absence, as well as the nature, of previously associated species, and of associated herbivores, seed predators etc. This may well impart sets of characteristics that are not easily seen as of value to ecological success in a particular climate, or to productivity in an agricultural system. A knowledge of evolutionary history and the species' exposure to past environments may sometimes help to explain how a particular plant genotype responds to a new environment (see *Acacia auriculiformis*, Boland *et al.*, 1990), but matching current climates and soils will not always enlighten us fully about the plant's phenotypic capability to adapt to new environments.

In general, it is the traits that enhance competitive ability ('aggressivity') that most concern agriculturists dealing with herbaceous crops. Mechanisms for ensuring survival (e.g. seed dormancy and other forms of growth regulation) are usually seen as unhelpful in a modern, highly intensive, sole cropping context. In agroforestry, however, we are often dealing with plant components, the crops and the multipurpose trees, that have received very different levels of genetic selection.

The degree to which genotypic characteristics are established that confer the ability for the species to maintain itself in one particular environmental niche (species 'fitness'), or that allow it to do so in several (its genetic 'flexibility', see Stebbins, 1950), will also be conditioned by the extent of phenotypic plasticity that it can exhibit (Bradshaw, 1965; Crick and Grime, 1987), that is, the capability of the *individual genotype* to respond to short term environmental changes. Plasticity is clearly of considerable significance with respect to within-season changes in managed systems. This issue is important when selecting plants for new sites, as the genotype × environment (G × E) interaction needs to be considered as well as the responses attributable to genotype (i.e. heritability) and the main effects of environment. The nature of the G × E responses in yield in relation to the changes incurred across a range of environments, i.e. whether or not a cultivar's yield response is especially favoured by improved environmental conditions, or especially sustained in poor ones, is an important selection issue.

Phenotypic plasticity

At present, there is little useful information about the G × E responses of the woody species that are now being utilized in agroforestry in tropical regions, although we know from the changes in yield responses of agricultural crop cultivars to changes in site and ecozone just how important this can be. Especially for trees the investment in this type of research is costly, long term and still may not provide all the information required (Stewart *et al.*, 1992). Irregular between-year phenology can also add to the difficulties. Many of these so called multipurpose tree species are outcrossing, highly heterozygous and, therefore, phenotypically variable in both morphological and, presumably, physiological characteristics. So, again, we might expect variation not only in the extent of their capability to adapt in form and function to climatic variations, but also in the way that they do so in time, i.e. in their individual phenological responses.

All this becomes important when plant species are moved to new regions. For example, the spread of tree and shrub germplasm is often done by first undertaking wide-ranging species and provenance trials. Notable successes have been obtained with genera such as *Eucalyptus* and *Pinus*, for example, where selections in new regions have done much better than in their indigenous environments; *Eucalyptus camaldulensis* and *Pinus radiata* are two good examples. With outcrossing species tree 'provenance', i.e. the concept of a relation between tree performance and exact geographical location, has proved to be essential; but there is usually little correlation with any particular set of morphological, and hence formal taxonomic, classification (Turnbull and Griffin, 1985). Where provenances are selected with the help of precise computerized matching of climate and soils (Booth *et al.*, 1987; Booth, 1991), this still leaves the extent of G × E responses unknown. A very large number of species of so-called 'multipurpose trees' are currently being introduced, tested and tried for various agroforestry practices all around the world in an unprecedented exchange of germplasm (e.g. Hughes, 1994; ICRAF, 1994). For many of these phenology in their place of origin may not have been studied in detail, and that in any new site is unpredictable.

Phenotypic or 'plastic' responses to environmental change, including management and those brought about by associated plants, can be of critical importance in overcoming short term environmental variations or perturbations and will tend to optimize resource capture, although such responses incur some metabolic cost. In some cases constancy of phenotypic response (stability) is of adaptive significance, e.g. as with seed size, and deviations might make the plant less fit. Maximum fitness does not require the same degree of plasticity in all characteristics, of course, and the *costs* of plastic responses (e.g. leaf deciduousness) play a part, as do any implications for survival (e.g. limits to partitioning imposed by storage features). Although plasticity can offer a 'buffering' capacity against the environmental changes encountered, there will be cases where it can serve to make best use of the environment by allowing diverse genotypes to assume a near-similar, optimal phenotype (as in maize composites). If the original environmental situation is restored a reversal of the previous changes may be needed. Again, the cost of achieving this has to be accounted for, especially where new, adapted organs have to be replaced.

The plant's capacity to effect a successful plastic response depends not only on the magnitude of the change imposed but also the timescale over which it occurs. Rapid response to a short term environmental change will often depend on several plant traits (e.g. the capacity of the leaves to respond morphologically and physiologically to shade, the present growth potential of the apical meristems, the ability of root systems to grow rapidly into new areas) or conditions (e.g. the current supply of immediately available nutrient resources). For example, for trees the nature, form and position of individual growth units, the numbers of vegetative and flowering buds and their condition, the physiological state of the root system, the levels of stored organic and inorganic materials and overall growth regulatory states can all contribute to the plant's overall capacity for responsive changes. Periods of very severe climatic stress can evoke an extreme plastic response in woody plants, i.e. death of some organs, from which the plants themselves can recover. In dry regions this can be an important aspect of woody plant behaviour. For example, dwarf shrubs are an important resource for pastoralists in northern Kenya (< 300 mm annual rainfall), where they provide a substantial part of the feed intake for both domesticated and wild animals (44% of total pastoral food energy overall for sheep, goats and camels). With one of them, *Indigofera spinosa*, dry season tissue mortality was shown to be more severe in larger (i.e. ungrazed) plants (Coughenour *et al.*, 1990a), thus tending to equalize for 'grazing stress'. Another abrupt form of environmental change occurs when a tree is coppiced or pollarded, or when a hedge is pruned, and it is beyond the capacity of some woody species to respond to such manipulations satisfactorily. Clearly, intrinsic growth rate is an issue, and one that is closely correlated with the length of the life cycle. Not all of the plant's organs will grow at the same rate, but those of vegetatively fast-growing species often do so, in part because they are able to allocate a high proportion of assimilates to investment in rapidly expanding new leaf, e.g. they are likely to be successful 'competitors' (see Grime, 1979). 'Stress-tolerators' on the other hand assign more to storage and, hence, are slower growing. Fast growth allows certain plastic responses to occur within the time span of relatively short duration environmental changes; plants with slower growth rates will have to endure such changes and will be physiologically and morphologically suited to do so. Long term environmental changes are likely to be accommodated through genotypic modification, but this will be dependent on the time needed effectively to establish successive mature generations. Woody perennials are likely, therefore, to exhibit relatively more short term plastic responses relative to their overall genetic fitness than annuals.

If the components of tree–crop mixtures are operating interactively in a complex way in time they are also doing so in a distinctly 'patchy' environment spatially and temporally. This will be due both to (i) initial spatial differences in the site (e.g. changes in soil characteristics over short lateral distances and depths), and also (ii) the ways in which any agroforestry system, itself, creates differences in the above- and below-ground environments with time, both within and between seasons. Sampling difficulties in agroforestry systems are greatly exacerbated by this. Plasticity has a critical part to play in such situations also, especially with regard to resource acquisition, as it is a relatively short term mechanism by which the 'foraging ability' of plants can be quickly adapted to fluctuating circumstances, and roots and leaves modified or new organs projected into resource-rich zones (see 'Below ground', p. 263). This capacity of plants to 'make the best', in a dynamic way, of the patchy environment that they invariably find themselves in is one reason why the accuracy of physical models may sometimes be limited.

Certainly, because of their longevity as individuals, woody perennial plants have the requirement to endure rather than avoid any relatively rare adverse climatic events and, equally important, to be able to respond and make use of favourable climatic sequences, as and when these occur. It might be argued that such phenological strategies are likely to optimize both resource capture and resource use efficiency through adaptation to appropriate space–time niches but, in some circumstances, this may depend on what other trade-offs have been reached (e.g. for slow growth and longevity so as to out-last competitors, or for survival benefits), and how much phenotypic plasticity is still available. Unwanted or unnecessary characteristics, such as branchiness, indeterminacy, fruit shedding, tend to have been eliminated from crop plants by modern day breeders. Because so little selection and breeding have as yet occurred in most multipurpose tree species, many will still retain their full complement of morphological and physiological traits relating to both adaptive *and* survival capabilities, and therefore their full capacity for phenotypic plasticity. Before beginning to narrow this through selection for as yet somewhat ill-defined improvement goals it may be as well to understand more about how form relates to function, as well as the extent and significance of their phenotypic responses.

Some Aspects of Growth and Form

Cultivated crop species have undergone varying levels of selection for sole or intercrop situations. With the growing interest in agroforestry they are used as they are but, eventually, there will be a need for plant breeders to select crop species, at least the major ones, with cohabitation with woody plants in mind. One example of reselection for agroforestry is found with pigeonpea (*Cajanus cajan*) where original, perennial branching types have been reinstated in place of the dwarf, seasonal types produced by breeders in recent years (Daniel and Ong, 1990).

To design and manage an agroforestry system optimally we need to understand what controls the growth and development of the various plant components in it. Indirectly, the form of any plant can be modified by practices such as changes in planting density and nutrition, water supply, shade, or combinations of these (e.g. shelter). Directly, it is regulated by choice of genotype and, with trees, by training and subsequently by different forms of pruning. The ways in which individual tree species respond to factors that can change the allocation of resources within the plant still requires a great deal of attention, but our perception of how this may come about is slowly becoming clearer. Cannell and Dewar (1994) have recently comprehensively reviewed the underlying concepts from a modelling point of view. However, functional imbalances will occur 'within-season' even if, in the longer term, carbon and nitrogen flows equilibrate in an understandable way and through mechanisms that can be examined and tested.

Environmental requirements for the satisfactory growth and development of our main temperate crop species and pasture grasses have been well studied (Loomis and Connor, 1992), but many tropical crops less so (Squire, 1990). In any case, this knowledge will not help predict what will happen at a tree–crop interface because this is a result of an interaction with a woody component about which we know rather little, and which will be highly site and management dependent. Indeed, other contributors to this volume attempt to explore various ways, ultimately, to resolve this problem. What is also needed is an equivalent understanding of environmental effects on the development of commonly used multipurpose tree species; initially through knowledge of how environment changes form and structure, and then how this relates to function.

Plant organs and growth control

For seasonal plants growth may be determinate or indeterminate and this has many implications both for form and function (Squire, 1990). Plants, or plant organs, are determinate when growth is terminated after a definite period by, for example, the abortion or permanent dormancy of an apical bud (for stems), or by the occurrence of a terminal flower or inflorescence. In lateral branches growth may be sylleptic, i.e. contemporaneous with the main axis, or not. Indeterminate growth on the other hand will occur over an undefined period terminated, ultimately, by outside influences. The rate and duration of development of a seasonal herbaceous plant will be largely determined by the particular set of current environmental conditions that it is subjected to, but the outcome will often be different depending whether the plant is determinate or indeterminate. This is because indeterminate plants can continue growth until the farmer chooses to harvest them, or adverse climatic conditions interrupt all growth and further development. Determinate crop plants, on the other hand, are chosen to be able to complete their life cycle within the expected growing season, so that finishing any phase in a shorter time, e.g. if temperatures are higher, will usually result in limiting the overall capture of environmental resources. These 'principles' will apply to woody plants but more pertinently to organs rather than the tree as a whole.

Temperature affects the rate as well as the duration of the processes occurring in any particular phenophase. Temperature and nutrition are also likely to influence the number of active meristems that are formed and/or that survive and, in some cases, temperature may affect the efficiency with which captured resources are utilized by crops, although this has not been generally established. Growth or development is often expressed not against time but 'thermal time', i.e. the temperature experienced above a base temperature and below an optimum, in units of degree days (°C day). There are few woody species that have been studied in this way, a useful exception being tea (*Camellia sinensis*) where yield is the outcome of the number and growth of young leafy shoots that lend themselves to such an investigation (Stephens and Carr, 1990).

Continuing rapid development may depend on whether growth is 'supply' or 'demand' driven for nutrients at any particular site, and over any part of a season. 'Fast-growing' tree species are likely to have high levels of nutrient demand. Internal source–sink relationships may limit development also. For example, woody species characteristically have erratic and sometimes profligate flowering, or there can be an early onset of fruiting. Both of these can impose internal stress for resources which may alter not only current partitioning but subsequent phenological patterns also (Fig. 7.7).

The above-ground structural form of trees

The principles for establishing and managing an effective canopy are well understood (Chapter 4). With agricultural crops these characteristics are 'streamlined' to fit the expected season with crops by manipulating sowing time, planting density and crop duration (genotype). Key issues are to (i) maximize light interception, (ii) maximize the transpiration through the crop during the main period of growth, and within the limits of available water in the system, and (iii) try to encourage flowering (if this is required) at a time which allows a suitable climatic period for subsequent fruit development and maturation. Thus carbon gains have to be maximized, carbon losses, through respiration, wasteful investment and the shedding of parts, limited. Woody and herbaceous plants will achieve these compromises differently. For example, apart from sometimes growing at apparently less than optimum times (from an agriculturist's point of view), trees also incur a carbon penalty for the competitive benefits an elevated perennial structure brings them, even though they create this structure with great efficiency (Cannell and Morgan, 1988), and they may shed parts both above and below ground either at regular or irregular intervals.

The disposition of parts in many tree species can be less regular than one finds in our commonly grown sole crops. For example, tree canopies can often be 'clumped' (Fig. 7.8), a feature that is likely to affect resource

Fig. 7.7. *Leucaena leucocephala* cv. K8 (at Machakos, Kenya). Heavily overbearing. There may sometimes be a conflict for plant breeders of multipurpose trees between the need to provide genotypes that seed easily, so that farmers can produce young plants cheaply, and the desirability of restricting dry matter allocation to non-fruiting parts, i.e. to limit fruiting/seeding capacity.

capture and, possibly, resource use efficiency (Baldocchi and Hutchinson, 1986). however, the effects of spatial irregularity in plant form may be difficult to evaluate. For example, tree root systems can also be clumped (Fig. 10.3), a situation which may permit separate root systems to occupy different microspaces. In agroforestry, this has yet to be shown to affect below-ground competition positively, and the initial suggestion to this effect (Huxley, 1994), based on visual appraisal of the joint fine root data from a *Grevillea robusta* hedgerow experiment with contiguous maize (Huxley *et al.*, 1994; also see Fig. 7.12), has not been substantiated by subsequent detailed hierarchical pattern and spectral analysis, which showed only weak negative correlations confined to the soil volume near to the hedge (Floyd, 1993).

We are certainly often dealing, in agroforestry, with spatially irregular canopies and a confusing juxtaposition of parts with potentially different levels of activity. In general, in plant associations the vertical distribution of leaves is likely to be most uniform in diverse and most clumped in species-poor ecosystems (Ewel *et al.*, 1982). Sole crop canopies can be stratified in terms of structure and function (expanding leaves, sun leaves, shade leaves, the carbon 'overdraft' layer, senescing/abscinding leaves (Thomas, 1992)), and the linearity of the relationship between accumulated light interception and dry matter accumulation (Chapter 4) implies that the proportion of these layers remains roughly constant over the period under study. However, few multipurpose tree canopies are of this form, unless they are grown in dense stands (e.g. as woodlots or fodder banks), especially in leaf-retaining species where any information on the duration of leaf activity, or even of leaf longevity, is yet to be obtained.

The final above-ground form of a plant depends on how, when and how many component branch structures develop, and how many are lost or shed. The generation of form is a dynamic process involving numerous different plant organs that make up the various structures (e.g. see Bell,

Fig. 7.8. *Croton megalocarpus.* An evergreen (leaf-retaining) species with clumped foliage. Each clump contains leaves with a range of different age classes.

1991, for an excellent illustrated account) and is conditioned, for any single genotype, by all the environmental factors that affect growth and morphogenesis (including responses to 'patchy' environments (Sutherland, 1990)). A major morphogenic modifier is light and, in agroforestry systems, this is something that should not be neglected, particularly for crop responses at the tree–crop interfaces.

For all tropical and temperate woody species some 23 'architectural models' serve to describe the arrangement of main stems and branches as they occur in young plants (Hallé et al., 1978; Barthélémy et al., 1989a) and, building on these general rules for branching, can explain later development through 'reiteration' (the repetition of the basic model), 'metamorphosis' (a change in plagiotropy to orthotropy, i.e. a more acute growing angle, for some of the new branches), and 'intercalation' of new branches into the mid-crown region of the canopy. Thus, a natural canopy will develop an accumulation of shoots into a form that is typical and recognizable for any one species depending on the growth characteristics of the trunk and branches, the position of the flowering sites, and the incorporation of additional determinate or indeterminate units in a predetermined manner. An additional complexity is that branch position in the tree in relation to distance from the root can affect growth vigour ('topophysis', Maggs and Alexander, 1967).

As woody plants grow older irreversible changes can take place ('phase changes', Wareing, 1987), for example a transition from a juvenile to an adult state (i.e. with a potential to flower), and sometimes with concomitant morphological differences (e.g. leaf characteristics as in *Eucalyptus* spp.). These changes will have effects on resource capture/resource use efficiency (see 'Reproductive processes', p. 259). 'Ageing' also occurs as trees grow older, but this process is not well defined although it can have varying effects on form (Barthélémy et al., 1989b), as well as usually detrimentally affecting growth and flowering (Wareing, 1959). Ageing *is* reversible, for example by appropriate pruning technology.

An understanding of these processes is important for physiologists if they are fully to comprehend the *changing* nature of the canopies they are considering, and the spatial and temporal complexities that can arise over time.

For many agricultural crops, studying the development of the whole crop canopy within the season has provided significant insights into various aspects of environmental resource capture and resource use efficiency. It may, sometimes, also be helpful to distinguish different within-canopy levels of physiological activity, as mentioned above, but analysing the structure and function of individual components in such a canopy may not be a major consideration except for special studies. However, because tree canopies can be so complex, studies of the development of woody plant canopies may often be facilitated by investigating samples of the individual growth units that go to make up the whole. For example, where leaf-retaining species have 'clumped' canopies, each leaf cohort consists of leaves of one of a series of ages arising along the individual branches (growth units) making up the clump. If fruiting takes place in parts of such a canopy (or even overlapping fruiting cycles) it may be impossible to disentangle what is happening without descending to a growth unit level. When we consider the additional consequences of the phenological time displacements of successive growth phases that can occur in tropical trees, any analysis of whole-canopy function of trees in agroforestry systems is certainly going to be difficult, although detailed investigations of parts of the canopy may still be feasible. With multipurpose trees the growth units are, indeed, the parts that provide various products that determine yield (e.g. leafy shoots for fodder or mulch, woody twigs for fuelwood, etc.), and these units form the basis on which it is possible to manipulate plant form precisely. In practice, this means identifying suitable units and observing and/or measuring their incremental growth and phenology (Fig. 7.9).

Understanding the nature of growth units and their development patterns will not only improve our comprehension of how overall tree shape comes about, but help provide an understanding of the likely branching responses to pruning, and an awareness of the physiological implications of its results. Recently, architectural models combined with an understanding of the formation and developmental sequences of individual 'modules', or 'growth units', have been used to model the growth of trees and to compare the efficiency of different canopy forms (Reffye, personal communication). Strictly speaking a 'module' is the leafy axis on which the entire sequence of aerial differentiation is carried out, i.e. it involves the sexual differentiation of the axis. A 'growth unit' (unit of morphogenesis) is usually delimited by the extent of the extension that has occurred in one season (Hallé, 1986).

These aspects, together with whether the growing axillary shoots grow synchronously with their main axis (sylleptic, and chronologically the same age; Tomlinson, 1978) or have a dormant period (proleptic), will all have physiological implications with regard to the timing and duration of, for example, the canopy development of individual trees. Thomas (1992) states that a tree is just 'A colony of autotrophs living on the surface of a large deposit of excreted carbon', although we do know that lateral transfer of assimilates, etc. within different parts of tree canopies can occur (e.g. Hansen, 1969). The growth unit approach may, nevertheless, provide a way in which canopy form can be related to functional aspects of growth and development, and it has been used to study resource allocation, e.g. Cannell and Huxley (1969) with ^{14}C tracer, and Maggs (1960) Cannell (1971a) Dick et al. (1990) and Stephens and Carr (1990) using growth analysis techniques.

Especially in the more intensively managed forms of agroforestry the woody components commonly have their form regulated through pruning,

Fig. 7.9. An example of a seasonally determinate 'growth unit' (*Bridelia micrantha*). As an alternative to the study of the whole canopy an analysis of the numbers and growth rates of representative samples of such units in response to resource availability and temperature can be made. However, as studies in tea (*Camellia sinensis*) have shown, the growth of individual shoots is conditioned not just by environment resource availability but by position on the tree and original age (Stephens and Carr, 1990).

coppicing or pollarding. By analogy we can learn a great deal about how best to do this from the abundant work done on more intensive forms of temperate top fruit production (e.g. Atkinson, 1980) and some tropical plantation crops (e.g. coffee, Huxley, 1969). Although, as discussed in Chapter 4, the difference in agroforestry is not only the effects of canopy geometry on growth, which need to be investigated *per se*, but the effects on resource capture in the system as a whole.

An immediate effect of changing canopy structure through pruning, etc. will be to modify water use, the outcome depending on how much canopy is removed, the time of season, the state of growth of the understorey and the status of the soil water profile (Chapter 6). Again, there is information available from other tree–crop systems (including associations with grass, e.g. Howard, 1925). For example, Chootummatat *et al.* (1990) in Western Australia looked into the effects on water use of four different ways of intensively training plum trees (*Prunus salicifolia*). With these systems seasonal water use changed by only some 10–12%, but it depended on the amount of water supplied. In much more water-limiting conditions some unexpected results can ensue: the interaction between level of water stress and defoliation through browsing, for example. Coughenour *et al.* (1990b) investigated this aspect (in a clipping experiment) with one of the fodder shrubs (*Indigofera spinosa*) that is an important constituent of the vegetation in northwest Kenya. As expected, unclipped plants were most negatively affected by water stress, but well-watered plants subjected to heavy clipping produced less biomass by the end of the season than water-stressed plants indicating that grazed plants will, in fact, be less affected by drought under these arid conditions, as mentioned on p. 250.

Finally, it is impossible to understand how branching and tree form arise without making observations on the position, condition and behavioural responses of buds (Longman, 1978, and see Fig. 7.10). This applies with even greater force with woody plants because of the delay that may often occur between bud initiation and any kind of endogenous or exogenously induced responses. Axillary buds can form complex series in which the behaviour of different members is genetically controlled by hormonal mechanisms. Ancillary buds can arise in a variety of locations (stipular, from stem and root phloem tissue, etc.) and be pre-formed or not (Bell, 1991). Species differ greatly, and information for multipurpose tree species seems to be almost entirely absent. The same applies to floral sites (do they occur terminally on shoots, on new or older wood, or on spurs?), as well as the time of flower initiation. As the history of research with temperate fruits has shown, this kind of information is indispensable if we are to relate growth and development to practical management issues such as pruning, and so to be able to maximize the physiological efficiency of these systems.

Many multipurpose tree species are thought to be outbreeding, and therefore to segregate genetically when raised from seed (Hughes and Styles, 1984). Vegetative propagation is, therefore, an important issue and there are many aspects of growth and form that need to be taken into account when selecting material for this purpose (Leakey *et al.*, 1992). The extent of genetic variation in woody plant form, and the opportunities it offers for plant improvement are beyond the scope of this chapter, however. Some comments on below-ground structure are found later, below, and dealt with in more detail in Chapter 9.

Reproductive Processes

Flowering

In many seasonal crop species a pre-condition for a satisfactory yield is the establishment of a suitable vegetative structure followed, if relevant, by an adequate number of suitably located fruiting points, 'adequate' being determined by sink : source considerations and, often, quality aspects such as

Fig. 7.10. *Vitex keniensis*. An example of a plant with a complex bud series. Although less obvious, many woody species have buds of different kinds in their leaf axils which may exhibit several kinds of response in terms of growth and/or development (e.g. *Coffea* spp.). Little is known about this, as yet, for multipurpose tree species.

fruit size. At any particular level of resources the regulation of fruit and seed numbers is usually through choice of genotype and planting density. Thus, the effects of plant form on the yield of the main agricultural crops are relatively well understood. Cereal crops may flower terminally (millets) or laterally (the female inflorescences of maize), and herbaceous crops can be determinate or indeterminate, with various implications for canopy development, the precise foliar sources of assimilated carbon for fruit and root growth, and nutrient and carbohydrate storage and utilization. For cereals, tillering is an important determinant of yield potential. Furthermore, crop plants have been selected so that the different development sequences, and the provision of an adequate branching structure (e.g. in legumes) take place within the 'normal' (i.e. average) cropping season. Indeed, the variations in development that occur from year to year as a consequence of departures from the 'normal' season are a major contributing factor to yield differences. In some woody species, apical meristems develop into fruiting structures but this, itself, can markedly regulate form. A plastic response of seasonal herbaceous species to adverse environmental conditions is often to advance flowering and seed formation. Woody perennials tend to defer flowering under these circumstances so improving the chances of survival of the parent plant (Grime *et al.*, 1986).

Flower or inflorescence bud structures arise in mature crop plants in suitable environmental conditions, usually after a particular number of leaves have been formed. Breeding has ensured that no interruption to this process normally takes place, and its occurrence is carefully matched to environmental conditions so as to ensure that both flowering and fruit growth and maturation take place at suitable times, depending on the expected length and conditions of the cropping season. After proceeding through a juvenile phase, which can vary from one season to many depending on species, woody plants can be classified according to the periods over which they are likely to flower and fruit. For example, they may flower continuously, seasonally, intermittently, only after a number of years, or once in a lifetime. Even in the lowland humid forest a range of flowering patterns can be observed (e.g. Newstrom *et al.*, 1991). Woody plants may initiate flower/inflorescence buds on the current or last season's growth, on older lateral branches, stems or even the trunk. If on the current season's shoots, the plants may be 'tip-bearers' or not, depending on the flower initiation sites. All these kinds of flowering and fruiting behaviour may be found in multipurpose tree species used in agroforestry systems, and they clearly have functional implications. For example, where fruit yield rather than vegetative biomass is the final output, irregular bearing is likely to impede attempts at accurate assessment of water use efficiency in terms of yield. And this will increase the need to understand the biological links between growth and the flowering/fruiting processes in multipurpose trees. Little or no such information is to be found for the multipurpose tree species used in agroforestry, either in the scientific literature or in more observational accounts.

Some aspects of fruiting

Where woody species are cultivated for their fruits and/or seeds we need to be able to control fruiting. Even if they are grown for vegetative parts, we cannot ignore the flowering/fruiting process and the ways in which it can modify plant functions, e.g. growth rates, dry matter distribution, rate of capture of soil nutrients, etc.; that is, unless the plants are to be grown entirely vegetatively in hedgerow intercropping just for mulch materials. For individual species management interventions can help if there is sufficient understanding of the way apical dominance operates, i.e. of the factors serving to distribute flowering points in space and time. And also of what controls, if any, exist for regulating fruit load. Practical management operations, for example, fertilization, irrigation and especially pruning, help to establish a suitable vegetative structure, and regulate fruiting load but, if the likely responses are first identified, they can be manipulated so as to help entrain growth and flowering to suitable environmental periods or, in agroforestry, to periods that offer better resource-sharing opportunities with the adjacent crops or pastures.

Trees and shrubs are subjected to environmental and/or competitive stresses over relatively long periods of time, so they have to build up a vegetative structure that will serve to accommodate such stresses while still ensuring succession through seed dispersal. Thus it is crucial to establish a long term balance between the growth of the vegetative meristems, which may be indeterminate or determinate, and the fruiting structures, which are usually determinate. Unlike seasonal crop plants, therefore, woody perennials need to be able to adjust the fruiting potential afforded by their previously accumulated structure according to the constraints imposed by future successive seasons. Different woody species resolve this by exploiting a wide range of fruiting arrangements. These can occur on woody stems originating in the current or past seasons, in the axils of leaves on main stems, or on laterals of different orders, on established spurs or short shoots, even on main stems, so that the growing fruits will draw on assimilate sources and stored nutrients in a greater variety of ways than is found with seasonal crop plants. The time at which flowering occurs in relation to the maturation of leaves can also play a part in resource allocation and, in the extreme case with a deciduous tree, flowering will occur before the emergence of the new leaves (in *Schizolobium excelsum*, for example).

The nutritional status of a tree at the time of flower initiation can influence fruitfulness. For example, Hill-Cottingham (1968) found that a nitrogen application to apple trees in late summer during the differentiation phase of the floral primordia resulted in 'stronger' blossom the following spring because the longevity of the embryo sac is increased, leading, in some cultivars, to an increase in self-fertility.

Many woody species bear fruit only late in their life but fast-growing multipurpose trees are usually early flowerers. Because it helps to avert overall sink limitation, fruiting is known to enhance carbon assimilation (Neales and Incoll, 1968), and fruiting plants usually achieve greater amounts of biomass than non-fruiting ones (e.g. Cannell, 1971b). However, there is usually a high degree of hormonally regulated, local, autonomous control, which can exert constraints on fruiting to some extent or another. In some species this occurs through apical dominance limiting flower initiation on new wood (but not with 'tip-bearers'), or current fruiting sites limiting flower initiation proximally, in a similar way. These negatively correlative mechanisms result in spatial distancing of competing meristems and, because of the degree of control they exert in heavily fruiting seasons, a marked amount of temporal phasing (i.e. 'on' and 'off' periods of fruiting, followed by vegetative growth), which is a notable feature of most woody species (Browning, 1985).

Both woody species and determinate herbaceous crops can adjust their fruit loads by restricting, in the appropriate development phases, one or more of the components of yield. However, woody species commonly 'overproduce' flowers and then serially adjust the subsequent fruiting load through a variety of hormonally induced fruit shedding mechanisms; a condition that has arisen because fruit maturation may extend over long periods, often 6–9 months or more. Excessive fruit load may impose severe stresses for carbohydrates and/or nutrients which can result in leaf senescence and abscission, so that both the amount and the location of fruits with respect to active carbon-fixing sources, i.e. young, fully expanded, well-illuminated leaves (Huxley, 1985) are factors that may well affect the functional well-being of a woody plant canopy, as well as its persistence. Certainly, it can have an immediate effect on fruit set (e.g. in citrus, Moss et al., 1972).

In a self-regulatory way, overproduction of flowers with the possibility for fruit shedding allows for a balance to be struck between canopy size and fruit load so as to avoid internal competitive stress, but it is also a process of control that may help offset flower, fruit or seed predation. Achieving and maintaining a balance between vegetative and fruiting parts is especially important for woody perennials (Luckwill, 1970), particularly in highly variable environments, and it is one which helps maximize genetic fitness over long life spans as it works by reducing weak or unfavourably sited sinks, thus economizing on resource utilization. In some woody species flower initiation is often reduced by shade (Cannell, 1983), again, a regulatory mechanism to ensure that fruits develop near to parts of the canopy best able to support them. Fruiting spurs show severe restriction of vegetative development, often together with a high level of flowering, and such leaf clustering enhances light penetration into the canopy. Trees with spur-fruiting, or species that can be made to behave this way (e.g. guava, *Psidium guajava*, see Fig. 7.11), can often be grown intensively as productive, small-statured plants (Huxley, 1975), and there is a good deal of scope for this kind of intensification in several agroforestry practices.

Below Ground

Rates of resource capture by roots will be affected by the plant's rate of growth, and hence demand for nutrients, the amount (root length density,

Fig. 7.11. Guava (*Psidium guajava*) although usually grown as a small spreading tree can easily be trained to produce fruit intensively as a hedgerow through suitable early training and forcing the plant to produce short 'fruiting spurs' by pinching out the tips of young lateral branches.

disposition and kind (thickness) of fine roots, and by factors that control the supply of water and nutrients in the soil (see Chapter 9).

As crop plants grow their root systems steadily extend, at first, both laterally and vertically, then, after root systems of individual plants mingle, a descending 'root front' is often identified. This usually continues until around the time of flowering or early fruit maturation (Gregory, 1994), after which roots senesce and die. Crop growth models function well by assuming a continuous increase in rooting depth, sometimes with provision for irregularities at the rooting front (and see Chapter 10).

The architecture of the main, structural roots of trees and shrubs can vary considerably depending on species, site and soil management, and, perhaps, on whether they were propagated vegetatively or from seed. Information on these aspects for multipurpose trees is still scarce. When young, and more easily studied, tree roots are often seen to extend rapidly through the production of 'long' or 'pioneer' (i.e. little branched) roots (e.g. Reynolds, 1970). A few of these will eventually form the main structures. Other long roots become 'mother roots' on which other roots branch to provide finer roots of different orders on which, eventually, 'short' or 'fine' roots (defined by no consistent diameter) are found, but this complex of roots of different categories and branching patterns (see Fitter, 1991) will often not fill the available soil volume. The *overall* volume of soil explored will double if the rooting radius is extended by only the cube root of 2, so that, in order to capture sufficient water and nutrients to maintain a steady rate of growth, the main root system of a free-standing tree will need to expand radially only at an ever diminishing rate. Furthermore, if the root system is in any way clumped, and clusters of fine roots in different parts of it are dying, there is plenty of space free of external competition *within* the overall rooting volume for new long roots to explore, or re-explore, and new fine root clusters to form and exploit (Fig. 7.12). Harper *et al.* (1991) have suggested that an ideal architecture ensures a spacing of resource-depletion zones that maximizes access to the most limiting resources and minimizes the overlap of resource-depletion zones. This overlooks the possibilities of the movement of soluble nutrients along macropores (e.g. old root channels) and the longer term trade-offs where fine root turnover is a major feature. In agroforestry the latter might be an important feature contributing to below-ground complementarity.

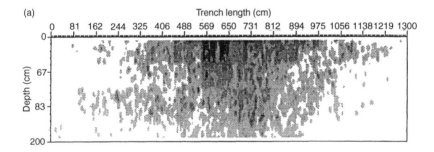

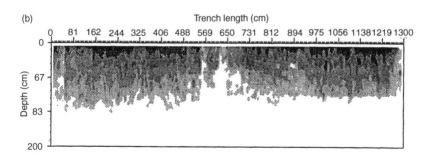

Fig. 7.12. The rooting profiles (fine root counts) in a *Grevillea robusta*–maize hedgerow intercropping experiment at Machakos, Kenya (Huxley *et al.*, 1994): (a) *Grevillea robusta* roots and (b) maize roots. The hedgerow was in the centre of the plot.

The ability to produce new fine roots rapidly ('root growth potential') is especially important to help secure the successful establishment of young, outplanted trees (Deans *et al.*, 1990). For mature trees, it may well be a useful characteristic if they suffer from below-ground herbivores, and it could possibly be investigated more widely with young plants to identify differences in mature rooting characteristics of different multipurpose tree species.

Growth regulation within tree root systems is not yet sufficiently well understood. An alternation of root and shoot growth may sometimes occur, and models of *overall* allocation of C and N (and other nutrients) between different sinks can allow for this (see Cannell and Dewar, 1994). We know that this may occur from studies on temperate woody species, e.g. apple root growth occurs in spring and autumn. However, even short term changes in root–shoot growth phases will be of some considerable practical importance. Internal regulation will be interrelated with canopy activity, and vice versa, but short and long term effects need to be distinguished. The former are likely to be affected by hormonal signals. We know, for example, that cytokinin production in roots can stimulate bud activity in shoots (Luckwill, 1970; Jacobs, 1979). When considering the diverse above-ground phenological patterns shown by tropical woody plants it is as well not to forget that we know nothing, as yet, about the part that roots might play. Indeed, one might speculate whether the pre-rains canopy activity of many African dryland trees and shrubs (e.g. *Faidherbia albida*) might be a response to early root activity (offering a competitive advantage over the associated grasses), rather than the other way round. The growth phases of patches of fine roots may not always be coincident throughout the root system but vary according to position, i.e. as arising from different mother roots (e.g. Huxley and Turk, 1975). Stress of one kind or another can evoke root responses in terms of physiological activity and/or morphological plasticity (root shedding). One might expect that, in terms of the carbon economy of the plant, fine roots could beneficially be shed when resource capture levels fall below a minimum level, but this ignores possible competitive and/or survival advantages of not doing so. Thus the root systems of trees, especially in the tropics, may well have evolved as wide a range of growth strategies as the above-ground parts (see 'Phenology', p. 275).

King (1993) has modelled forest growth, nitrogen uptake and biomass allocation and he points out that in a stand wood production is maximized by quite a low allocation to roots for any specified amounts of plant-available nitrogen. Wood production of the individual tree, however, will be maximized by a higher allocation to roots because this will enable the tree better to compete below ground, a feature that may also enhance long term productivity by decreasing nutrient losses from the system. Such considerations are likely to be meaningful in agroforestry systems.

The root system fulfils vital functions with regard to storage and numerous metabolic changes. Organic compounds that have been conducted through the phloem, and that have arisen in leafy shoots may be re-transferred back again. For example, in studying the source–sink relationships of individual leaves of tomato using ^{14}C-labelled CO_2, Khan and Sagar (1969) discovered that large amounts of labelled assimilate that had been transferred to the roots overnight were returned to the shoot system the following day. Similarly, Cooper and Clarkson (1989) have shown that there is recycling of N from root to shoot to root, and back again. Whether such transfers and transformations affect short term resource capture efficiency is uncertain, but they certainly will be implicated in longer term growth patterns. Sap analysis techniques can be used to determine the movements of hormones as well as organic and inorganic substances moving from the roots (Jones, 1971; Dambrine *et al.*, 1995).

Except with regard to aspects of dry matter distribution and the flow of N and mineral nutrients it is clear that the links between above- and below-ground activities have, as yet, not been at all well explored for mature trees other than through a few observational studies (e.g. Howard, 1925; Rogers, 1969; Huxley and Turk, 1975). Such links are clearly of considerable importance to any functional analysis of agroforestry systems.

Many other aspects of roots and rooting in agroforestry are discussed in Chapter 9.

Organ Replacement in Trees

To shed and replace ageing parts with new ones would seem an inevitable process for a long term perennial plant, and woody perennials have evolved a whole range of strategies that can increase the overall efficiency of resource use by this means, and hence confer competitive advantage and/or increase survival prospects. For example, above ground with leaves, and depending how often and under what circumstances it occurs, replacement offers possible strategies to conserve carbon. Below ground with fine roots, it provides opportunities to re-explore soil volumes within the rooting zone.

If shedding of parts is a significant part of the reiterative growth pattern of a tree, in agroforestry it will also have implications for tree–crop interface effects because of the residues provided (Chapter 1), and hence play an important part in the recycling of nutrients and the maintenance of good physical conditions in the topsoil in agroforestry systems in general (Glover and Beer, 1986). Some species, e.g. the thorny acacias, shed a good deal of branch litter, but we are concerned here only with leaves and fine roots.

Leaf longevity

The number of occasions on which trees replace their leaves, and the precise times within seasons when they do this can vary according to species and the degree of external and internal stress experienced (see 'Phenology', p. 275), periods of drought and overbearing being especially influential. A number of commonly used leaf-retaining multipurpose tree species are sensitive to stress and may become only partially clad over some of the year, e.g. *Leucaena leucocephala*, *Sesbania sesban*, *Senna* spp., others seldom lose much leaf under these circumstances, e.g. *Grevillea robusta*, *Gliricidia sepium*. Such species may shed old leaves over long periods but, often, more predominantly at the same time as deciduous species (cf. Huxley and Van Eck, 1974). Very obviously, retaining a canopy throughout the year, even of older leaves, is likely to enhance water loss from the system considerably compared with one utilizing deciduous species. Surprisingly, this is seldom given much consideration when selecting multipurpose trees.

Leaf longevity of woody plants can vary between a few months (e.g. a deciduous species such as African nutmeg, *Monodora myristica*), through several years (many of the palms), to 40 years or more (*Pinus longaeva*), but invariably there is a functional decline with age, both for stomatal sensitivity (Reich and Borchert, 1988) and photosynthetic efficiency. For the canopies of deciduous tree species this may roughly parallel what is happening to an agricultural crop (e.g. maize, Dwyer and Stewart, 1986, Fig. 7.2) but this is unlikely to be the case for leaf retainers where the age structure of the canopy can be complex (Fig. 7.13). So, for individual species, when does photosynthetic decline commence, and how does it progress with time? In addition, nutrient deficiencies (and other stresses) can greatly affect canopy age structure (Harper, 1989; see Fig. 7.14). Overall, the choice of deciduous or leaf-retaining species and, within the latter, the canopy demography are important factors in considering the type of woody species best suited for any particular agroforestry practice, as such traits as are involved will affect the functional nature of the canopy with consequences for dry matter accumulation, water loss and tree–crop interface interactions. There is so far only anecdotal information about leaf longevity under any particular set of conditions for multipurpose tree species, and almost nothing about the relationships between leaf age and functional decline. As leaf longevity is seriously affected by pests, pathogens and even seemingly innocuous microorganisms living in the phyllosphere (e.g. in coffee, Hollies, 1967), ascertaining the 'normal' life of leaves on trees is not necessarily easy.

Any particular degree of leaf longevity implies a trade-off in energy loss versus maintenance and renewal energy costs. Taken alone this is too simplistic a view, however, because nutrient storage aspects, competitive advantages of canopy height (even senescent leaves shade out lower storey plants), and the maintenance of a functional hormonal balance are all part

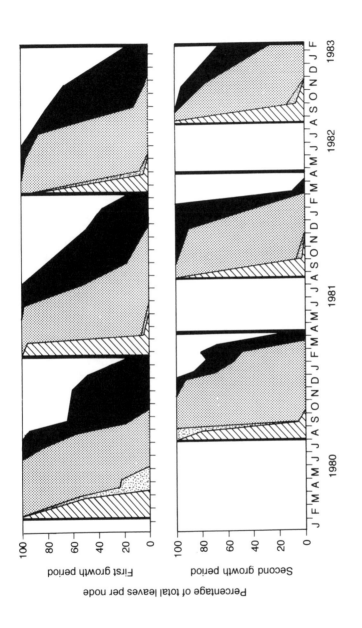

Fig. 7.13. Leaf development for two cohorts of leaves on a population of *Prosopis glandulosa* in the Sonoran desert of California. Each point is a mean of 100 + branches and error bars represents LSDs at $P > 0.05$. I = Initiated; J = juvenile; M = mature; S = senescent. (Source: Nilsen *et al.*, 1987.)

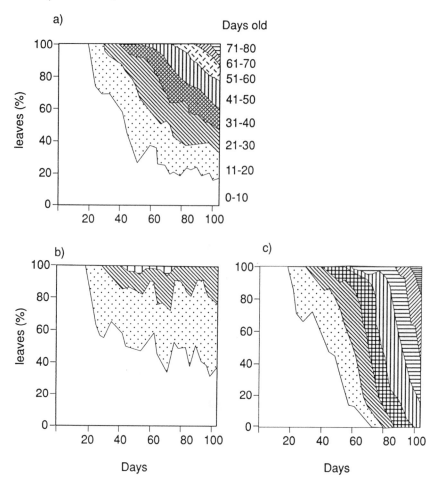

Fig. 7.14. The changing age structure of the leaf populations of *Linum usitatissimum* grown in sand culture receiving (a) full Long Ashton nutrient solution, (b) nutrient solution omitting potassium, and (c) nutrient solution omitting calcium. (Source: Harper, 1989.)

of the compromise that has to be reached if a species is to attain a balance between competitive ability and a capacity for survival. Leaf-retaining species tend to have higher LAIs than deciduous species, perhaps because they can spread the energy cost of smaller increments of canopy renewal over a longer period of time. The older, shaded leaves on such species may also become light acclimatized, i.e. possess lower leaf light saturation rates (F_{max}) (Cannell, 1989), and so maintain better carbon balances, although this, apparently, is not always the case (e.g. lodgepole pine, *Pinus contorta,*

Schoettle, 1990). Generally, deciduous woody species have mesomorphic leaves (Walter, 1971), which lower the replacement costs.

Nutrient sufficiency increases dry matter partitioning to above-ground parts, including leaves (Ledig, 1983; Cannell, 1989), and it helps foliage persist (e.g. Priestley, 1972; Thomas, 1992). Nutrient demands of forest stands are usually greatest just before canopy closure. Infertile habitats may favour evergreen leaf-retaining species as these tend to conserve nutrients (Grime, 1979). Stress tolerators, i.e. relatively slow-growing, low nutrient demanding plants that favour the allocation of assimilates to storage rather than leaf (Grime, 1979), will be able to offset at least short periods of nutrient depletion, and they may achieve higher nutrient resource capture rates under longer duration pulses of nutrient release (Grime et al., 1991). These are all considerations to be taken into account in the choice of multipurpose species, and the management of agroforestry systems to make best use of available environmental resources.

$^{14}CO_2$ labelling of leaves of different ages in many plants has shown that old leaves do not normally import newly acquired assimilates from younger leaves. Whether older leaves are shed when they no longer maintain a positive carbon balance, and they have used up all carbohydrate reserves (Koslowski, 1971; Thomas, 1992), or not, may depend on whether they still have a function to fulfil, e.g. as a reserve for mineral nutrients, and/ or to help shade out competitors. Shedding may be brought about prematurely by several forms of stress: either environmental, e.g. dense shade, particularly high radiation and leaf temperatures and/or severe vapour pressure deficits and drought, and nutrient insufficiency; or endogenous factors such as depletion of carbon and nutrients to a strong sink (e.g. a heavy fruiting load), or hormone 'shock' (through removal of a shoot apex, or removal of root apices by pests or pathogens).

Fine root longevity

Like the crown, the root system of a tree can be regarded as a modular structure adapted to maximize resource acquisition per unit energy of cost. Fine roots die (but do not abscise) because of internal cues (Harper et al., 1991), environmental stress (e.g. drought, which within a few days will also result in the sudden loss of all root hairs) and depletion of carbohydrate reserves. We already know that the longevity of fine roots can vary between species considerably – in Pinus, for example, from 0.2 years to over 5 years (Schoettle and Fahey, 1994) – and it can be influenced by mycorrhizal associations (Fahey, 1992). Such differences might be based on the carbon economy of the system. However, as these authors point out, it is impossible to ascertain whether fine roots are shed only when soil depletion lowers the efficiency of resource capture below some critical level, as may occur with leaf longevity. This is because unlike leaves, where the gradients for light are largely unidirectional, and lower leaves shaded by upper ones thus senesce, spatial and temporal patterns in water and nutrient availability in the soil are multi-directional and vary greatly for all kinds of reasons. In addition, the presence of neighbouring plants almost invariably induces light gradients, but nutrient gradients can occur in their absence (Keddy, 1990). Thus the conditions under which fine roots become a net cost would be difficult to establish. Herbivorous organisms are also responsible for considerable root loss in the field, as are plant pathogens.

Physiological and ecological explanations about the effects of soil fertility on fine root longevity differ. There is plenty of experimental and observational evidence to support the conclusion that an increase in soil water availability and/or nutrient supply reduces the proportion of the total dry matter that is allocated to the root system (e.g. Cannell, 1985). The investigations by Linder and Axelsson (1982) that he quotes on 20-year-old *Pinus sylvestris* provide experimental evidence that additional nutrients greatly reduced fine root turnover in absolute terms, even though the trees were bigger (Fig. 7.15). On the other hand, Grime's proposals for primary plant strategies ('ruderals', 'competitors' and 'stress tolerators'; Grime, 1979) visualize active foraging by competitors in fertile soils by rapid exploitation through a morphologically dynamic root system, in which the life span of individual roots is short (Grime *et al.*, 1991). Stress tolerators will grow on less fertile soils in harsher environments and have to capture and *retain* resources, so that their root systems are supposedly comparatively long-lived structures in which plasticity is expressed mainly through reversible physiological changes. Ruderals are characterized by rapid early root growth and early flowering (but this view is simplistic for trees, see Oldeman and van Dijk, 1992), and they will forage actively in response to localized resource depletion. These primary strategies are modified depending on whether constituent species are 'dominants' (i.e. tending to monopolize resource capture) or 'subordinates' (plants with specializations of various kinds that can restrict their ability to monopolize the environment). The response of dominants to resource depletion is considered to be 'coarse grained' and will thus tend to maximize resource capture from the patchy environment as a whole. Subordinates, contrastingly, forage in a 'fine-grained' way which involves a high level of within-root system plasticity enabling the plant to maximize capture from specific parts of the environmental resource. In these circumstances not only 'patches' but also 'pulses' (i.e. the regular or spasmodic release of nutrients within microspaces in the soil) are important in providing resources and evoking plant responses. Here, again, stress tolerators and competitors are predicted to differ in that the former, being less successful in unproductive habitats, should be able to exploit briefer pulses of mineral nutrient availability than would competitors. These below-ground attributes are accompanied by above-ground traits related to 'foraging' for light. Some experimental evidence has been obtained to support these concepts (e.g. Grime *et al.*, 1991, see Figs 7.16 and 7.17) and, although seen as controversial (Grace, 1990), these ways of viewing the plants' abilities to capture resources could help with a better understanding of the types of multipurpose tree to be selected under different agroforestry conditions.

Schoettle and Fahey (1994) mention a positive effect of increased soil heterogeneity on fine root longevity, because access to favoured sites would be maximized (McKay and Coutts, 1989), and they quote evidence for pines suggesting that fine roots live longer in the organic rather than the mineral soil horizon in forests (Persson, 1983; Gholz *et al.*, 1986; with additional information from Vogt *et al.*, 1986, correlating leaf litter fall N flux and apparent fine root longevity). Schoettle and Fahey (1994) also suggest that (wide-spaced) temporal patterns in the nutrient-supplying power of soils will adversely affect fine root longevity, as this implies a heavier maintenance respiration load (a reduction of uptake per unit maintenance cost). The timing of seasonal leaf litterfall or, in agroforestry, the intermittent use of plant residues as mulch, is clearly a factor that may have to be considered. Thicker roots live longer (Gholz *et al.*, 1986), so this

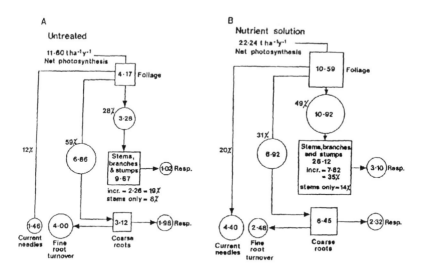

Fig. 7.15. Current and annual dry matter budgets of 20-year-old *Pinus sylvestris* trees, (a) untreated and (b) supplied for 6 years with a complete nutrient solution each growing season. The squares show the standing biomass (t ha^{-1}); the circles show the fluxes of dry matter (t ha^{-1} year^{-1}). The areas of the squares and circles are proportional to the values. (Source: Cannell, 1989, data from Linder and Axelsson, 1982.)

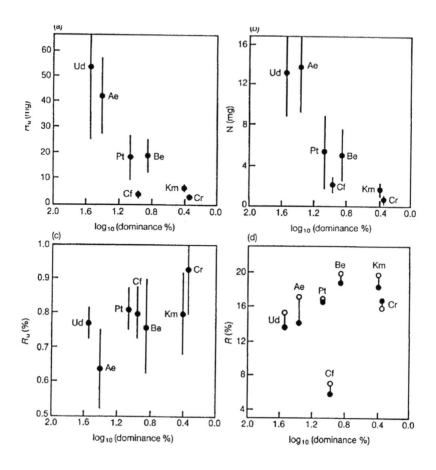

Fig. 7.16. The relationship between species ranking ('dominants' to left, 'subordinates' to right) in an experimental community and the ability to modify root distribution (a and c) and maintain nitrogen capture (b and d) in a nutritionally heterogeneous rooting volume (i.e. sand culture with segments alternately free of or supplied with nutrients). In each figure species are ranked according to the shoot biomass attained in the experimental community after 16 weeks' growth in a productive greenhouse environment. R_u = root dry weight increment to undepleted segments. N = total increment of nitrogen during the depletion phase. (R(%)) = proportion of dry weight increment allocated to roots in depleted (○) and control (●) treatments. Ae = *Arrhenatherum elatius*; Be = *Bromus erectus*; Cr = *Campanula rotundifolius*; Cf = *Cerastium fontanum*; Km = *Koeleria macrantha*; Pt = *Poa trivia*; Ud = *Urtica dioica*. Confidence limits of 95% are indicated by the vertical lines. See p. 272. (Source: Grime *et al.*, 1991.)

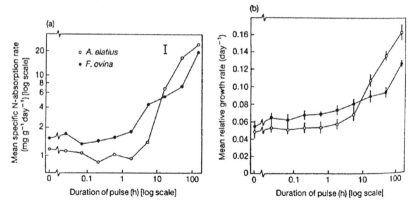

Fig. 7.17. (a) Mean specific N-absorption rate of *Arrhenatherum elatius* (from productive habitat, potentially fast-growing) and *Festuca ovina* plants (from productive habitat, potentially slow-growing) exposed once every 6 days to nutrient pulse treatments of differing duration. Vertical bar is LSD (*P* > 0.05) for comparing means on a logarithm scale. (b) Mean relative growth rate of same two species under the same conditions as before. Means are shown ±95% confidence limits. (Source: Grime *et al.*, 1991.)

will also contribute to differences between species and the duration over which 'clumps' of fine roots persist.

We clearly have more to learn about this whole topic, which needs to be explored both in terms of contributions that agroforestry systems may or may not make to plant residue effects on soil fertility, and also in respect of below-ground resource capture and resource sharing potentials.

Phenology

Climates can be specified as suitable for the complete phenological development of particular agricultural crops (FAO, 1978; Higgins *et al.*, 1987), but this categorization may often not adequately describe the potentials for development of woody species because of their propensity for 'irregular' behaviour (Fig. 7.18). Significant departures are less often apparent in many of the plantation crops (coffee, tea, coconut, rubber etc.) probably because of selection or, in the case of *Coffea* spp., an inherent control that tends to entrain subsequent development sequences (flower bud dormancy broken by rain, Alvim, 1960). The degree to which different multipurpose tree species show irregular behaviour is so far documented only fragmentally (e.g. in the ecological or range management literature). However, with the very large number of trials now proceeding worldwide there is every opportunity to make quantitative phenological comparisons (Akunda and Huxley, 1990) between species, seasons and sites.

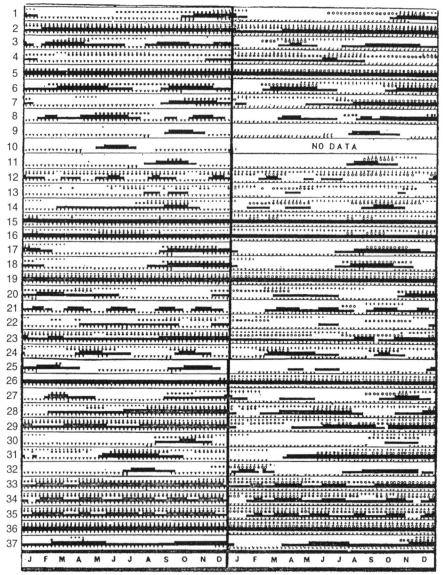

1967   1968

Although woody plants will be selected to be generally adapted to particular climatic regions (climate matching, see Booth, 1991) they will still have to endure local weather patterns. All phenological events must commence, continue and conclude through concomitant endogenous or exogenous factors that first in woody plants 'trigger' and then support it. Phenological patterns may well have evolved, in part, so as to optimize environmental resource use over a long life span (i.e. taking into account the stochastic nature of climatic events). How they do so in any one season, however, is due in part to the consequences of previous seasons' events on the current season's growth and development. Two common causes of changes in the phenological pattern of the current season are a previous severe depletion of nutrient and carbohydrate resources caused by overbearing (Fig. 7.7), often exacerbated by poor environmental conditions, and any factor that prolongs vegetative growth unduly. For example, in the drier parts of East Africa, after an unusual continuation of a rainy period, it is not unusual to find that woody vegetation 'flushes' late (sometimes very late) in the following rains.

We also know that some woody species in semi-arid regions can flush vegetatively before a rainy season, and others at some time during it, or even after it has finished; and not all the individuals of any one species will, necessarily, be doing the same thing at the same time. Patterns of leaf flushing can be complex (Fig. 7.19), and may have been evolved in response to many factors unrelated to resource capture – herbivory, for example (Nilsen et al., 1987; Coley, 1988) – or they may be coupled through branch morphology with flowering requirements or factors related

---

Fig. 7.18. Diagram of part of a series of phenological records of woody species obtained over 8 years (1962–69) shown here for 37 specimens (32 species) at Kabanyolo (near Kampala, Uganda) with bimodal rainfall, i.e. long rains from March to June, short rains from September to December. Thin and thick lines, leafy shoots growing moderately or fast; below line, v = senescent leaves falling; above line, ○ = flowering and ● = flowering prolifically; + = fruits ripening. Note the diversity of behaviour. (1) *Acrocarpous fraxinifolius*, (2) *Aleurites moluccana*, (3) *Averrhoa carambola*, (4) *Bauhinia alba*, (5) *Bougainvillea glabra*, (6) *Cassia grandis*, (7) *Cassia javanica*, (8) *Senna splendida*, (9) *Chlorophora excelsa* 1, (10) *Chlorophora excelsa* 2, (11) *Chlorophora excelsa* 3, (12) *Cinnamomum zeylanicum*, (13) *Citrus limon*, (14) *Citrus sinensis*, (15) *Coffea arabica*, (16) *Coffea canephora*, (17) *Delonix regia* 1, (18) *Delonix regia* 2, (19) *Eucalyptus saligna*, (20) *Jacaranda mimosifolia*, (21) *Litchi chinensis*, (22) *Mangifera indica*, (23) *Markhamia platycalyx*, (24) *Milletia dura*, (25) *Monodora myristica*, (26) *Musanga cercropioides*, (27) *Persea americana*, (28) *Psidium guajava*, (29) *Punica granatum*, (30) *Schizolobium excelsum*, (31) *Spathodea campanulata*, (32) *Sterculia dawei*, (33) *Theobroma cacao* 1, (34) *Theobroma cacao* 2, (35) *Theobroma cacao* 3, (36) *Tecoma stans*, (37) *Vitis vinifera*. (Source: Huxley and Van Eck, 1974.)

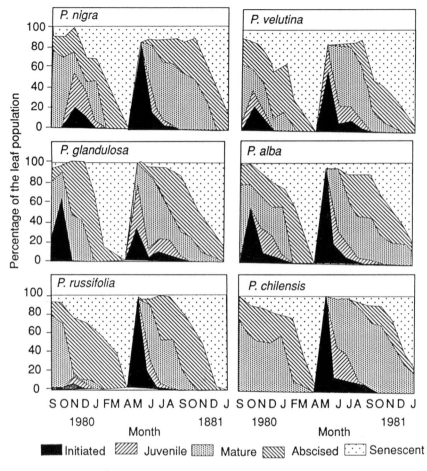

Fig. 7.19. The relative proportion of the leaf population in each of five developmental classes for six *Prosopis* species. (Source: Nilsen *et al.*, 1991.)

to competitive root growth (Huxley, 1985). Some of the 'curious' behavioural patterns of tropical woody plants may also have evolved as a stratagem to avoid seed predators.

In deciduous, temperate, plantation forest species the time of bud break and the rate of foliation in spring has a major effect on the amount of light intercepted over the growing season (Cannell, 1989), and hence yearly wood increment. Again, the early canopy development (Watson, 1958) and the maintenance of green leaf area (Wolfe *et al.*, 1988) of crops is essential for high yields. In the tropics, late sowing after the start of the rains decreases crop yields considerably (e.g. Hawkins and Cooper, 1981, for maize). These are relatively simple examples of the effects of time

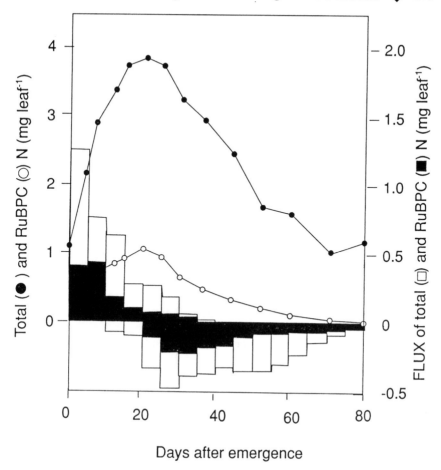

Fig. 7.20. Levels of nitrogen in, and fluxes through, the total N and ribulose-1,5-biphosphate carboxylase fractions of *Oryza* leaves during ontogeny. (Source: Thomas, 1987, data from Makino *et al.*).

displacements on growth and yield compared with the range of within- and between-season variability often experienced with freely grown multi-purpose tree species in many parts of the tropics.

   As well as the transfer of mobile nutrients such as K and Mg a leaf becomes an active net exporter of N during senescence (Fig. 7.20, from Thomas, 1987), but the time at which leaf abscission occurs in woody, leaf-retaining species may not represent the conclusion of such transfers, as this may have occurred previously. In drier regions leaf shedding is often associated with the onset of drought (Frankie *et al.*, 1974), but elsewhere it can occur in both wet and dry seasons (Reich and Borchert, 1988), and not even be affected by soil moisture status (Joseph-Wright and Cornejo,

1990), presumably due to various internal controls. There is some evidence that leaf fall in the tropics may be related to the onset of cooler environmental periods (Huxley and Van Eck, 1974; see also Fig. 7.24).

Although we have yet to explore the effects of such temporal displacements of growth and development in multipurpose tree species, studies on seasonal plant species may well provide an indication of its relevance, particularly if we consider a model tree as an assemblage of growth units acting similarly. For example, Fig. 7.21 shows the large changes in the dry mass of parts, and in biomass allocation, of a small South African ephemeral subjected, in its desert fringe habitat, to unpredictable rainfall and, hence, a very variable date on which germination might occur in any one year, resulting in growing seasons of very different durations. In this experiment (Van Rooyen *et al.*, 1992) seeds were sown in pots outside at intervals from March to September (i.e. early to late growing season), resulting in life spans of 25 to 9 weeks, respectively. The increasingly hot dry conditions in and after October normally terminated active growth. Plants having the benefit of a long, partly growth-favourable growing season invested a high proportion of their biomass to vegetative structures which increased their potential reproductive output. The detrimental effect of a short, and harsher, growing season on reproduction was to some extent offset by a greater dry matter allocation to reproductive organs but at the expense of leaves, stems and roots. These results are predictable because all seasonal plants have to conserve their reproductive processes in order to survive; woody plants need not do so each season. But to what extent does such adaptive plasticity to shifts in growth periods occur in woody plants from semi-arid and seasonally arid regions? Where flowering occurs the magnitude of the trends exhibited could be similar, but there are also other priority sinks (perennating stem and root meristems and storage sites to ensure resources for early growth next season). In order to consolidate our understanding of the outcome of the different phenological patterns often exhibited by multipurpose trees, particularly for sequential effects of one set of phenophases on the next, we need to assess not only the results of the duration and time of displacements of growth and developmental phases outside the 'normal' growing season for resource capture and resource use efficiency, but for dry matter distribution also.

As yet the regulatory mechanisms that evoke phenological responses are poorly understood, although there is probably a simple underlying control that initiates shoot growth phases. In climatically variable tropical regions this will be most flexibly controlled though the physiological decline of the canopy past a threshold, either by ageing and/or the imposition of various environmental stresses. When this happens the growth regulation within the shoot will be changed. For example, apical dominance of existing lateral vegetative buds may be removed so that a new canopy, or cohort of

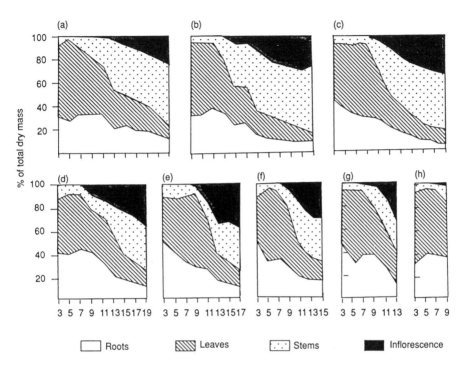

□ Roots   ▧ Leaves   ⋯ Stems   ■ Inflorescence

Fig. 7.21. Changes in the biomass allocation of *Dimorphotheca sinuata* sown on different dates: (a) early March; (b) late March; (c) April; (d) May; (e) June; (f) July; (g) August and (h) September. (Source: Van Rooyen *et al.*, 1992.)

leaves, can be produced at the next environmentally favourable time (Huxley and Van Eck, 1974). Such a mechanism would be much less rigid than control through daylength, for example, although the latter has been claimed to affect shoot growth rates in some tropical tree species (Longman, 1978). Termination of growth phases may be through some form of 'feedback' control (Borchert, 1978). In leaf-retaining trees the endogenous factors that control vegetative 'flushing' are obscured because of the old leaves that are retained in the canopy, but the same patterns of growth followed by leaf fall followed by growth are still to be observed (Huxley and Van Eck, 1974), especially if growth on individual branches is monitored (Akunda and Huxley, 1990; Fig. 7.22).

Once a phenophase has started its continuation will depend on the availability of environmental resources and, for woody plants especially, stored resources may initially be important. Indeterminate development is dependent on factors other than temperature, e.g. an overall restriction of the growing period, but processes that can be considered determinate

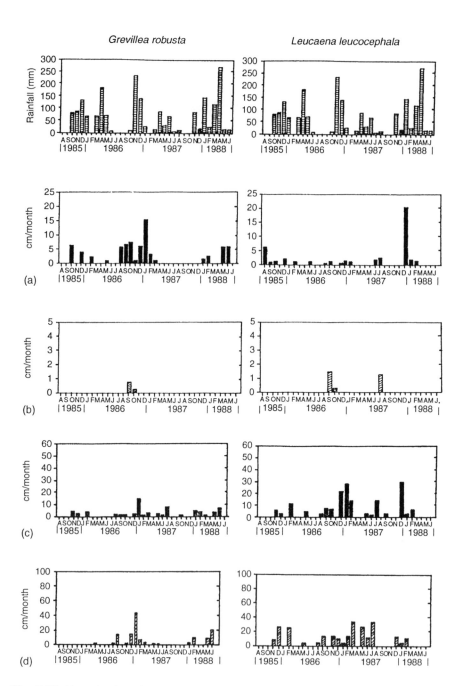

Fig. 7.22. Branch phenology data for *Grevillea robusta* and *Leucaena leucocephala* at Machakos, Kenya. (a) Extension growth of uncut primary shoots (cm month^{-1}), and of (b) primary shoots that had their apices removed in August 1985. Similarly, total extension growth of lateral branches on the same primaries either (c) uncut or (d) with apex removed as before. Rainfall data at top of diagram. (Source: Akunda and Huxley, 1990.)

(flowering, the extension of individual growth units) will be regulated by thermal time. As Squire (1990) points out, this can have important implications for the production and distribution of dry matter in crop plants because temperature strongly affects the expansion of leaves and the extension of stems. Where there is only a single cohort of leaves produced in a season, i.e. on a fully deciduous tree, then individual growth units will probably behave in the same way as a determinate crop species and the final size will be independent of temperature, but highly dependent on water and nutrient stress. However, the relative proportions of leaves of different ages can vary quite considerably even between fully deciduous species of the same genus as Nilsen *et al.* (1987) have shown in a detailed growth comparison of six *Prosopis* species in California. In leaf retainers, where one or several leaf cohorts may be produced in a season, the mean size of the canopy will depend on the difference between leaf production and leaf loss, and canopy efficiency will be highly dependent on changes of leaf efficiency with age and position in the tree. Where leaves are long-lived *and* remain reasonably photosynthetically efficient for some time, the production of new leaf cohorts in peripheral positions may shade older still-effective ones, although the leaves of such woody plants often show adaptation to shade (Cannell, 1989). Clumped foliage may enhance the photosynthetic capability of the canopy as a whole. Because they are usually well ventilated, the aerodynamic resistance of spaced trees to water loss will, in any case, be small and a major factor regulating water use in such a highly coupled situation will be stomatal sensitivity, and the period over which this is maintained by ageing leaves. The rate and time of production of successive leaf cohorts may well affect overall water use, therefore, with implications for dry matter production and, hence, transpiration efficiency (see Chapter 6).

Perhaps the most obvious way in which phenological events will have an impact is in determining the trees' ability to capture resources successfully. In tropical regions mean daily solar radiation fluxes are often highest in dry periods (e.g. because of sun elevations, as in East Africa, or cloudiness during the monsoons as in India), so that new, i.e. young photosynthetically efficient, leaves may often be produced outside the main rainy seasons, and in times when the saturation deficit is high. Indeed, this is commonly seen in African acacias and other dryland shrub species. Thus, where phenological displacements to less favourable environmental conditions occur, ε, the dry matter : intercepted radiation ratio, and e_{w}, the water use ratio, are likely to diminish in a way that is species characteristic (as Monteith *et al.* (1991) points out, the product of e_{w} and D, where D is the saturation vapour pressure deficit of the air, is relatively conservative for any one species/cultivar). Furthermore, the availability of soil water and nutrients, especially in relation to the use of these resources by associated crop plants, may vary considerably from season to season, as will the trees'

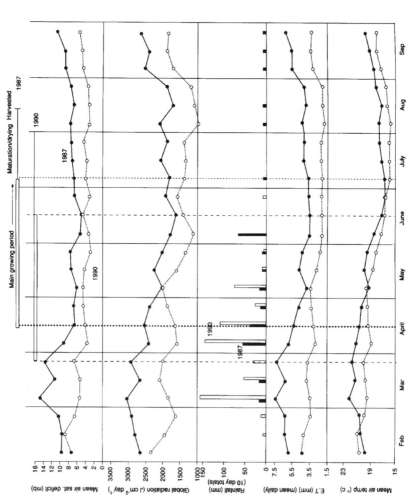

Fig. 7.23. Meteorological data for ICRAF's Experimental Station, Machakos, Kenya during and on either side of the normal 'first rains' cropping season for 1987 and 1990 which were, respectively, 'dry' and 'wet' years and showing the extent of the within- and between-year differences occurring. The precise effects of these on woody species that are maintaining an active canopy, or flowering or fruiting at times before or after the cropping season, and that can reach a source of soil water will be complex (see Chapters 4, 5 and 6); apparent benefits for e.g. dry matter accumulation such as high global radiation in February/March may be offset by disadvantageously high saturation deficit. Prolonging growth beyond the cropping season can incur a period of lower temperatures in some years, and so on. Maintaining the flexibility to be both 'opportunistic' and to extend the overall duration of growth are important attributes of many tropical woody perennials.

effects on loss of water from the soil surface, and topsoil temperatures and nutrient supplies. In such complex situations a formal analysis of growth may be difficult to achieve and growth and yield prediction over the short term impossible. Another consequence of different phenological patterns for woody and non-woody species is that they will not be exposed to exactly the same periods of environmental stress. Figure 7.23 shows the extent of differences in some climatic parameters in and on either side of the 'cropping season' for a wet and dry year at Machakos, Kenya.

The time of development and duration of the canopies of taller deciduous trees and shrubs will clearly affect the environment, and hence the resource capture of lower storey crops or grasses in many ways (see Chapter 4). They can also diminish water loss from the soil surface (e.g. Belsky *et al.*, 1989), which, especially in arid and seasonally arid regions, may represent a large fraction of rainfall (Chapter 6). The choice of leaf-retaining or deciduous trees for windbreaks (Chapter 5) will undoubtedly need careful thought so as to balance annual water use against provision of shelter at the times it is needed. Even though hedgerows have been shown not markedly to affect the microclimate nearby (at Hyderabad, India; Monteith *et al.*, 1991), under some circumstances small barriers can provide shelter that benefits the adjacent crop (Huxley *et al.*, 1994). Also, trees among crops or in grasslands do markedly affect the microclimate in their vicinity (e.g. Vandenbelt and Williams, 1992). Thus there will be many interactive effects as a result of microclimatic changes brought about by an upper storey canopy the growth of which, itself, may be subject to within- and between-season variations.

Because the joint canopies in most agroforestry practices are open and rough, and so are well ventilated, we might expect them to be atmospherically 'well coupled' (Jarvis, 1985). Thus, within the physical limits imposed on water loss from the system, plant characteristics (e.g. species characteristic stomatal dynamics) will also have a significant part to play, and this aspect becomes even more important when growth is occurring in less favourable conditions. Will the woody plants and the agricultural crops/ grasses behave differently in this respect? Of perhaps equal or greater significance will be differences in rooting characteristics between plant components that will affect their phenology. Hesla *et al.* (1985) investigated leaf conductances and dawn leaf water potentials of shrubs and grasses in two rangeland habitats in Kenya where predictable seasonal droughts occur (Nairobi National Park and the Masai Mara). Although during the wet seasons both plant types showed similar patterns of diurnal water relations, in this case shrub stomatal conductances and leaf water potentials were always higher than the comparative grass values. Unlike the grasses, three of the five shrub species studied showed progressive stomatal closure during the day, not due to increased water potential. Dawn water potentials for the grasses mirrored seasonal soil surface soil moisture changes, but the shrubs showed little or no change between wet and dry seasons (suggesting that their access to available soil water was not greatly affected by seasonal rainfall patterns). The greater use of water by these rangeland shrubs compared with the grasses was assumed to result in the larger and more extended dry matter production that was observed (but not measured). Interestingly, the grass species maintained near-constant stomatal conductances in both wet and dry seasons but three of the shrubs did not, suggesting different water use strategies. The authors suggest that the grasses competed strongly for water and did not reduce water loss to conserve limited soil water supplies, on the other hand the shrubs curbed water loss, keeping leaf water potentials from increasing and thereby extending activity into the dry seasons. The onset of drought affects growth processes more severely than it does photosynthesis. Leaf water potentials need to fall above 0.5 MPa before photosynthesis is seriously affected, but leaf expansion and protein synthesis are curtailed at < 0.3 MPa. So that short periods of water stress will tend to influence growth by limiting LAI.

Unlike the agricultural crops, trees and shrubs may often be tapping relatively deep ground water sources ('phreophytes'), or even be unable to grow successfully without them (and note the isolated trees or palms that one occasionally sees in sand desert, presumably growing on a perched water table). Clearly, the phenology of such species may well differ from that of plants that are rain-dependent. Nilsen *et al.* (1987) in a detailed study illustrated this for *Prosopis glandulosa* var. Torreyana grown in the Sonoran desert in California where the growth of most other plants is coupled to and dependent on rainfall availability. Rapid vegetative growth spurts occurred under extremely different climatic conditions mainly in the 'second' season each year and following a growth phase in each 'first' season (Fig. 7.13). Second-season leaves were retained for about 240 days compared with an average of 352 for those produced in the first season, and second-season leaves were smaller in area and weight. Shoot elongation was rapid and seasonally consistent and from newly developed laterals, by extension of the previous year's meristems. The maximum canopy size occurred from May through to October during the hottest and driest part of the year, and woody increment accrued over some 8 months of the year. Low minimum temperatures brought about leaf abscission (see Huxley and Van Eck, 1974; Fig. 7.24).

It has been suggested that the rapid synchronous emergence of off-season growth is an adaptation to counter herbivory (for example, *P. glandulosa* is attacked by a whole range of seed, pod and stem eating animals). Certainly, woody species differ greatly in both their susceptibility to and capacity to recover from herbivore attack (a subject that is too large to cover in this volume!). For example, in an investigation of 21 different tree species Coley (1982) found that rapidly growing pioneers were 'grazed' by insects four to ten times more in both wet and dry seasons than trees

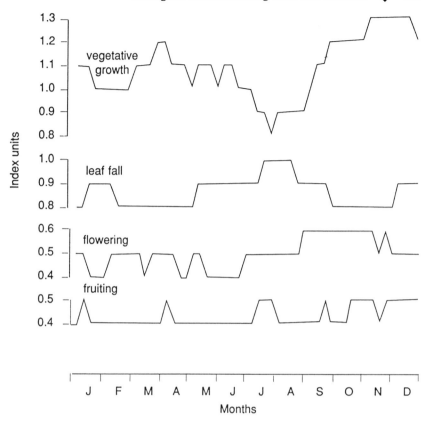

Fig. 7.24. Overall seasonal trends in vegetative growth, leaf fall, flowering and fruiting of the 38 woody specimens growing at Kabanyolo (as in Fig. 7.18) and averaged over 8 years. The first part of the year (January–February) is usually hot and dry, the first rains occur from late March to May followed by a cool cloudy period mid-year, with the second rains mainly in October–November. Source: Huxley and Van Eck, 1974.

with persistent foliage. Because quite commonly very large amounts of the canopy can be taken and/or leaf efficiency greatly reduced by herbivores of one kind or another (and some not very obvious, e.g. leaf miners), it is important to monitor this carefully, particularly when dealing with trials involving raised (i.e. unobserved) canopies and several different species.

Because tropical woody species are thought to be only rarely subjected to daylength controlled dormancies we are able to manipulate the phenophases of multipurpose tree species through practices that suitably modify the environment and/or by managing the tree. As with crop plants, providing irrigation and/or nitrogen fertilizer can prolong the activity of leafy shoots and delay flowering, whereas conditions that impose stress (drought, high temperatures) act conversely. Shade may be beneficial in this respect, mainly by alleviating stress (e.g. for water and nutrients). Direct plant management can be by practices that alter the carbohydrate source : sink relationships (e.g. fruit thinning) and/or the numbers of potentially active meristems (e.g. pruning roots or leafy shoots). The effects of pruning can be drastic, for example, in tea, because the removal of most stem apices at plucking time leaves the next harvest dependent upon the numbers and the rate of growth of the next set of lateral buds on the plucking table. Thus regular cycles of growth are set up (Fig. 7.25) as long as the environment favours growth and until internally induced growth regulation restricts the supply of buds available. As well as changing the period over which new leafy shoots will capture light, pruning in different seasons can affect the time at which new lateral shoots will start to grow out (Maggs, 1965; Akunda and Huxley, 1990), probably because the state of bud inhibition varies throughout the season.

Pruning is thus a powerful management tool for 'entraining' growth and flowering phases to periods at which they would otherwise not occur, as, to a lesser extent (and where it is applicable in agroforestry), is irrigation. Perhaps, also, other forms of plant management, e.g. leaf removal has been shown to delay flowering in *Piper arieianum* (Marquis, 1988), and by flower removal farmers on the East African coast alter the timing of fruit yields of mango (*Mangifera indica*). However, the adoption of an agroforestry practice may often depend on the availability of management resources, so that an increase in labour requirements or skills, as we have with pruning, may be counter effective.

Although some general principles can be outlined the precise outcome of environmental or managemental changes on phenology cannot always be predicted without prior investigation because they depend on complex interactions between growth regulatory processes, that is, growth allometry and source : sink relationships (e.g. root : shoot interactions, and the effects of changes in assimilate transfers due to the removal of leaves or of developing fruits), and hormonal regulation (e.g. the removal of apical dominance and the disturbance of hormone production from different plant parts). Thus there may be both short and long term phenological effects which will depend very much on species, and to some extent on the season at which any management operation is carried out. As an example, the time taken for any short term vegetative response to occur to pruning (in the form of a growth flush) will depend on the number and position of units responding, as well as their subsequent rate and duration of growth. Furthermore, because both the total annual duration of growing periods, and the potential they offer for growth, can vary year by year, a reappraisal of the 'growing season' is needed for agroforestry tree–crop mixtures in order to take into account the growth phases of the woody components. Unfortunately, it may not always be easy to predict, precisely, any expected

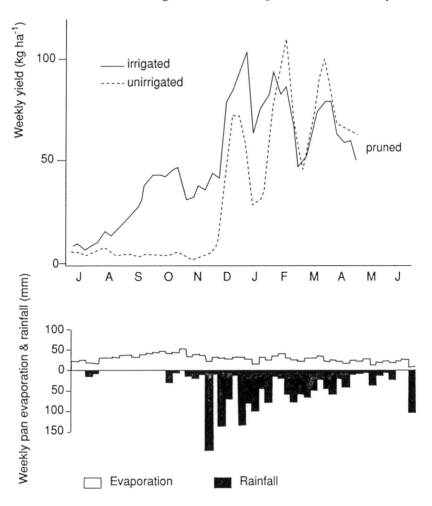

Fig. 7.25. Effects of ±dry season irrigation on the weekly yield of made tea (kg ha^{-1}) in Malawi. Although irrigation resulted in greater amounts of tea being picked during the dry season, successive yields for unirrigated tea were larger later. Removing one set of young shoots entrains the bush so that 'flushes' of new shoots occur. These will ultimately diminish in magnitude in tea, irrespective of environmental resource availability, because of the onset of bud dormancy. (Source: Dale, 1974.)

growing periods for all woody tropical species, especially in seasonally arid environments. Clearly, and especially for trees that shed their whole canopy of leaves, the timing of growth events is crucial, and the phenological patterns of the woody components will influence contiguous crop species.

Phenology has been a sadly overlooked aspect of agroforestry research, but the value of characterizing plant behaviour in order to explore and explain competitive and complementary interactions between plant components is clear. Phenological studies are now becoming more widely incorporated in experiments, at ICRAF's Machakos Experiment Station, for example, to help relate tree growth and competitiveness of multipurpose tree species such as *Melia volkensii*, which is fast-growing but deciduous in the crop grain-filling period, with *Grevillea robusta*, a slower-growing, leaf retaining species (Chin Ong, personal communication).

References

Akunda, E. and Huxley, P.A. (1990) *The Application of Phenology to Agroforestry Research*. ICRAF Working Paper No. 63. ICRAF, Nairobi, 50 pp.

Allison, J.C.S. and Watson, D.J. (1966) The production and distribution of dry matter in maize after flowering. *Annals of Botany N.S.* 30, 365–381.

Alvim, P. de T. (1960) Fisiologia del crecimiento y de la floración del cafeto. (The physiology of growth and flowering in coffee.) *Coffee* 2, 57–64.

Atkinson, D. (1980) Water use and the control of water stress in high density plantings. *Acta Horticulturae* 114, 445–456.

Baldocchi, D.D. and Hutchinson, B.A. (1986) On estimating canopy photosynthesis and stomatal conductance in a deciduous forest with clumped foliage. *Tree Physiology* 2, 155–168.

Barthélémy, D., Edelin, C. and Hallé, F. (1989a) Architectural concepts for tropical trees. In: Holme-Nielso, L.B. and Baslev, H. (eds) *Tropical Forests: Botanical Dynamics, Speciation and Diversity*. Academic Press, London, pp. 89–109.

Barthélémy, D., Edelin, C. and Hallé, F. (1989b) Some architectural aspects of tree ageing. In: Dreyer, E., Ausserac, G., Bonnet-Masimbert, M., Dizengrenel, P., Farre, J.M., Garrec, J.P., Le Tacon, F. and Marta, F. (eds) *Annales des Sciences Forestières* 46 Suppl. *Forest Tree Physiology*. Elsevier/INRA, Amsterdam, pp. 194s–198s.

Bell, A.D. (1991) *Plant Form: An Illustrated Guide to Flowering Plant Morphology*. Oxford University Press, Oxford, 341 pp.

Belsky, A.J., Amundsen, R.G., Duxbury, J.M., Riha, S.J., Ali, A.R. and Mwonga, S.M. (1989) The effects of trees on their physical, chemical and biological environments in a semi-arid savanna in Kenya. *Journal of Applied Ecology* 26, 1005–1024.

Boland, D.J., Pinyopusarerk, K., McDonald, M.W., Jovanovic, T. and Booth, T.H. (1990) The habitat of *Acacia auriculiformis* and probable factors associated with its distribution. *Journal of Tropical Forest Science* 3, 159–180.

Booth, T.H. (1991) Where in the world? New climatic analysis to assist species and provenance selection for trials. *Unasylva* 42, 51–57.

Booth, T.H., Nix, H.A. and Hutchinson, M.F. (1987) Grid matching: a new method for homocline analysis. *Agricultural and Forest Meteorology* 39, 241–255.

Borchert, R. (1978) Feedback control and age-related changes of shoot growth in seasonal and non-seasonal climates. In: Tomlinson, P.B. and Zimmerman, M.H. (eds) *Tropical Trees as Living Systems*. Cambridge University Press, Cambridge, pp. 497–515.

Bradshaw, A.D. (1965) Evolutionary significance of phenotypic plasticity in plants. *Advances in Genetics* 13, 115–151.

Browning, G. (1985) Reproductive behaviour of fruit tree crops and its implications for the manipulation of fruit set. In: Cannell, M.G.R. and Jackson, J.E. (eds) *Attributes of Trees as Crop Plants*. National Environmental Research Council/Institute of Terrestrial Ecology, Abbots Ripton, pp. 409–425.

Cannell, M.G.R. (1971a) Effects of fruiting, defoliation and ring-barking on the accumulation and distribution of dry matter in branches of *Coffea arabica* L. in Kenya. *Experimental Agriculture* 7, 63–64.

Cannell, M.G.R. (1971b) Production and distribution of dry matter in trees of *Coffea arabica* L. in Kenya as affected by seasonal climatic differences and the presence of fruits. *Annals of Applied Biology* 67, 99–120.

Cannell, M.G.R. (1983) Plant management in agroforestry. In: Huxley, P.A. (ed.) *Plant Research and Agroforestry*. ICRAF, Nairobi, pp. 455–487.

Cannell, M.G.R. (1985) Dry matter partitioning in tree crops. In: Cannell, M.G.R. and Jackson, J.E. (eds) *Attributes of Trees as Crop Plants*. National Environmental Research Council/Institute of Terrestrial Ecology, Abbots Ripton, pp. 160–193.

Cannell, M.G.R. (1989) Physiological basis of wood production: a review. *Scandinavian Journal of Forest Research* 4, 459–490.

Cannell, M.G.R. and Dewar, R.C. (1994) Carbon allocation in trees: a review of the concepts for modelling. *Advances in Ecological Research* 25, 59–104.

Cannell, M.G.R. and Huxley, P.A. (1969) Seasonal differences in the pattern of assimilate movement in branches of *Coffea arabica* L. *Annals of Applied Biology* 64, 345–357.

Cannell, M.G.R. and Morgan, J. (1988) Support costs of different branch designs: effects of position, number, angle and deflection of laterals. *Tree Physiology* 4, 219–231.

Chootummatat, V., Turner, D.W. and Cripps, J.E.L. (1990) Water use of plum trees (*Prunus salicina*) trained to four canopy arrangements. *Scientia Horticulturae* 43, 255–271.

Coley, P.D. (1982) Rates of herbivory on different tropical trees. In: Leigh Jr., E.G., Rand, R.S. and Windsor, D.M. (eds) *The Ecology of a Tropical Forest: Seasonal Rhythms and Long-Term Changes*. Smithsonian Institute Press, Washington, pp. 123–132.

Coley, P.D. (1988) Effects of plant growth rate and leaf lifetime on the amount of and type of anti-herbivore defence. *Oecologia* 74, 531–536.

Cooper, D. and Clarkson, D.T. (1989) Cycling of amino-nitrogen and other nutrients between roots and shoots in cereals – a possible mechanism integrating root and shoot in the regulation of nutrient uptake. *Journal of Experimental Botany* 40, 753–762.

Coughenour, M.B., Coppoch, D.C., Rowland, M. and Ellis, J.E. (1990a) Dwarf shrub ecology in Kenya's arid zone: *Indigofera spinosa* as a key forage resource. *Journal of Arid Environments* 18, 301–312.

Coughenour, M.B., Detling, J.K., Bamberg, I.E. and Mugambi, M.M. (1990b)

Production and nitrogen responses of the dwarf shrub *Indigofera spinosa* to defoliation and water limitation. *Oecologia* 83, 546–552.

Crick, J.C. and Grime, J.P. (1987) Morphological plasticity and mineral nutrient capture in two herbaceous species of contrasting ecology. *New Phytologist* 107, 403–414.

Dale, M.O. (1974) Irrigation of mature tea. *Tea Research Foundation of Central Africa, Quarterly Newsletter* 35, 10–15.

Dambrine, E.F., Martin, F., Carisey, N., Granier, A., Hallgren, J.E. and Bishop, K. (1995) Xylem sap composition: a tool for investigating mineral uptake and cycling in adult spruce. *Plant and Soil* 168/169, 233–241.

Daniel, J.N. and Ong, C.K. (1990) Perennial pigeonpea: a multipurpose species for agroforestry systems. *Agroforestry Systems* 10, 113–119.

Deans, J.D., Lundberg, C., Cannell, M.G.R., Murray, M.B. and Sheppard, L.J. (1990) Root system fibrosity of Sitka spruce transplants: relationship with root growth potential. *Forestry* 63, 1–7.

Dick, J.McP., Jarvis, P.G. and Barton, C.V.M. (1990) Influence of male and female cones on the assimilate production of *Pinus contorta* trees within a forest stand. *Tree Physiology* 7, 49–63.

Dickman, D.I. (1985) The ideotype concept applied to forest trees. In: Cannell, M.G.R. and Jackson, J.E. (eds) *The Attributes of Trees as Crop Plants*. National Environment Research Council/Institute of Terrestrial Ecology, Abbots Ripton, pp. 89–101.

Donald, C.M. (1968) The breeding of crop ideotypes. *Euphytica* 17, 385–403.

Dwyer, L.M. and Stewart, D.W. (1986) The effect of leaf age and position on net photosynthetic rates in maize. *Agricultural and Forest Meteorology* 37, 29–46.

Ewel, J., Benedict, F., Berish, C. and Brown, B. (1982) Leaf area, light transmission, root and leaf damage in nine tropical plant communities. *Agro-Ecosystems* 7, 305–326.

Fahey, T.J. (1992) Mycorrhizal and forest ecosystems. *Mycorrhiza* 1, 83–89.

FAO (1978) *Report to the Agroecological Project, Vol. 1. Methodology and Results for Africa*. FAO, Rome, 158 pp.

Fitter, A.H. (1991) The ecological significance of root system architecture: an economic approach. In: Atkinson, D. (ed.) *Plant Root Growth: An Ecological Perspective*. Blackwell Scientific, Oxford, pp. 229–243.

Floyd, S. (1993) The spatial distribution of roots. M.Sc. thesis, University of Reading, Department of Applied Statistics, 94 pp.

Frankie, G.W., Baker, H.G. and Opler, P.A. (1974) Comparative phenological studies of trees in tropical wet and dry forest in the lowlands of Costa Rica. *Journal of Ecology* 62, 881–919.

Gholz, H.C., Hendry, L.C. and Cropper, W.P., Jr. (1986) Organic matter dynamics of fine roots in a plantation of slash pine (*Pinus elliotti*) in northern Florida. *Canadian Journal of Forest Research* 16, 529–538.

Ghosh, S.P. (1973) Internal structure and photosynthetic activity of different leaves of apple *Journal of Horticultural Science* 48, 1–9.

Glover, N. and Beer, J. (1986) Nutrient cycling in two traditional Central American agroforestry systems. *Agroforestry Systems* 4, 77–87.

Grace, J.B. (1990) On the relationship between plant traits and competitive ability. In: Grace, J.B. and Tilman, D. (eds) *Perspectives in Plant Competition*. Academic

Press, San Diego, pp. 51–65.

Gregory, P.H. (1994) Resource capture by root networks. In: Monteith, J.L., Scott, R.K. and Unsworth, M.H. (eds) *Resource Capture by Crops*. Proceedings of the 52nd Easter School in Agricultural Science. Nottingham University Press, Loughborough, pp. 77–97.

Grime, J.P. (1979) *Plant Strategies and Vegetative Processes*. John Wiley, Chichester, 222 pp.

Grime, J.P., Crick, J.C. and Rincon, J.E. (1986) The ecological significance of plasticity. In: Jennings, D.H. and Trewavas, A.J. (eds) *Plasticity in Plants*. Proceedings of the Symposium of the Society of Experimental Biology No. 30. Symposia of the Society of Experimental Biology, Cambridge, pp. 5–29.

Grime, J.P., Campbell, B.D., Mackey, J.M.L. and Crick, J.C. (1991) Root plasticity, nitrogen capture and competitive ability. In: Atkinson, D. (ed.) *Plant Root Growth: An Ecological Perspective*. Blackwell Scientific, Oxford, pp. 381–397.

Hallé, F. (1986) Modular growth in seed plants. In: Harper, J.L., Rosen, B.R. and White, J. (eds) *The Growth and Form of Modular Organisms*. Royal Society, London, pp. 77–87.

Hallé, F., Oldeman, R.A.A. and Tomlinson, P.B. (1978) *Tropical Trees and Forests: An Architectural Analysis*. Springer-Verlag, Berlin, 441 pp.

Hansen, P. (1969) 14C studies in apple trees. IV. Photosynthate consumption in fruits in relation to the leaf–fruit ratio and the leaf–fruit position. *Physiologia Plantarum* 22, 186–198.

Harper, J.L. (1989) Canopies as populations. In: Russell, G., Marshall, B. and Jarvis, P.G. (eds) *Plant Canopies: Their Growth, Form and Function*. Cambridge University Press, Cambridge, pp. 107–128.

Harper, J.L., Jones, M. and Sackville-Hamilton, N.R. (1991) The evolution of roots and the problems of analyzing their behaviour. In: Atkinson, D. (ed.) *Plant Root Growth: An Ecological Perspective*. Blackwell Scientific, London, pp. 3–32.

Hawkins, R.C. and Cooper, P.J.M. (1981) Growth, development and grain yield of maize. *Experimental Agriculture* 17, 203–208.

Hesla, B.I., Tiessen, L. L. and Boutton, T.W. (1985) Seasonal water relations of savanna shrubs and grasses in Kenya, East Africa. *Journal of Arid Environments*, 8, 15–31.

Higgins, G.M., Kassam, A.H., Van Velthuizen, H.T. and Purnell, M.F. (1987) Methods used by FAO to estimate environmental resources, potential outputs of crops and population supporting components in developing nations. In: Bunting, A.H. (ed.) *Agricultural Environments: Characteristics, Classes and Mapping*. FAO/CGIAR/CAB, Wallingford, pp. 171–183.

Hill-Cottingham, D.G. (1968) The effect of climate and time of application of fertilisers on the development and crop performance of fruit trees. In: Hewitt, R.J. and Cutting, C.V. (eds) *Recent Aspects of Nitrogen Metabolism in Plants*. Academic Press, London, pp. 243–253.

Hollies, M.A. (1967) Chronic leaf fall in Arabica coffee in Tanzania. *East African Agricultural and Forestry Journal* 32, 404–410.

Howard, A. (1925) The effect of grass on trees. *Royal Society Proceedings B* 97, 284–321.

Hughes, C.E. (1994) Risks of species introductions in tropical forestry. *Commonwealth Forestry Review* 73, 243–252.

Hughes, C.E. and Styles, B.T. (1984) Exploration and seed collection of dry zone multipurpose trees from Central America. *International Tree Crops Journal* 1, 1–31.

Huxley, P.A. (ed.) (1969) *Proceedings of a Seminar on Intensification of Coffee Growing in Kenya.* Coffee Research Foundation, Ruiru, 231 pp.

Huxley, P.A. (1975) Potential new systems for perennial tropical fruit crops. In: Proceedings Third African Symposium Horticultural Crops, Nairobi, Sept. 18–21, 1973. *Acta Horticulturae* 49, 33–42.

Huxley, P.A. (1983) Phenology of tropical woody perennials and seasonal crop plants with reference to their management in agroforestry systems. In: Huxley, P.A. (ed.) *Plant Research and Agroforestry.* ICRAF, Nairobi, pp. 501–525.

Huxley, P.A. (1985) The basis of selection, management and evaluation of multipurpose trees: an overview. In: Cannell, M.G.R. and Jackson, J.E. (eds) *Attributes of Trees as Crop Plants.* National Environmental Research Council/ Institute of Terrestrial Ecology, Abbots Ripton, pp. 13–35.

Huxley, P.A. (1994) Roots systems of some tree species and implications for resource capture in agroforestry. In: Monteith, J.L., Scott, R.K. and Unsworth, M.H. (eds) *Resource Capture by Crops.* Proceedings of the 52nd Easter School in Agricultural Science. Nottingham University Press, Loughborough, pp. 409–410.

Huxley, P.A. (1996) Multipurpose trees: biological and ecological aspects relevant to their selection and use. In: Last, F. (ed.) *Tree Crop Ecosystems.* Elsevier, Amsterdam (in press).

Huxley, P.A. and Turk, A. (1975) Some preliminary observations on root growth of *Coffea arabica* in a root observation laboratory. *East African Agricultural and Forestry Journal* 40, 300–312.

Huxley, P.A. and Van Eck, W.A. (1974) Seasonal changes in growth and development of some woody perennials near Kampala. *Journal of Ecology* 62, 579–592.

Huxley, P.A., Pinney, A. and Gatamah, D. (1989) *Final Report on Development of Agroforestry Research Methodology Aimed at Simplifying the Study of Potential Tree/ Crop Mixtures.* (Tree/Crop Interface Project No. 1-432-6005613 for BMZ/ GTZ.) (Limited circulation) ICRAF, Nairobi, 67 pp. + 5 appendices.

Huxley, P.A., Pinney, A., Akunda, E. and Muraya, P. (1994) A tree/crop interface orientation experiment with a *Grevillea robusta* hedgerow and maize. *Agroforestry Systems* 26, 23–45.

ICRAF (1994) Programme 2: Multipurpose Tree Improvement and Management. In: *Annual Report of the International Center for Research in Agroforestry for 1993.* ICRAF, Nairobi, pp. 42–61.

Jackson, J.E. (1980) Light interception and utilisation by orchard systems. *Horticultural Reviews* 2, 208–267.

Jackson, J.E. and Palmer, J.W. (1977) Effects of shade on the growth and cropping of apple trees. II, Effects of components of yield. *Journal of Horticultural Science* 52, 253–266.

Jacobs, W.P. (1979) *Plant Hormones and Plant Development.* Cambridge University Press, Cambridge, 339 pp.

Jarvis, P.G. (1985) Transpiration and assimilation of tree and agricultural crops: the omega factor. In: Cannell, M.G.R. and Jackson, J.E. (eds) *Attributes of Trees as*

Crop Plants. National Environment Research Council/Institute of Terrestrial Ecology, Abbots Ripton, pp. 460–480.

Jarvis, P.G. and Sandford, A.P. (1986) Temperate forests. In: Baker, N.R. and Long, S.P. (eds) *Photosynthesis in Contrasting Environments*. Elsevier, Amsterdam.

Jones, O.P. (1971) Effects of rootstocks and interstocks on the xylem sap composition in apple trees: effects on nitrogen, phosphorus and potassium content. *Annals of Botany* 35, 825–836.

Joseph-Wright, S. and Cornejo, F.H. (1990) Seasonal drought and the timing of flowering and leaf fall in a neotropical forest. In: Bawa, K.S. and Hadley, M. (eds) *Reproductive Ecology of Tropical Forest Plants*. Man and the Biosphere Series, No. 7. UNESCO/Parthenon, Paris, pp. 49–61.

Keddy, P.A. (1990) Competitive hierarchies and centrifugal organisation in plant communities. In: Grace, J.B. and Tilman, D. (eds) *Perspectives in Plant Competition*. Academic Press, San Diego, pp. 265–290.

Khan, A.A. and Sagar, G.R. (1969) Changing patterns of distribution of the products of photosynthesis in the tomato plant with respect to the time and to the age of leaf. *Annals of Botany* 33, 763–779.

King, D.A. (1993) A model analysis of the influence of root and foliage allocation on forest production and competition between trees. *Tree Physiology* 12, 119–135.

Koslowski, T.T. (1971) Control of shoot growth. In: *Growth and Development of Trees*. Academic Press, New York, pp. 296–386.

Leakey, R.B., Mesen, F., Shiembo, P.N., Ofori, D., Nketiah, T., Hamzah, A., Tchoundjeu, Z., Njoya, C., Odoul, P., Newton, A.C., Dick, J.McP. and Longman, A. (1992) *Low Technology Propagation of Tropical Trees: Appropriate Technology for Rural Development*. Institute of Terrestrial Ecology, Edinburgh.

Ledig, F.T. (1983) The influence of genotype and environment on the dry matter distribution in plants. In: Huxley, P.A. (ed.) *Plant Research and Agroforestry*. ICRAF, Nairobi, pp. 427–454.

Leverenz, J., Deans, J.D., Ford, E.D., Jarvis, P.G., Milne, R. and Whitehead, D. (1982) Systematic spatial variation of stomatal conductance in a Sitka spruce plantation. *Journal of Applied Ecology* 19, 835–851.

Linder, S. and Axelsson, B. (1982) Changes in carbon uptake and allocation patterns as a result of irrigation and fertiliser in a young *Pinus sylvestris* stand. In: Waring, R.H. (ed), *Carbon Uptake and Allocation in Sub-alpine Ecosystems as a Key to Management*. Oregon State University Forestry Research Laboratories, Corvallis, pp. 38–44.

Longman, K.A. (1978) Control of shoot extension and dormancy: external and ianternal factors. In: P.B. and Zimmerman, M.H. (eds) *Tropical Trees as Living Systems*. Cambridge University Press, Cambridge, pp. 465–495.

Longman, K.A. and Jenik, J. (1974) *Tropical Forest and its Environment*. Longman, London, 400 pp.

Loomis, R.S. and Connor, D.J. (1992) *Crop Ecology: Productivity and Management in Agricultural Systems*. Cambridge University Press, Cambridge, 538 pp.

Luckwill, L.C. (1970) The control of growth and fruitfulness in apple trees. In: Luckwill, L.C. and Cutting, C.V. (eds) *Physiology of Tree Crops, 2nd Long Ashton Symposium 1969*. Academic Press; London, pp. 237–254.

McKay, H. and Coutts, M.P. (1989) Limitations placed on forestry production by the root system. In: *Roots and the Soil Environment*. Aspects of Applied Biology, No. 22. pp. 245–254.

Maggs, D.H. (1960) The effect of number of shoots on the quantity and distribution of increment in young apple-trees. *Annals of Botany N.S.*, 24, 345–355.

Maggs, D.H. (1965) Dormant and summer pruning compared by pruning young apple trees once on a succession of dates. *Journal of Horticultural Science* 40, 249–265.

Maggs, D.H. and Alexander, D.McE. (1967) A topophysic relation between regrowth and pruning in *Eucalyptus cladocalyx* F. Muell. *Australian Journal of Botany* 15, 1–9.

Marquis, R.J. (1988) Phenological variation in the neo-tropical understorey shrub *Piper arieianum*: causes and consequences. *Ecology* 69, 1552–1565.

Monteith, J.L. (1991) Steps in crop climatology. In: Unger, P.W., Sneed, T.V., Jordan, W.R. and Jensen, R. (eds) *Challenges in Dryland Agriculture: A Global Perspective*. Conference on Dryland Farming. Texas Agricultural Experimental Station, Amarillo, pp. 273–282.

Monteith, J.L., Ong, C.K. and Corlett, J.E. (1991) Microclimatic interactions in agroforestry systems. *Forest Ecology and Management* 45, 31–44.

Moss, G.I., Steer, B.T. and Kriedermann, P.E. (1972) The regulatory role of inflorescence leaves in fruit setting by sweet orange (*Citrus sinensis*). *Physiologia Plantarum* 27, 432–438.

Nair, P.K.R. (ed.) (1989) *Agroforestry Systems in the Tropics*. Kluwer Academic/ICRAF, Dordrecht/Nairobi, 664 pp.

Neales, T.F. and Incoll, L.D. (1968) The control of leaf photosynthetic rate by the level of assimilate concentration in the leaf: a review of the hypothesis. *Botanical Review* 34, 107–125.

Newstrom, L.E., Frankie, G.W. and Baker, H.G. (1991) Survey of the long term flowering patterns in lowland tropical rainforest trees at La Selva, Costa Rica. In: Edelin, C. (ed.) *L'Arbre: Biologie et Développement, Naturalia Monspeliensia, Series A7*. Institut de Botanique, Montpellier, pp. 345–366.

Nilsen, E.T., Sharifi, M.R., Virginia, R.A. and Rundel, P.W. (1987) Phenology of warm desert phreophytes: seasonal growth and herbivory in *Prosopis glandulosa* var. Torreyana (honey mesquite). *Journal of Arid Environments* 13, 217–229.

Nilsen, E.T., Sharifi, M.R. and Rundel, P.W. (1991) Quantitative phenology of warm desert legumes: seasonal growth of six *Prosopis* species at the same site. *Journal of Arid Environments* 20, 299–311.

Oldeman, R.A.A. and van Dijk, J. (1992) Diagnosis of the temperament of tropical rainforest trees. In: Gòmez-Pompa, A., Whitmore, T.C. and Handley, M. (eds) *Rainforest Regeneration and Management*. Man in the Biosphere Series No. 6. UNESCO/Parthenon, Paris, pp. 21–65.

Ong, C. and Monteith, J.L. (1993) Canopy establishment: light capture and loss by crop canopies. In: Baker, N.R. (ed.) *Crop Photosynthesis: Spatial and Canopy Determinants*. Elsevier, Amsterdam, pp. 1–10.

Persson, A. (1983) The distribution and productivity of fine roots in boreal forests. *Plant and Soil* 71, 87–101.

Priestley, C.A. (1969) Some aspects of the physiology of apple rootstock varieties under reduced illumination. *Annals of Botany* 33, 967–980.

Priestley, C.A. (1970) Carbohydrate storage and utilisation. In: Luckwill, L.C. and Cutting, C.V. (eds) *Physiology of Tree Crops* pp. 113–127.

Priestley, C.A. (1972) The responses of young apple trees to supplementary nitrogen and their relation to carbohydrate resources. *Annals of Botany N.S.* 36, 513–524.

Reich, P.B. and Borchert, R. (1988) Changes with leaf age in stomatal function and water status of several tropical tree species. *Biotropica* 20, 60–69.

Reynolds, E.R.C. (1970) Root distribution and the cause of spatial variability in *Pseudotsuga taxifolia* (Poir.) Britt. *Plant and Soil* 32, 501–507.

Richards, P.W. (1952) *The Tropical Rainforest: An Ecological Study.* Oxford University Press, Oxford, 450 pp.

Roberts, J., Cabral, Osvaldo, M.R. and de Aguiar, L.F. (1990) Stomatal and boundary layer conductances in Amazonian Terra Firme forest. *Journal of Applied Ecology* 27, 336–353.

Rogers, W.S. (1969) The East Malling root laboratories. In: Whittington, W.J. (ed.) *Root Growth.* Butterworths, London, pp. 361–376.

Schoettle, A.W. (1990) The interactions between leaf longevity, shoot growth and foliar biomass per shoot in *Pinus contorta* at two elevations. *Tree Physiology* 7, 209–214.

Schoettle, A. and Fahey, T.J. (1994) Foliage and fine root longevity in pines. *Ecological Bulletin* 43, 136–153.

Squire, G.R. (1990) *The Physiology of Tropical Crop Production.* CAB International, Wallingford, 236 pp.

Stebbins, G.L. (1950) *Variation and Evolution in Plants.* Columbia University Press, New York.

Stephens, W. and Carr, M.K.V. (1990) Seasonal and clonal differences in shoot extension rates and numbers in tea (*Camellia sinensis*). *Experimental Agriculture* 26, 83–98.

Stewart, J.L., Dunsden, A.J., Hellin, J.J. and Hughes, C.E. (1992) *Wood Biomass Estimation of Central American Dry Zone Species.* Oxford Forestry Institute, Oxford, 83 pp.

Sutherland, W.J. (1990) The response of plants to patchy environments. In: Shorrocks, B. and Swingland, I.R. (eds) *Living in a Patchy Environment.* Oxford University Press, Oxford, pp. 45–61.

Thomas, H. (1987) Foliar senescence mutants and other genetic variants. In: Thomas, H. and Grierson, D. (eds) *Developmental Mutants: Higher Plants.* Cambridge University Press, Cambridge, pp. 245–265.

Thomas, H. (1992) Canopy survival. In: Baker, N.R. and Thomas, H. (eds) *Topics in Photosynthesis and Crop Photosynthesis: Spatial and Temporal Determinants.* Elsevier, Amsterdam, pp. 11–41.

Tomlinson, P.B. (1978) Branching and axis differentiation in tropical trees. In: Tomlinson, P.B. and Zimmerman, M.H. (eds), *Tropical Trees as Living Systems.* Cambridge University Press, Cambridge, pp. 187–207.

Turnbull, J.W. and Griffin, A.R. (1985) The concept of provenance and its relationship to intraspecific classification in forest trees. In: Styles, B. (ed.) *Intraspecific Classification of Wild and Cultivated Plants.* Oxford University Press, Oxford, pp. 157–189.

Vandenbelt, R.J. and Williams, J.H. (1992) The effect of soil surface temperature on the growth of millet in relation to the effect of *Faidherbia albida* trees. *Agriculture and Forest Meteorology* 60, 93–100.

Van Rooyen, M.W., Grobbelaar, N., Theron, G.K. aand Van Rooyen, N. (1992) The ephemerals of Namaqualand: effect of germination date on development of three species. *Journal of Arid Environments* 22, 51–66.

Vogt, K.A., Grier, C.C. and Vogt, D.J. (1986) Production, turnover and nutrient dynamics of above- and below-ground detritus of world forests. *Advances in Ecological Research* 15, 303–377.

Walter, H. (1971) *Ecology of Tropical and Subtropical Vegetation*. Oliver and Boyd, Edinburgh, 539 pp.

Wareing, P.F. (1959) Problems of juvenility and flowering in trees. *Journal of the Linnean Society of London, Botany* 56, 282–289.

Wareing, P.F. (1987) Phase change and vegetative propagation. In: Abbott, A.J. and Atkin, R.K. (eds) *Improving Vegetatively Propagated Crops*. Academic Press, London, pp. 263–270.

Watson, D.J. (1958) The dependence of net assimilation rate on leaf area index. *Annals of Botany N.S.* 22, 37–54.

Wolfe, D.W., Henderson, D.W., Hsiao, T.C. and Alvino, A. (1988) Interactive water and nitrogen effects on senescence of maize: Photosynthetic decline and longevity of individual leaves. *Agronomy Journal* 80, 865–870.

8

◆

Tree–Soil–Crop Interactions on Slopes

D.P. GARRITY

ICRAF, SE Asia Regional Program, PO Box 161, Bogor, Indonesia

Tree–crop interactions are fundamentally altered on slopes by the dramatic soil spatial changes that occur as terraces develop behind contour vegetative barriers. Tree–crop interactions on flat land are complex phenomena in their own right. But on sloping land the introduction of trees (or other vegetation) in contour hedgerows adds a new dimension. In time the hedgerows result in dramatic spatial changes in the properties of the soils of the entire system.

Hedgerow systems alter the length of slope, degree of slope, soil profile depth, and soil chemical and hydrological properties. These changes may be quite drastic over short distances. This review emphasizes 'tree–*soil*–crop' interactions, for it is the major soil spatial changes that transform the nature of the species interactions on slopes.

This chapter reviews current understanding of these soil changes and how they interact with tree–crop competition in production systems. First, the conventional row-wise pattern of alley-crop response on flat land is contrasted with the patterns observed in contour hedgerow systems on slopes. Second, the factors underlying the uniquely skewed response function on slopes are discussed in terms of a tillage-induced 'scouring' effect, with emphasis on the spatial soil fertility changes that occur. The flat-land row-wise crop response pattern across the alleyway (convex parabolic) shifts on slopes to a skewed response surface with (often) drastically lower yields in the upper alley, and maximum yields in the lower-middle alley zone. The process of natural terracing behind vegetative barriers is conceptualized as the creation of a series of microlandscapes. Aggregate and local effects of vegetative barriers on soil loss are examined. Third, a model integrating the two major processes, species competition and soil scouring, is presented. The chapter concludes with a discussion of the serious sustainability implications of these processes, and the urgent research needed to address them.

Species Interactions on Slopes

On slopes, as well as on flat land, a common row-wise pattern of crop response across the alleyways of an alley-cropping system during the early years is a convex parabola. This is true whether the hedgerow is a pruned leguminous tree or a grass (Fig. 8.1). The progressive decline in crop productivity as rows approach the hedgerow is attributed to a combination of above-ground and below-ground competition. The dome-shaped spatial response pattern is characteristic for upland rice (Salazar *et al.*, 1993). It is less frequently observed in the case of maize, which often has a more neutral response, with less evidence of the hedgerow-proximity yield depression observed by rice (Garrity *et al.*, 1995).

A third pattern, that of a concave parabola, has been observed in limited cases (e.g. dry season cowpeas in Samzussaman, 1994). This result implies a stimulatory effect of proximity to the hedgerow (perhaps a drought-mitigating microclimate effect). Thus, care is necessary in generalizing about the expected pattern of crop spatial response to hedgerows. Similarly there are major differences among hedgerow species

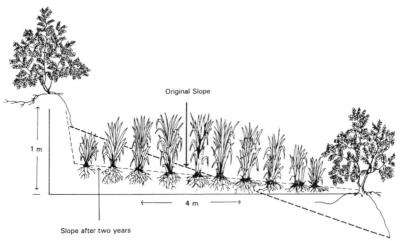

Fig. 8.1. Terrace formation and crop growth in a contour hedgerows system of upland rice and leguminous trees on a strongly acid oxisol. (Adapted from Basri *et al.*, 1990.)

in their tendency to reduce crop yields in adjoining alley zones (Garrity *et al.*, 1995). Also, the patterns may change as the trees age.

Because agroforestry is very often practised on sloping lands, a key issue is how slope may alter these relationships. First, there are several ways in which slope may *directly* impinge upon and alter tree–crop interactions. Three of these phenomena will be discussed briefly, before we turn to the main issue of how within-alley soil redistribution changes the system.

Competition for light

On flat land, intercrops of dissimilar height are ideally laid out in a north–south row direction (i.e. perpendicular to the solar path). This maximizes direct sunlight penetration down the row. The principle applies equally on slopes, but if contour planting is practised the control of row direction is rarely possible. Indeed, the aspect of the land (i.e. the direction the slope faces) is often highly variable. Slopes that either face east or west, with the perennial component laid out in contour lines, will tend to optimize sunlight penetration to the crop because the rows will be more-or-less in a north–south pattern, the configuration being analogous to north–south planting on flat land. With slopes facing away from the sun (e.g. north-facing fields in the northern hemisphere) light penetration to the dominated species is most severely reduced.

Tree root growth

This may be asymmetrical on slopes. Solera (1993) found that the density of roots was considerably higher in the soil upslope above 3-year-old pruned hedges of *Senna spectabilis* than in the soil downslope below the hedges. This was particularly true for the density of fine roots. This study was done on a site with approximately 20% slope in which a 50-cm high terrace riser had formed.

Field hydrology

This is directly affected by the presence of hedgerows. Kiepe (1995) observed that effective steady infiltration rates within the hedgerow were three to seven times greater than in the adjoining alleyways. Agus (1993) also observed a major contrast in infiltration rates between the upper alley and the hedgerows. Clearly, hedgerows exhibit the potential to capture and store in the profile much more water than occurs in their absence, particularly in climates with serious water deficits. This advantage is countered by the tendency of the perennial root systems to explore and exploit a large soil volume, in potential competition with a shallower-rooted annual crop

(Hauser, 1993). When terraces develop, further major changes in field hydrology are induced.

Skewed Distribution of Crop Yields in Sloping Alleys

There have been few long term observations of crop yields in contour hedgerow systems on slopes. But where these are reported, it is commonly observed that the yield response tends to become skewed after a few years: a distinct pattern of low yields on the upper rows of the alley and higher yields on the lower rows. Solera (1993) found that, after 5 years of upland rice cropping between *Gliricidia sepium* hedgerows on a typic hapludox with 20% slope in the Philippines, yields on the upper rows were 50% lower than those on the lower rows within the alley (Fig. 8.2). Without added phosphorus fertilizer the differential between the upper and lower alley zones was even more pronounced. A similar response was recorded with *Senna*

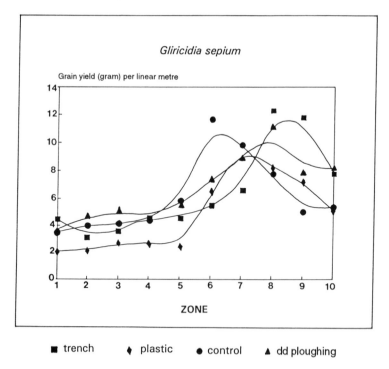

Fig. 8.2. Yield response of upland rice across the alley between hedgerows of *Gliricidia sepium* with and without root barriers (plastic), trenching or double-depth ploughing (dd ploughing), Claveria, Misamis Oriental, 1990. (Solera, 1993.)

spectabilis and *Pennisetum purpureum* (Napier grass) hedgerows. At a nearby location Agus (1993) obtained a yield depression in the upper alley rows with both rice and maize associated with *Gliricidia* after 4 years.

In 5-year-old hedgerow systems of either leguminous trees (*Leucaena leucocephala + Cajanus cajan*) or grass (*Paspalum conjugatum*) on an ustic kandihumult of 21–35% slope in Thailand, Turkelboom *et al.* (1993) reported reductions in rice yields of greater than 50% in the upper alley zones, compared with the middle and lower areas. Peden (personal communication, 1993) has observed a comparable yield depression in hedgerow trials in Uganda.

The pattern of response that is typically found does not indicate that the phenomenon can be explained as a manifestation of competition by the hedgerow component. Solera (1993) excluded both above-ground and below-ground hedgerow competition by intensive pruning and installation of 50-cm deep plastic barriers installed 30 cm from the outer of the hedgerows. He found that the spatial pattern of yields was unchanged (Fig. 8.2). Ramiaramanana (1993) and Turkelboom *et al.* (1993) investigated this issue with contour hedgerows composed of uncompetitive low-biomass producing grass strips. They could not account for the large yield depressions they observed in the upper rows of the alley compared with the lower rows as a competition effect. Clearly, processes other than hedgerow–crop competition were operating.

Contour Hedgerows: The Conventional Paradigm

The early literature on contour hedgerow systems extolled the advantages of vegetative barriers in reducing soil erosion on sloping fields (Young, 1989). The natural development of terraces was observed. This was highlighted as evidence that soil loss was being effectively controlled.

Figure 8.3 shows a widely held but erroneous view of a model of terrace development occurring with pruned tree hedgerows. Similar schematic drawings of hedgerow systems are found in the agroforestry literature. Note the emphasis on the buildup of soil behind the barrier. In this scheme the 'new' soil layer is deepest at the hedge, and overlies the old surface to a progressively thinner extent upslope. But there is no indication of soil redistribution within the alley from the upper part to the lower part. The process of within-alley soil redistribution and its serious implications were not appreciated until quite recently.

Soil Spatial Fertility Gradients

Soil may be redistributed across the alley quite rapidly after hedgerow installation on moderate to steep slopes. Basri *et al.* (1990) reported a 60-cm

The silvicultural challenge of inhospitable sites

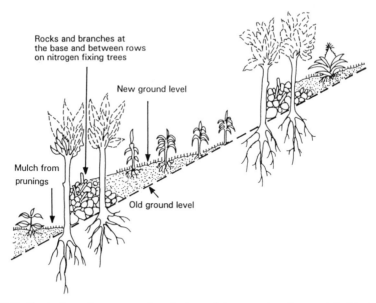

Fig. 8.3. Illustration of a conventional view of terrace development behind vegetative barriers as the accumulation in 'new' soil, without soil redistribution from the upper alley zones to the lower zones. Recent studies show this model to be unrealistic. (Source: MBRLC, 1988.)

drop in soil level between the alleys after 2.5 years on a field with 25% slope. Fujisaka (1992) observed similar rates of terrace development on several farmers' fields in Claveria, Philippines. A tendency toward rapid terrace riser development was frequently observed in trials on a range of other soil types (Sajjapongse, 1992).

Rapid terrace development was associated with a reduction in crop yield in the upper alley zone in many cases (Agus, 1993; Solera, 1993; Garrity *et al.*, 1995). Suspecting that soil fertility was a factor in producing the skewed grain yields, researchers began to monitor more carefully the pattern of soil properties across the alleyways. In a *Senna spectabilis* hedgerow experiment (Garrity *et al.*, 1995), soil organic carbon was found to vary from 1.7% near the hedgerow in the upper part of the alley to 2.8% near the lower hedgerow. Available P was twice as high in the lower zone compared with the upper. Soil pH was unchanged but exchangeable aluminium increased. These patterns were observed in the subsoil (15–30 cm) as well as in the surface soil, and occurred within 4 years of hedgerow establishment.

Agus (1993) sampled five points across the alleyways between *Gliricidia* contour hedgerows, and observed a quite linear soil fertility gradient. Exchangeable calcium, organic C, pH, and Bray-2 extractable P increased from the upper to the lower part of the alleyway, exchangeable Al decreased, while K and Mg were unchanged. Samzussaman's (1994) row-wise study of soil properties elucidated linear increases in organic carbon (from 2% to 3%) and nitrogen (0.20% to 0.27%) on an oxic palehumult (Fig. 8.4). Turkelboom *et al.* (1993) monitored soil organic carbon on an entire slope transect. They observed a 'saw-tooth' pattern, with a tendency for SOM differences across the alleys to accentuate on the lower slope terraces (Fig. 8.5).

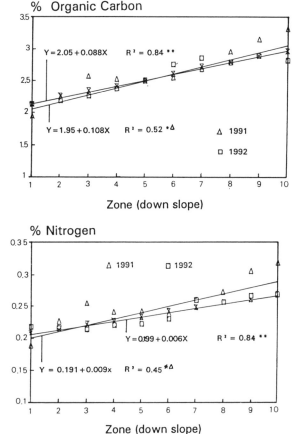

Fig. 8.4. Variation in soil organic carbon and nitrogen row by row across the alleyways in a contour hedgerow system with *Senna* (*Cassia*) *spectabilis* and upland rice on an oxic palehumult (Samzussaman, 1994).

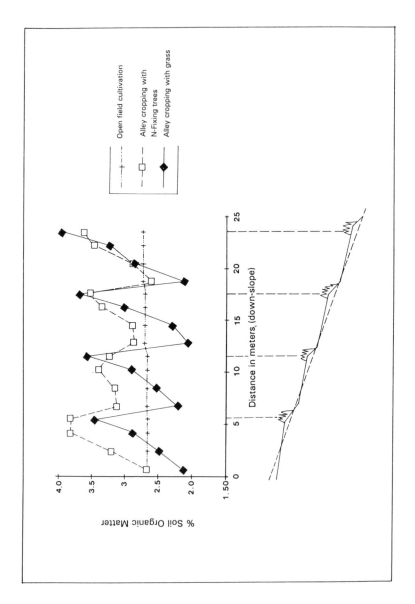

Fig. 8.5. The variation in soil organic matter gradient within alleyways and down the landscape in three cropping treatments (source: Turkelboom et al., 1993).

Why does such fertility accumulation occur towards the barrier strips, and why does it happen so rapidly?

Tillage Erosion Dominates Water-induced Soil Transport

Terracing tends to reduce the slope length (a primary effect) and slope angle (a secondary effect). This reduces the energy of water to move soil particles downslope. The upper alleyway zone nearest the hedgerow is the area that theory predicts would have the least expectation of soil loss by water erosion (separate direct splash impact from surface flow). Yet this is where maximum soil removal actually occurs. Water gains kinetic energy by flow downhill. Water in the upper alley zone would therefore tend to have picked up least energy. Runoff that may have passed through the bund would have presumably lost much of its energy and deposited soil particles in the hedgerow before emerging in the upper alley zone below.

The vegetative barrier above an alley ensures that very little enrichment of soil can occur from above. Nevertheless, it is in the upper alley zone where maximum soil losses occur. The enormous rates of profile scouring observed in many hedgerow systems led to the presumption that tillage transport must account for the overwhelming amount of soil movement (Turkelboom *et al.*, 1993; Garrity *et al.*, 1995).

Upper alley scouring and rapid soil displacement to the lower alley seem particularly prominent where draught animal plough systems are used. The plough efficiently turns soil downhill, and makes frequent tillage practicable. Philippine hedgerow farmers normally plough four to six times per year for two successive annual crops. The rate and type of scouring appears to be related to the way in which ploughing operations are done. This is not usually done with the express objective of accelerating the flattening of terraces. Rather, it is a by-product of 'normal' soil management operations, although some farmers have been known to increase ploughing frequency to flatten their terraces faster (A. Mercado, 1993, personal communication).

In Claveria, farmers conduct ploughing operations in the alleyways following three different patterns. Each has distinct effects on the pattern of soil redistribution. The scouring process also occurs with hand hoe farming in Thailand on steep slopes (Turkelboom *et al.*, 1993). However, scouring is not always observed. It is conspicuously absent when no-tillage is practised continuously, or when hand-hoeing is done superficially (as in shifting cultivation).

The Gap in Erosion Modelling

Much work has been done to develop models of water-induced erosion, but these have little utility in situations in which tillage transport is the dominant process. The modelling of soil erosion in contour hedgerow systems should be reoriented to give more attention to tillage processes: a fresh challenge to researchers.

In such modelling efforts it will be important to separate the water and tillage effects on natural terrace formation. Collecting data for this will require innovative approaches, because soil movement will have to be investigated over small spatial areas within the alleys. The mesh-bag method proposed by Hsieh (1992) may be useful in examining water-induced transport over such small areas, and for mapping the degradation/ aggradation at the soil surface over the face of the alleyway. Another approach is measuring soil loss from erosion experiments with collectors embedded at the lower side of a single alleyway (isolated from run-on water by physical barriers) compared with losses from an entire slope of hedgerows. Monitoring tillage-induced microshifts in surface soil can also be attempted by uniformly embedding small pellets in the soil and later examining their distribution (e.g. with a metal detector).

Landscapes, Hedgerows and Erosion: A Reappraisal

There is widespread evidence that vegetative barriers composed of trees, grasses or other species tend to reduce aggregate soil loss from sloping agricultural fields. Agus (1993) observed a 70–80% decline in soil loss with hedgerows of Gliricidia + Napier grass with rainfall of 1200–1500 mm; over 90% of this was bedload sediment, the rest was suspended. On a range of sites in Indonesia, Philippines, Thailand and Vietnam, IBSRAM researchers found that contour hedgerows reduced soil loss 49% to 89% compared with conventional farming, which had incurred erosion levels ranging from 5 to 413 t ha^{-1} in 3 years (Sajjapongse, 1992).

Results of this nature have impressed scientists and policy-makers and contour hedgerows are a superb way to control accelerated erosion on sloping agricultural fields (Young, 1989; Garrity, 1994). Thus, it is puzzling to discover that, in many cases, the annual crop yields in the alleys do not show any advantage compared with open-field results, even on a per-planted area basis, i.e. ignoring the loss of land to the hedgerows (Ramiaramanana, 1993; Turkelboom et al., 1993). This is despite the fact that soil erosion is dramatically reduced from rates that would seem unsustainable (100–300 t ha^{-1} year^{-1}), and substantial amounts of biomass may be added to improve soil fertility. This occurs even on fields with relatively non-competitive grass strips (e.g. bahia grass) or natural vegetative strips (Turkelboom *et al.*, 1993; ICRAF, 1994).

Why does effective soil conservation seem to have such apparently small effects on crop performance? The explanation may lie in the nature of erosion on open slopes. Recent work that monitored patterns of soil movement across entire slopes and watersheds has confirmed new insights on the patterns of soil loss in a landscape (Brown *et al.*, 1991a, b; Busacca *et al.*, 1993). These workers found that soil loss was not clearly or simply related to the slope gradient or elevation on the slope.

In a landscape, most of the soil loss occurs only on the upper summit positions of slopes, and on the tops of 'knobs' (see Fig. 8.6). The great majority of the area, located in mid-slope positions, experiences little or no net soil loss, although the slope gradient in those landscape positions is often the steepest (Fig. 8.6). Surprisingly, at this site net erosion was confined to just two areas, a small portion of the landscape.

On a generalized slope model such as that in Fig. 8.7, the 'shoulder' area exhibits maximum erosion. The footslope may receive substantial deposition. But little or no net soil loss may occur on the upper linear and lower linear landscape positions, although they have much the greatest slope angle.

The implication for contour hedgerow systems is profound: hedgerow experiments are most often laid out on the linear slope positions, and results are compared with adjacent open field treatments on the same slope positions. Although the open field treatments may exhibit very high total soil loss, little net loss may actually be occurring on the mid-slope positions. Losses are coming from above, and being deposited below. Monitoring of [137]caesium, which was deposited naturally as radioactive fallout, shows that

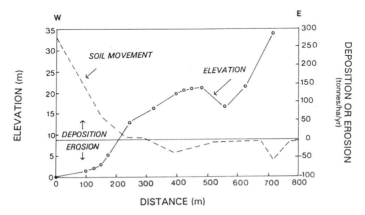

Fig. 8.6. Comparison of relative elevation and rates of soil movement estimated from ^{137}Cs tracer along an east–west transect across a study watershed in the Palouse region of Washington State, USA (from Busacca *et al.*, 1993).

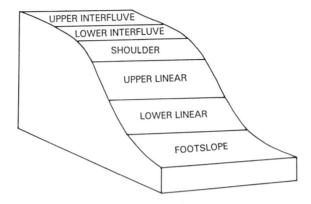

Fig. 8.7. Generalized physiographic representation of landscape positions (after Jones *et al.*, 1989).

in reality nearly every ton of soil loss at the mid-slope is replaced by soil from above (Busacca *et al.*, 1993).

This net loss applies to soil movement due to tillage as well as to water erosion. Tillage gradually removes the topsoil from the ridges, knobs and shoulders of a landscape. Soil from all points tends to slump downward on disturbance, but these are the only areas that do not receive a replenishment of soil from above.

The Alleyway as a Microlandscape

In contour hedgerow trials, it is the mid-slope areas on adjacent open fields that are often compared with the vegetative barrier treatments. We may now suspect that the open field mid-slope may, in many case, not be experiencing a net soil loss, even though the landscape as a whole is doing so.

What may be the case where the mid-slope area is cordoned off to water in-flow by perimeter barriers? It is considered 'correct' procedure to so isolate the area from which erosion measurements are to be interpreted. However, this procedure may introduce an artefact into the measurement: the area of the erosion plot experiencing net soil loss may be confined to areas where deposition from above is blocked, either by an artificial perimeter barrier or an exposed topographic position. Thus, a zone of degradation may be created on the upper side of any sloping field or experimental run-off plot, simply by creating an effective barrier to soil movement on the upslope boundary. Scouring will occur in the zone immediately below, due to lack of potential for soil enrichment from upslope.

The contoured sloping field can, therefore, be regarded as broken into a series of alley–hedge repeating units: a series of microlandscapes with erosion/deposition behaviour analogous to that detected at the macro level on whole landscapes. The alleyway may be seen to exhibit the same soil movement properties as the landscape, albeit with a simplified and repeating morphology.

Modelling Crop Performance Across the Alleys

The row-wise performance of crops in alleyways subjected to scouring (e.g. Fig. 8.3) suggests a response surface that is determined by two phenomena: the classical hedge–crop interspecies competition for growth resources, and the soil scouring during terrace formation. The conventional response function of crop yield or dry matter may be represented as the integrated effect of these two independent sets of processes. Figure 8.8 illustrates the generalized response surfaces that tend to be observed in situations where both processes are operating, compared with the response surfaces observed where the independent processes are desegregated. In future, as more work focuses on the disaggregation of the integrated effect we should be able to model the interaction between these effects as part of a more general understanding of the entire tree–crop system.

Figure 8.8 illustrates the utility of dividing the alleyway into homogeneous zones in which the species competition effect and the scouring effect are differentially active. A minimum of five zones, rather than three, as used in other tree–crop interaction studies, seems necessary to capture the response surface and analyse contributing processes to this skewed distribution. The scouring–deposition effect creates a radically more favourable crop environment above, as opposed to below the hedgerow, but species competition in the area of direct interface (the upper and lower zones) tends to deflate the differences among their integrated effects on performance.

It is useful to subdivide the central area of the alleyway into three subzones: (i) the mid-alley with (theoretically) no net soil gain/loss; (ii) the area just upslope (soil removed, without major tree interference); and (iii) the area just downslope (soil gain, without major tree interference). The lower-middle zone tends to reveal maximum response from the topsoil deposition (Figs 8.2 and 8.8). Exceptions to this general model will be expected. For example, A. Mercado and D.P. Garrity (unpublished) have observed situations where the middle-alley zone has a distinct yield depression compared with the adjoining upper-middle or lower-middle zones. This happens when ploughing consistently leaves a back-furrow in the alley centre.

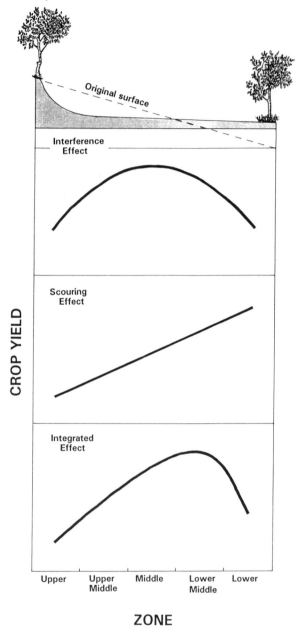

Fig. 8.8. Models of row-wise yield of alley crops as an integrated function of species interference and scouring.

As a first approach to modelling the crop effects, van Noordwijk and Garrity (1995) proposed a parabolic distribution of relative crop yields across the alley (see curve a on right side of Fig. 8.9). With an average yield of 1 (no net benefit or loss due to the hedgerows) and the distance (X) scaled from 0 to 1, the yield (Y) may be described as:

$$Y = pX(1 - X) + 1 - p/6 \tag{8.1}$$

where the parameter p determines the shape of the curve. If the topsoil fertility (F) is redistributed, without net losses or gains, it may be represented by:

$$F = qX + 1 - 0.5q \tag{8.2}$$

where the parameter q determines the gradient. By multiplying Equations 8.1 and 8.2 we obtain a yield curve of the form:

$$Y = -pqX^3 + pX^2a(1.5q - 1) + X\left(p\left(1 - \frac{2q}{3}\right) + q\right) + 1 - \frac{p}{6} - \frac{q}{2} + \frac{pq}{12}$$

$$\tag{8.3}$$

This is a two-parameter cubic equation in X, which might be fitted to data sets as shown in Fig. 8.9 to attempt to unravel the interactions. The figure gives examples for three values of q reflecting three stages of terrace formation by soil redistribution. In this multiplicative model the effects of topsoil depth and tree–crop interactions act independently on crop yield. A

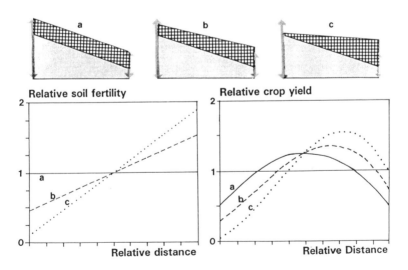

Fig. 8.9. Three stages in the process of terrace formation by soil redistribution and effects on crop yields (van Noordwijk and Garrity, 1995).

consequence of the linear soil fertility gradient and the symmetric shade and mulch interaction curve is that the effect on average yield is neutral (which is shown by integrating Equation 8.3 over the interval of 0 to 1).

Samzussaman (1994) used the SCUAF model to examine long term trends in soil properties of contour hedgerow systems with data validation from Philippine trials in Luzon (oxid palehumult) and Mindanao (typic hapludox). It was apparent that, without compartmentalizing the model by alleyway zones, predictions may be unrealistic. It will be necessary to include hedgerow effects on soil spatial properties in models of soil and crop response to contour hedgerow farming systems. In fact, modelling contour hedgerow systems more comprehensively presents an important and crucial challenge in the next generation of research. It is because these systems are so complex that simulation will prove useful in addressing a host of new concerns about their viability and management.

Is Farming on Natural Terraces Sustainable?

It is generally assumed that scouring effects will dissipate with time as a terrace surface stabilizes and more organic matter can be retained in the surface soil in the upper zones. However, it is not known how long this process may take on different sites and under different management regimes. It is not surprising to expect that any short term losses due to scouring may appear to make the investment in contour hedgerow installation and maintenance unattractive to farmers. For example, some farmers in Leyte Province in the Philippines who established natural vegetative strips for soil conservation nevertheless fallowed them after a few years. Soil loss control was apparently effective, but it could not prevent nutrient depletion due to the export of nutrients by continuous cropping, as fertilization was not practised.

Thus, we urgently need to know:

1. Are scouring effects a short term or long term phenomenon?
2. How can they be effectively avoided or alleviated?

It may be argued that scouring is only a temporary phenomenon and, as has been observed with scraped soils in other contexts (e.g. in creating lowland rice fields), the productivity differential between the scraped area and the depositional area will gradually equilibrate. On deep soil with reasonable soil organic carbon levels, the scoured areas may no doubt recover in a few years. The evidence indicates that the effect may be relatively short term on narrower terraces, and on certain soils; but on wider terraces and other soils the yield depression may last for many years. On strongly acid ultisols, oxisols and inceptisols, the solum may be quite deep physically, but these soils are often 'chemically shallow' due to excessive subsoil acidity due to soluble aluminium. Scouring scrapes off the more fertile topsoil, thus restricting crop rooting depth.

In situations such as reported by Agus (1993), the aggregate grain yields were still higher with tree legume hedgerows than on open slopes, although a scouring effect was evident. But in other cases they were similar or lower (Ramiaramanana, 1993; Turkelboom et al., 1993). The absence of short term benefits to crop yield is certainly evident in such cases, and a deterrent to adoption. (We are currently monitoring long term hedgerow trials to observe when and under what species combinations and management practices the upper alley zone yield depression may disappear.)

On calcareous soils with a shallow solum over limestone parent material (a common upland soil in parts of Southeast Asia) a contrasting problem arises: scouring removes the entire topsoil down to unconsolidated rock. The loss of productivity is permanent and drastic.

Some ecologists have also questioned the terracing of landscapes as an alternative to shifting cultivation (Ramakrishnan, 1990). In high rainfall climates, flat terraces tend to stimulate much greater leaching losses than on sloping fields due to greater vertical infiltration. The additional leaching losses may partially offset the advantages in soil conservation. The permanent terraces also tend to lead to continuous cultivation, creating severe soil nutrient depletion. Without nutrient importation from off-field through significant quantities of fertilizers or manures, crop yields cannot be maintained. In such cases, terracing may lead to a less sustainable agricultural system than conventional rotation fallowing.

Considering the degree of uncertainty that still prevails concerning the effects of vegetative barriers on landscape processes, and on system sustainability, research must bring to bear more analytical and critical tools to understand them. Within the subset of landscapes where vegetative barriers are useful, we need to know how to cope with the scouring process. The last section addresses this issue.

Coping with the 'Scouring' Effect

There are three basic ways to try to cope with scouring: to avoid it, alleviate it or change the cropping system to live with it. The research agenda for each is briefly addressed.

Reducing the intensity of tillage

To avoid scouring a farmer must not practise primary tillage too frequently. In zero or minimum tillage systems there is less tendency for hedgerow risers to develop or for soil scouring to occur. There are many variations of reduced tillage that smallholders can adopt. Ridge tillage appears to be one of the most promising ones, in which alternating unploughed strips are maintained indefinitely. Timeliness of field operations, e.g. harvesting and planting, can reduce the number of tillage operations required. Adapting the ridge-till concept to small-scale farmers has received almost no attention to date.

Adjusting biomass and nutrient inputs

To alleviate scouring it is logical to consider changing the spatial pattern of nutrient inputs. The application of crop residues, hedgerow prunings, animal manure and fertilizers may be biased to the upper alley zones. Since scouring tends to redistribute topsoil nutrients downward naturally, application to the upper alley may ensure a more even distribution in the long run. But it may also be argued that application to the lower zones with more favourable soil may make more efficient use of limited nutrient inputs. This depends on the comparative nutrient utilization efficiencies of crops growing in the different zones. Practical answers to these fertility management issues, and an underlying predictive knowledge about them, should be at the top of the agenda for immediate research.

Changing the cropping (or hedgerow) system

The final strategy is to accept the scouring effect and to alter the cropping system to adjust to the changes (particularly if they are drastic). Substitution of perennials (e.g. fruit or timber trees) for annual crops, or more tolerant annuals (e.g. cassava), is a practice that farmers have tried. Researchers and farmers must learn how to manage the degraded zone better so that a conservation technology may be created.

Cognizant that the highest soil fertility is found inside the hedgerow strip, Turkelboom et al. (1993) have even suggested that this 'sleeping' fertility be exploited by periodically moving the hedgerows downslope and cultivating the former hedgerow area. The idea is more practical with a grass strip but not trees. Such species as pigeonpeas (Cajanus cajan) may be suitable for this; they naturally die off after a couple of years, making it easy to relocate the hedgerow.

Conclusion

Contour hedgerow systems attempt to exploit and conserve the available soil fertility on slopes in more efficient ways. They may dramatically reduce off-field soil erosion, but can accelerate soil degradation through soil redistribution downslope within the alleyways. In developing natural terraces, crop production is an integrated function of two processes: hedge and crop species interference, and a soil scouring effect. The scouring effect creates further complexity in the management of such systems.

The contour hedgerow concept has now been widely recommended. Unfortunately, it is often indiscriminately applied through standard extension packages. Superficial expectations about hedgerow intercropping performance have often been wrong. A concerted research effort is now urgently needed to elucidate the processes more fully, model the interactions and gain a predictive understanding of when, where and how contour hedgerow systems embody a superior technology for hillslope agriculture. In particular we need to understand the three basic ways to cope with the scouring effect: avoiding it, alleviating it or changing the cropping system to live with it.

References

Agus, F. (1993) Soil processes and crop production under contour hedgerow systems on sloping Oxisols. PhD dissertation, North Carolina State University, Raleigh, NC, USA, 141 pp.

Basri, I.H., Mercado Jr, A.R. and Garrity, D.P. (1990) *Upland Rice Cultivation Using Leguminous Tree Hedgerows on Strongly Acid Soils.* IRRI Saturday Seminar Paper, March 31. Agronomy, Plant Physiology, and Agroecology Division, International Rice Research Institute, Los Baños, Philippines, 32 pp.

Brown, R.B., Cutshall, N.H. and Kling, G.F. (1991a) Agricultural erosion indicated by Cs-137 redistribution: I. Level and distribution of Cs-137 activity in soils. *Soil Science Society of America Journal* 45, 1184–1190.

Brown, R.B., Kling, G.F. and Cutshall, N.H. (1991b) Agricultural erosion indicated by Cs-137 redistribution: II. Estimates of erosion rates. *Soil Science Society of America Journal* 45, 1191–1197.

Busacca, A.J., Cook, C.A. and Mulla, D.J. (1993) Comparing landscape-scale estimation of soil erosion in the Palouse using Cs-137 and RUSLE. *Journal of Soil and Water Conservation* 48, 361–367.

Fujisaka, S. (1992) *A Case of Farmer Adaptation and Adoption of Contour Hedgerows for Soil Conservation.* International Rice Research Institute, Los Baños, Laguna, Philippines.

Garrity, D.P. (1994) Sustainable land use systems for the sloping uplands of Southeast Asia. In: Ragland, J. and Lal, R. (eds) *Technologies for Sustainable Agriculture in the Tropics.* American Society of Agronomy, Madison, Wisconsin, pp. 41–66.

Garrity, D.P., Mercado Jr, A. and Solera, C. (1995) The nature of species interference and soil changes in contour hedgerow systems on sloping acidic lands. In: Kang, B.T. (ed.) *Alley Farming.* International Institute of Tropical Agriculture, Ibadan, Nigeria.

Hauser, S. (1993) Root distribution of *Dactyladenia (Acioa) barteri* and *Senna*

(*Cassia*) *siamea* in alley cropping on Ultisol. I. Implication for field experimentation. *Agroforestry Systems* 24, 111–121.

Hsieh, Y.P. (1992) A mesh-bag method for field assessment of soil erosion. *Journal of Soil and Water Conservation* 47, 495–499.

ICRAF (1994) *Annual Report for 1993*. International Centre for Research in Agroforestry, Nairobi, Kenya.

Jones, A.J., Mielke, L.N., Batles, C.A. and Miller, C.A. (1989) Relationship of landscape position and properties to crop production. *Journal of Soil and Water Conservation* 44, 328–332.

Kiepe, P. (1995) *No Runoff, No Soil Loss: Soil and Water Conservation in Hedgerow Barrier Systems. Tropical Resource Management Paper 10. Wageningen Agricultural University, The Netherlands.*

MBRLC (1988) *A Manual on How to Farm your Hilly Land without Losing your Soil.* Mindanao Baptist Rural Life Centre, Davao del Sur, Philippines.

Ramakrishnan, P.S. (1990) Agricultural systems of the northeastern hill region of India. In: Gliessman, S.R. (ed.) *Agroecology.* Springer-Verlag, New York, pp. 251–274.

Ramiaramanana, D.M. (1993) Crop–hedgerow interactions with natural vegetative filter strips on slopic acidic land. MSc thesis, University of the Philippines at Los Baños, 141 pp.

Sajjapongse, A. (1992) Management of sloping lands for sustainable agriculture in Asia: An overview. In: *Technical Report on the Management of Sloping Lands for Sustainable Agriculture in Asia, Phase I (1988–1991)*. Network Document No 2, International Board for Soil Research and Management, Bangkok, Thailand, pp. 1–15.

Salazar, A., Szott, L.T. and Palm, C.A. (1993) Crop–tree interactions in alley cropping systems on alluvial soils of the Upper Amazon Basin. *Agroforestry Systems* 22, 67–82.

Samzussaman, S. (1994) Effectiveness of alternative management practices in different hedgerow-based alley cropping systems. PhD dissertation, University of the Philippines at Los Baños.

Solera, C.R. (1993) Determinants of competition between hedgerow and alley species in a contour hedgerow intercropping system. PhD dissertation, University of the Philippines at Los Baños, 135 pp.

Turkelboom, F., Ongprasert, S. and Taejajai, U. (1993) Alley cropping on steep slopes: Soil fertility gradients and sustainability. Paper presented at the *International Workshop on Sustainable Agricultural Development: Concepts and Measures.* Asian Institute of Technology, Bangkok, December 14–17, 16 pp.

van Noordwijk, M. and Garrity, D.P. (1995) Nutrient use in agroforestry systems. In: *Proceedings of the 24th International Colloquium of the International Potash Institute*, Basel, Switzerland.

Young, A. (1989) *Agroforestry for Soil Conservation.* CAB International, Wallingford, UK.

9

◆

Root Distribution of Trees and Crops: Competition and/or Complementarity

M. VAN NOORDWIJK[1], G. LAWSON[2], A. SOUMARÉ[3],
J.J.R. GROOT[4] AND K. HAIRIAH[5]

[1] ICRAF, SE Asia Regional Program, PO Box 161, Bogor 16001, Indonesia;
[2] Institute of Terrestrial Ecology (ITE), Edinburgh, UK; [3] Institut d'Economie
Rurale (IER), BP 258, Bamako, Mali; [4] AB-DLO, Haren, The Netherlands;
[5] Brawijaya University Malang, Indonesia

Introduction

This chapter takes a look at the 'hidden half', because interactions between trees and crops for below-ground resources are often as important as those for light and above-ground space (Anderson and Sinclair, 1993). The basic efficiency of utilization of water and nutrients for photosynthesis is probably not very different between trees and C_3 crops, so tree growth does not constitute a 'free lunch' in any agroforestry system. However, the below-ground resource base for tree growth, and thus the degree of potential competition or complementarity between trees and crops, is usually more difficult to assess than that for shorter-lived crops. Because of their longer lifetime, tree root systems have the chance to explore larger areas, both laterally and vertically, and exploit zones of rich localized supply of water and/or nutrients. If these zones are largely out of reach of annual crops, the lunch bill for the trees is provided for and trees can increase the total production of the system, although they may not improve crop growth as such. This is especially true for deep rooted trees, which can exploit deep soil water reserves, either stored or part of subsurface flow pathways, and weathering saprolite or bed-rock layers inaccessible to crops, or intercept leaching water and nutrients on the way down, below the crop root zone.

The general concept that all trees are deep rooted may be greatly over-stated, however, as there are large differences between species and sites and the horizontal scavenging ability of tree roots is often underestimated. Everyday examples are urban trees in front of a house, surrounded by pavements and stones, which thrive on roots passing under the house and exploiting the back garden, or the septic tank there.

Experiments with annual crops often involve closely-spaced small plots. If such experimental layouts are used for agroforestry experiments, the tree component may exploit the soil of the (fertilized) 'control' plots as well as their own; this is likely to reduce crop yield in the control plot and, in the long run, increase that in the agroforestry plots. Both effects lead to an overestimate of the positive yield effect of agroforestry. This situation has been found for the semi-perennial cassava plant and may invalidate many experiments leading to the conclusion that cassava is not responsive to N fertilizer (van Noordwijk *et al.*, 1992). It can be more pronounced for trees and many of the previous experiments on alley cropping and other agroforestry systems are difficult to interpret, because root interactions in the so-called no-tree control plots were not properly excluded (Hauser, 1993; Coe, 1994). A basic idea of root distribution of the various components in a system is thus needed to do valid agroforestry field experiments. On small farm plots, trees may mine neighbouring areas (the neighbour's plots?) and the farmers' perceptions of advantages of trees may be biased for this reason too.

Roots are often discussed and wild generalizations about deep-rooted or horizontal scavenger roots are common, but few people take the effort to observe roots under their particular conditions. Chapter 3 illustrated that agroforestry systems are only viable if the tree and crop components represent approximately equal direct value to the farmer relative to the area occupied, unless there is a strong complementarity in the use of above-and/or below-ground resources.

In this chapter some of the existing concepts for studying the below-ground functioning of both tree and crop component separately are reviewed, a number of hypotheses on below-ground interactions between trees and crops are developed which can be tested with currently available methods and the methods available for simple observations and detailed tests of hypotheses are briefly described.

Basic Root Ecological Concepts

Morphogenetic and functional shoot : root balance

Serious root observations in agricultural systems started more than a century ago. The relation between above- and below-ground growth has fascinated researchers for a long time and almost every layman expects that there is a clear relation between the height of a plant (tree) and the depth of its root system, or between the lateral spread of the shoot or crown and the lateral spread of the root system. The 'root atlases' published by Kutschera and Lichtenegger (1982) show that no generalizations of this nature are valid, however. Root characteristics across plant species vary apparently independently of shoot characteristics. Natural selection has led to a large number of root and shoot combinations, apparently adapted to different environmental conditions. There is thus ample scope for selection, breeding and biotechnology to modify the genetic determinants of root development, if only we knew in what direction they should be changed.

The relationship between above- and below-ground growth should first of all be investigated within a single genotype to be meaningful. Often researchers found that a better root development was correlated with a higher yielding crop, and a 'basic law' of agriculture was formulated that any restriction to root growth by adverse soil conditions would directly lead to a reduced yield (Hellriegel (1883) quoted in van Noordwijk and De Willigen, 1987). Although evidence contradicting this 'basic law' was gradually accumulating (ibid.), it took a long time before the law itself was refuted and replaced by the hypothesis of a 'functional equilibrium' between root and shoot growth (Brouwer, 1963, 1983).

Figure 9.1 shows a generalized form of the response of above- and below-ground parts to increased water and/or nutrient supply, based on Schuurman (1983). In a part of the range both shoot and root biomass increase with improved resource supply, but the maximum root biomass is generally obtained at a lower level of resource supply than maximum shoot biomass. The shoot:root ratio may increase gradually to the left of the maximum root biomass and sharply above that point. This scheme can be used to explain the conflicting evidence in literature about external factors 'increasing' or 'decreasing' root growth in experiments which cover only part of the range. Classical statements such as 'phosphorus stimulates root growth' are based on only part of the range; elsewhere one may find that 'lack of phosphorus stimulates root growth'.

Although primarily developed for annual plants, the functional equilibrium concept appears to be equally valid for perennial plants, when expressed as the ratio of leaves and fine roots. The large amounts of storage and stability tissue, both above and below ground, complicate comparisons of total biomass, however. Gower (1987), for example, reported that fine root biomass in tropical wet forests is inversely related to phosphorus and calcium availability. Vitousek and Sanford (1986) found that shoot:root ratios decrease with decreasing soil fertility. We here follow the convention of the older Dutch literature of expressing shoot:root, rather than root:shoot ratios, as common in the English literature; roots are a more elegant basis for this ratio as the roots of a seedling emerge before the shoot.

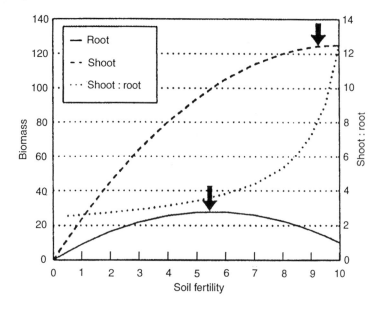

Fig. 9.1. Schematic relationship between shoot and root biomass production and soil fertility. The maximum root size is often obtained at intermediate shoot biomass and between the optimum conditions for root and those for shoot growth the uptake rate per unit root (reflected in the shoot:root ratio) increases rapidly (based on Schuurman, 1983).

Dhyani *et al.* (1990) found that root weight ranged from 27% (*Leucaena leucocephala*) to 72% (*Eucalyptus tereticornis*) of total tree biomass in a comparison of five tree species at 2 years age. Toky and Bisht (1992) found for 6-year-old trees (12 species) that root biomass ranged from 9% (*Acacia catechu*) to 27% (*Morus alba*) of total biomass, with a median value of 20.3%. These figures probably do not reflect the relative importance of roots in current carbon allocation in trees, as roots may have a higher turnover rate than above-ground tissues. Sanford (1985) estimated fine root turnover in Venezuela in the top 10 cm at 25% per month. Berish (1982) observed a fine root biomass under successional vegetation of around 40% of that in adjacent natural forest. Fine root biomass reached the undisturbed level after only 5 years, at the time that leaf area index reached the control level as well.

According to the 'functional equilibrium' concept (Brouwer, 1963) the allocation of growth resources in the plant to root and shoot meristems is modified by the major current environmental conditions. If water or nutrients are in short supply within the plant, the root system will get a larger share of the carbohydrate supply within the plant and will increase in size relative to the shoot (as measured in a shoot:root ratio) or even in an absolute sense. When light (or CO_2 supply) is limiting plant production, the shoot will increase in size relative to the root system.

Subsequent research (Lambers, 1983) has shown that the underlying mechanism is more complex than the direct resource limitation of shoot and root meristem activities envisaged by Brouwer (1963) and that there is more variation between plants in how rapidly and to what extent they adjust to modified conditions. The functional equilibrium is, however, still a source of inspiration of hypotheses about actual plant responses, as it explains their overall functionality. In the actual coordination of activities of shoot and root meristems plant hormones play a role as signals, but the mechanisms are not yet clearly understood by which they receive and transmit the information about the current environmental situation or internal conditions in the plant as a result of this.

Maximum plant production can be obtained with relatively small root systems, provided that the daily water and nutrient requirements are met by technical means (van Noordwijk and De Willigen, 1987). Better possibilities for uptake mean that a smaller root system is sufficient to supply the needs of a shoot. The answer to the question 'How many roots does a plant need?' thus depends on the environment in which the plant grows and its intrinsic growth rate. During agricultural intensification human control over the supply of water and nutrients has gradually increased; the endpoint of this development may be reached in horticulture based on soilless culture techniques. Reducing the size of the root system has a limit, however, where the physiological capacity for uptake is reached – this limit may be encountered first of all for water (De Willigen and van Noordwijk, 1987; van Noordwijk, 1990). A plant growing in free water still needs a considerable root surface area, as can be approximated (for non-saline conditions) by:

$$A_{r,w} = \frac{E_p}{L_p \Delta H_p} \tag{9.1}$$

where

$A_{r,w}$ = root surface area required for water uptake [m^2],
E_p = transpiration rate per plant [$cm^3\ s^{-1}$],
L_p = hydraulic conductance of roots for water entry [$cm^3\ m^{-2}\ MPa^{-1}\ s^{-1}$],
ΔH_p = maximum acceptable difference in plant water potential between root xylem and root environment [MPa].

Equation 9.1 and parameters for full grown tomato or cucumber plants predict that the required root surface area is approximately 1 or 2.4 m^2, respectively, or 50% of the leaf area in both species. Shoot:root ratios expressed on a dry weight basis may reach 20–30 in this situation. The actual root surface area formed under non-restrictive conditions was 50–

100% of the leaf surface area in a series of experiments (De Willigen and van Noordwijk, 1987). The specific root area (root surface area per unit dry weight) can be ten times higher than the specific leaf area (with 0.2 and 0.02 m^2 g^{-1} as order of magnitude, respectively). For oak and aspen trees grown in pots Wiersum and Harmanny (1983) observed a root surface area of approximately twice the leaf surface area.

In the field, the required size of the root system is not determined by the maximum physiological abilities of the roots, but rather by the transport rates of water and nutrients in the soil and hence by the need to reduce transport distances and the required gradients as determined by required uptake per unit root length in an extensively branched root system. The more restricted the water supply, the larger is the root system needed relative to the shoot.

Root densities and uptake efficiency

Model approaches

Although a large root system may not be needed for maximum growth rates, roots are of direct importance for the efficient use of available water and nutrient reserves in the soil, and hence in reducing negative side effects of agriculture. As a first estimate, we may still expect that 'the more extensive the root system is, the higher nutrient and water uptake efficiency may be' (van Noordwijk and De Willigen, 1991). The possibility of obtaining a higher resource uptake efficiency can only be realized if total supply of nutrients and water is regulated in accordance with the crop demands and the resource use efficiency attainable. On a field scale both resource supply and possible crop production show spatial variability and inadequate techniques for dealing with this variation may reduce the resource use efficiency much below what is possible in the normally small experimental units considered for research (van Noordwijk and Wadman, 1992).

In modelling nutrient and water uptake a number of levels of complexity can be distinguished:

1. 'Models without roots', based on measured or estimated 'uptake efficiencies' (ratio of uptake and amount of available resources); roots remain implicit in such models.
2. Models predicting uptake efficiency on the basis of measured root density and distribution; these models have to integrate the activities of single roots to the root system level.
3. Models based on descriptive curves fitted to root growth in space and time, e.g. negative exponential functions to describe root length density as a function of depth or deterministic root branching models, driven by time or cumulative temperature (Diggle, 1988; Pagès *et al.*, 1989).

4. Models based on functional equilibrium concepts, relating overall root growth to the internal water, nutrient and carbohydrate status in the plant.
5. Models that include differential response of root growth to zones with different environmental conditions (nutrient, water, oxygen supply, mechanical impedance).

Models at level 2 are a prerequisite for any of the levels following, and considerable efforts have been made to develop and test them (Nye and Tinker, 1977; Barber, 1984; De Willigen and van Noordwijk, 1987; Gillespie, 1989). Earlier models described the nutrient uptake rate of roots as determined by the external concentration, based on Michaelis–Menten kinetics or similar relationships between concentration and uptake rate. At times that the external supply exceeds the current crop demand, however, such models overestimate uptake, as internal feedback mechanisms down-regulate uptake in most plants under such circumstances. When demand exceeds supply, on the other hand, the affinity of the uptake mechanisms for nutrients is so high that roots can deplete the concentration at the soil solution–root interface to virtually zero. The model description of De Willigen and van Noordwijk (1987, 1994) therefore is based on a notion of crop nutrient demand, similar to potential transpiration rates, regulating uptake per unit root when supply is sufficient and a 'zero-sink' (actually an infinite sink strength leading to a concentration of zero) when supply is limiting. The amount of available nutrients left in the soil at the transition between these two situations is termed N_{res}. Figure 9.2 shows a concentration profile in the soil surrounding a single root; if the roots are regularly distributed, the soil 'belonging' to each root is approximately a cylinder of constant radius. N_{res} is defined as the integral of the concentration in this cylinder at the moment when transport towards the root just falls short of uptake demand. It determines the highest uptake efficiency that can be achieved without reducing crop growth:

$$\text{max. efficiency} = \frac{\text{crop demand}}{\text{crop demand} + N_{res}} = 1 - \frac{N_{res}}{\text{crop demand} + N_{res}}$$

(9.2)

When nutrient supply is less than the sum of crop demand and N_{res}, the uptake efficiency may be (slightly) higher. When supply becomes limiting, nutrient uptake can gradually deplete the N_{res} nutrient stock, asymptotically approaching complete depletion.

Model for simple root–soil geometry
De Willigen and van Noordwijk (1987, 1991, 1994) derived, under simplifying assumptions on root–soil geometry, an equation for N_{res} as function of root length density L_{rv} and root diameter, which can be used to predict uptake efficiency from a single homogeneous layer or which can be part of

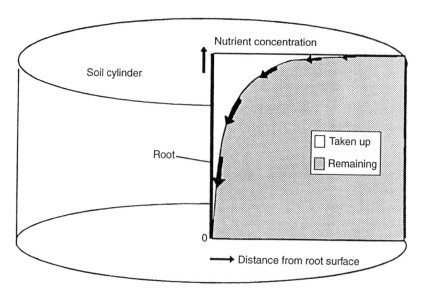

Fig. 9.2. Concentration profile in the soil surrounding a single root; if the roots are regularly distributed, the soil 'belonging' to each root is approximately a cylinder of constant radius.

dynamic uptake models from layered soils.

$$N_{res} = \frac{A(K_a + \Theta)D_m^2 G(\rho, v)}{4H(a_1\Theta + a_0)\Theta D_0} \quad (9.3)$$

where:

A = daily nutrient demand [kg ha^{-1} day^{-1}],
K_a = apparent adsorption constant [ml cm^{-3}],
Θ = soil water content [ml cm^{-3}],
a_1 and a_0 = parameters describing decrease of effective diffusion coefficient with decreasing Θ,
H = depth of soil zone considered [cm],
D_0 = diffusion coefficient of nutrient in free water [cm^2 day^{-1}],
D_m = root diameter used for model [cm].

with:

$$\rho = 2(\pi L_{rv}D_m^2)^{-0.5} \quad (9.4)$$

and

$$G(\rho, 0) = \frac{\rho^2}{8}\left[-3 + \frac{1}{\rho^2} + \frac{4\ln\rho}{\rho^2 - 1}\right] \quad (9.5)$$

and a slightly more complex definition if the dimensionless group based on transpiration rate, v, is not zero (De Willigen and van Noordwijk, 1987). As diffusion constants do not differ much between most solutes, the zero-sink concentration profile for all major nutrients NO_3^-, NH_4^+, K^+ and $H_2PO_4^-$ can be treated in a similar way. Only the demand parameter A and the adsorption parameter K_a (which relates the total available amount to the concentration in soil solution) will differ considerably between them; K_a for $H_2PO_4^-$ is 100–1000 ml cm^{-3}, while for NO_3^- adsorption may be negligible; thus the factor $(K_a + \Theta)$ is 300–5000 times larger for P than for N. N_{res}/A expresses the residual amount as a number of days with unconstrained uptake which would be possible for an infinitely dense root system ($N_{res} = 0$ for $L_{rv} = \infty$. For nitrate N_{res}/A may be only a few days, while for P it easily encompasses one or several growing seasons. So and Nye (1989) showed that for a tenfold decrease in effective diffusion constant $(a_0 + a_1\Theta) D_0$ from its value at field capacity ($pF = 2.0$) a sandy loam has to dry out until pF 3.3 and a silty clay until pF 4.5. Such a decrease in soil water content renders N_{res} for NO_3^- in a dry soil similar to that of K^+ at field capacity.

For water uptake a similar approach is possible if the factor A is replaced by the potential transpiration rate and the concentration is replaced by the matrix flux potential (De Willigen and van Noordwijk, 1987, 1991); for a more refined treatment of water uptake, however, the hydraulic conductance of roots, L_p, should be considered as well. Under wet conditions L_p will dominate the total soil–plant resistance and water uptake may be proportional to root length density; in drier soil the soil resistance gradually becomes more important (De Willigen and van Noordwijk, 1991).

Figure 9.3 shows N_{res} as a function of L_{rv}, A and Θ for a standard parameter set for NO_3 uptake (De Willigen and van Noordwijk, 1987). N_{res} becomes less than 10 kg ha^{-1} for L_{rv} values in the range 0.2–2 cm cm^{-3} (lower values for wetter soil and lower daily N demands); increasing root length density above this value will allow only a small amount of additional N uptake. Some of the simplifying assumptions, especially on the uniformity of root diameters and on the effects of root distribution pattern can now be avoided (van Noordwijk and Brouwer, 1995).

Mycorrhizal hyphae and heterogeneity in root diameter
If root systems of different diameter are compared at equal root length density (length × diameter^0), the larger the diameter is, the smaller N_{res} and thus the more efficient the uptake can be. If the comparison is made at equal surface area (length × diameter^0 × π), N_{res} decreases with decreasing root diameter (De Willigen and van Noordwijk, 1987). If the comparison is made at equal root volume (length × diameter^2 × π/4) or weight, the advantage of the smaller root diameters is even more pronounced. The most stable result is obtained for a comparison at equal length × diameter$^{0.5}$. Figure 9.4 shows the required P availability in the soil – indicated by the

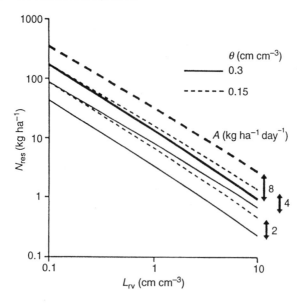

Fig. 9.3. The amount of mineral N, N_{res}, in the soil (at two water contents), required to maintain crop demand A (three values, representing high–normal demands for N), as a function of root length density L_{rv} (De Willigen and van Noordwijk, 1987).

(water-extractable) P_W index – when root systems of different diameter are compared on the basis of equal root length, root surface area, root volume or sum of root length × diameter$^{0.5}$. The more efficient the root system, the lower the required P level of the soil. Calculations were made with the P model of van Noordwijk *et al.* (1990), which is based on N_{res} and P adsorption isotherms, and parameters for the growth of the velvet bean *Mucuna* growth on an ultisol in Lampung, Indonesia (Hairiah *et al.*, 1995). With the length × diameter$^{0.5}$ index, calculation results are approximately independent of root diameter over at least one order of magnitude. We thus have a method to add hyphal length of mycorrhizal fungi (which are about a factor 25 smaller in diameter than the finest roots) to the crop root length: roughly 1/5 (or $25^{-0.5}$) of the hyphal length can be added to the root length density. If only 'infection percentage' data are available for the mycorrhiza, we have to assume a reasonable length of hyphae per unit infected root length (a value between 10 and 100 seems reasonable, say 50 as first estimate), and we thus obtain an increased root length density by a factor $1 + (0.5) \times$ %inf/5. For a normal infection percentage of 15%, this means that the effective root length density is 2.5 times the length of roots alone (van Noordwijk and Brouwer, 1995). The lack of adequate methods for quantifying hyphal length makes this a priority area for research, if mech-

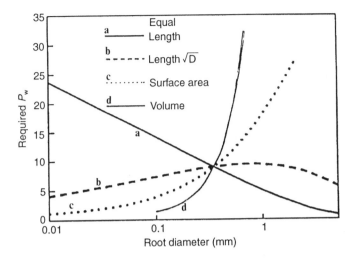

Fig. 9.4. Required P availability in the soil – indicated by the (water extractable) P_W index – when root systems of different diameter are compared on the basis of equal root length, root surface area, root volume or sum of root length × diameter$^{0.5}$ (van Noordwijk and Brouwer, 1995).

anistic P uptake models are to be used for mycorrhizal plants, i.e. for nearly all species found in agroforestry.

A similar method can be used to obtain a weighted average root diameter for a branched root system, with a diversity of root diameters.

Non-regular root distribution

With the 'root position effectivity ratio' R_{per}, the uptake efficiency for any actually observed root distribution pattern can be related to that for a theoretical, regular pattern. The effects of incomplete root–soil contact can be incorporated as well, in an approximate manner (van Noordwijk *et al.*, 1993a, b). R_{per} is defined as a reduction factor on the measured root length density, to account for the lower uptake efficiency of real-world root distributions, when compared with the theoretical, regular pattern assumed by most existing uptake models (based on a cylinder geometry of the root–soil system), including the model used for deriving Equation 9.1. For ran-dom root distributions, R_{per} is approximately 0.5 (that means root length density $X/2$ in a regular pattern has the same N_{res} as a random pattern at density X); the figures shown by van Noordwijk *et al.* (1993a) are based on a different and incorrect definition of R_{per}, which was later corrected (van Noordwijk *et al.*, 1993b). For clustered root distribution, as may be expected in structured soils, where roots grow mainly along cracks, R_{per}

values in the range 0.05–0.4 can be expected. R_{per} tends to decrease with higher absolute root length densities.

Dynamics of root growth and decay
Estimates of L_{rv} normally have a fairly wide confidence interval, because of the considerable spatial variability of root length density. If root growth and decay are estimated from a time series of destructive sampling, the results tend to have an unacceptably large uncertainty. If sequential non-destructive observations can be made on the same roots, e.g. those located next to a mini-rhizotron, and the resulting images are analysed for changes relative to the root length present, a much smaller sampling error can be obtained. The cost of this, however, is a potential bias, as the observation method may affect root behaviour (Gijsman *et al.*, 1991; Anderson and Ingram, 1993). Details are given by van Noordwijk *et al.* (1994a), who presented results of an analysis of sugarbeet and winter wheat root turnover.

Effective root length density as function of time and depth
Combining these elements (Table 9.1), we can derive an 'effective root length density' L_{rv}^* as a function of time and depth from (van Noordwijk and Brouwer, 1995):

$$L_{rv}^*(i,T) = R_{per}(i,T) \cdot \frac{\int_{t=0}^{T} (G(i,t) - D(i,t))\, dt}{\int_{t=0}^{s} (G(i,t) - D(i,t))\, dt} \cdot \frac{\sum_{j=0}^{n} L_{rv}(i,s,j)\sqrt{D_j}}{\sqrt{D_m}} \qquad (9.6)$$

where:

$L_{rv}^*(i,T)$ = effective root length density (cm cm^{-3}) in layer i at time T,
$L_{rv}(i,s,j)$ = measured root length density in layer i at time of sampling s for root diameter j,
$R_{per}(i,T)$ = root position effectivity ratio (procedure defined in van Noordwijk *et al.*, 1993b),
$G(i,t)$ = observed root growth along minirhizotrons as a function of time in zone i,
$D(i,t)$ = observed root decay along minirhizotrons as a function of time in zone i,
D_m = root diameter used for model calculations,
D_j = root diameter for diameter class j and observed root length density $L_{rv}(j)$.

If R_{per} is about 0.4 and the mycorrhizal correction factor 2.5, the two correction factors may, accidentally, cancel and the use of direct L_{rv} values can be correct in practice.

Table 9.1 Steps in describing root–soil geometry for uptake models (van Noordwijk and Brouwer, 1995).

1. Choose relevant sampling zones, based on depth, distance to crop rows, expected synlocation (spatial correlation) of roots and resources; measure the root length density, $L_{rv}(i,s)$, for each stratum i at sample time s close to the expected maximum root development).

2. Effective root length density for a root system with a known frequency distribution of root diameters (including hyphae):

$$L_{rv'} = \frac{\sum_{j=1}^{n} L_{rv,j}\sqrt{D_j}}{\sqrt{D_m}}$$ (9.9)

where D_m = diameter used for model calculations, D_j and $L_{rv,j}$ are root diameter and root length density of n diameter classes.

3. Extrapolation from sampling time s to any time t is based on:

$$L_{rv}(i, t) = L_{rv}(i, s)\frac{R_p(i, t)}{R_p(i, s)}$$ (9.10)

where $R_p(i,t)$ = relative root presence at zone i at time t on minirhizotron images.

$$R_p(i, t) = R_g(i, t) - R_d(i, t)$$ (9.11)

$R_g(i,t)$ = root growth in zone i till time t relative to year production,
$R_d(i,t)$ = root decay in zone i till time t relative to year production.

4. Measure effectiveness of the root distribution via the R_{per} method (van Noordwijk *et al.* 1993a, b) and derive the effective root length L_{rv}^* for time t at zone i:

$$L_{rv}^*(i, t) = R_{per}(i, t)L_{rv}(i, t)$$ (9.12)

where R_{per} = root position effectivity ratio, accounting for non-regular root distribution and incomplete root–soil contact; the sum of the G-functions (compare Equation 9.3) for the observed root pattern is the same as that for a regularly spaced pattern, where L_{rv} is reduced by the factor R_{per}.

Critical densities for various functions

Van Noordwijk (1983) gave an indication of the root length densities L_{rv} needed to meet the demands of an average crop for water and nutrients from a normal agricultural soil in northwest Europe: 0.1–1 cm cm^{-3} for NO_3^-, 1–10 cm cm^{-3} for $H_2PO_4^-$ and intermediate values for K^+ and water uptake. Root length densities beyond these ranges will have a relatively small effect on decreasing N_{res}, although for P uptake L_{rv} increases up to 30–50 cm cm^{-3} may still be meaningful. The carbon investments required for additional root growth can be balanced against the carbon fixation that is made possible by additional water uptake. In climatic conditions where re-wetting of dried soil is rare or in situations where fine roots will not survive a drying–wetting cycle, root length densities L_{rv} above 3–5 cm cm^{-3} may not be economical for a plant, in terms of its C economy. The values given here are no more than indications of the order of magnitude, as both soil (K_a, Θ, H) and crop parameters (A, D_m) affect their values.

Allocation of uptake in multilayer systems
In a stratified soil (by layer or any other division in internally relatively homogeneous zones), we need an algorithm for allocating total demand (A) over the various strata in those situations where total supply exceeds demand. Although there are insufficient physiological data to choose between them, a number of algorithms are possible. For example, the demand can be allocated proportional to:

1. relative root length density,
2. N_{res}, or
3. the external nutrient concentration in each stratum.

De Willigen and van Noordwijk (1989, 1991) used an algorithm that is based on allocation method 1 if total supply exceeds demand, but which will increase the demand allocation to zones where supply exceeds demand stepwise if certain zones cannot meet the originally allocated demand.

Allocation of uptake in multispecies systems
The simplest description of competition for water and nutrients is based on zero-sink uptake by both or all species competing for the same resource. The relative competitive strength will then be proportional to the N_{res} value for each component, based on its effective root length density in the zone or layer where competition occurs. For more refined descriptions differences in phenology (leading to different A values over time) and root development (different $L_{rv}^*(i,T)$) should be taken into account as well and a dynamic simulation model is needed. Below-ground competition is for resources that are stored in the soil and thus is affected by the recent history of uptake, in contrast to competition for light and CO_2.

Root growth and distribution patterns

Genotype × environment interactions
Although certain generalizations about deep/shallow or narrow/wide root distribution patterns can be made at a species or genotype level, the actual root pattern is based on genotype × environment interactions (Kerfoot, 1963). Van Noordwijk (1992) contrasted the results of root ecological studies at the single root, the whole plant and split-root level. For the root response to factors such as P supply, Al^{3+} concentration, soil compaction and O_2 concentration these three levels of complexity may lead to contrasting results. Of special interest here is the 'split-root' level, which can be used to analyse the local response of root systems to heterogeneities in the environment. The response of a root tip to its local environment depends in many ways on the conditions elsewhere in the root system (around other roots) as well as in the shoot. For example, branch root development is often

stimulated in zones of relatively high P supply; this response is absent, however, when P supply in the plant as a whole is adequate. Thus the often made generalization that 'phosphate stimulates root growth' is only partially true. The results can be explained by assuming that root meristems with direct access to P have a first choice in using it and may thus attract a larger share of the carbohydrates necessary for growth in a plant where P within the plant is a growth-limiting resource. Once the local roots' needs are met, P supply to the shoot will increase, and by internal redistribution in the phloem, also P supply to other roots. This phenomenon has been extensively studied for crop plants (De Jager, 1985), but also applies to wild species (Caldwell *et al.*, 1992). Similarly, Hairiah *et al.* (1993) showed that fewer *Mucuna* roots develop in a solution containing a relatively high Al^{3+} concentration if part of the root systems grows in a solution without Al^{3+}; yet, if this Al^{3+}-containing solution is used for the whole root system, it will stimulate root growth compared with a homogeneous control solution. The response of a root tip to Al^{3+} thus depends on the environment around other roots. No separate Al-signalling mechanism has to be invoked to explain these results, however, as the Al-avoidance response disappears at improved P supply to the plant and may be based on Al-induced P shortage in exposed roots.

In the local response of root growth, a distinction should be made between the growth of main axes and branch root development. Most of the responses appear to be based on stimulated branch root development and can also be described as a reduced degree of apical dominance, the mechanism by which the top meristem of shoots or roots reduces or delays the development of branch axes. In perennial root systems, a large proportion of the finer branch roots is relatively short-lived (Chapter 7) but new branch roots can develop annually from the surviving secondary thickened transport roots. Wiersum (1982) noted a pronounced branching response of coconut roots to local fertilizer application and proposed a simple soil nutrient test. Roots of mature, field grown trees can be induced to grow in a mini-basin with a nutrient solution of various composition. The intensity of the local stimulation of branch root development can be taken as an indicator of which nutrient is in short supply in the tree as a whole. A similar method, based on a modified in-growth core technique, was used by Hairiah *et al.* (1991).

Putz and Canham (1992 and personal communication) and Putz *et al.* (1994) found no differences between trees and shrubs in below-ground architectural plasticity or in root extension along a nutrient gradient. Species from a poor habitat, however, tended to have higher root plasticity (response to local nutrient supply) and root growth rates than species generally occurring in nutrient richer habitats. This finding was contrary to a prediction by Grime (1979), but is consistent with a higher relative spatial heterogeneity of nutrient availability on poor sites.

Deep rooting is common in xerophytic species like *Alhagi camelorum* (25 m recorded), *Glycyrrhiza glabra* (10–15 m), *Andina* sp. (18–19 m) (Daubenmire, 1959) and *Acacia senegal* (32 m) (Deans, 1984). Where there is no access to a ground water table, however, desert shrubs may have a very extensive horizontal root system to intercept rainfall from a large area. Roots of the small desert shrub *Tamarix* were found to extend up to 40 m (Ladover (1928) quoted in Daubenmire (1959)).

The considerable genotype × environment interactions mean that statements about the 'typical' root development of tree and crop species lose their validity when transferred to a substantially different soil type, soil fertility status or climatic zone. Simple observation methods are thus needed to 'ground-truth' generalizations about root patterns. Evidence of genotypic differences in tree root distribution is scarce, but this is probably due to a lack of suitable observation methods rather than to a true lack of genotypic variation in root characteristics. Vandenbeldt (1991) found clear differences in rooting depth of young plants of *Faidherbia albida* genotypes from western and southern Africa.

Considerable genotypic differences exist in tolerance to stress factors such as poor aeration, high soil bulk density and/or high Al^{3+} content. Coster (1932b) compared the waterlogging resistance of a large number of tropical trees and found considerable differences: some trees wilt and die back within one or a few days of waterlogging, other non-wetland trees can endure it for a half year or more. Waterlogging resistance is normally based on effective internal aeration structures (van Noordwijk and Brouwer, 1993), which are very well developed in mangrove trees, in addition to pneumatophores for 'breathing' above the soil or water surface. Many tree species, however, are able to develop at least some roots below the water table. The need for internal aeration structures above the water table (for short distance O_2 transport) depends on soil architecture and root–soil contact (Kooistra and van Noordwijk, 1995).

Horizontal and vertical distribution
Simplified curves fitted to actual root distributions can be used for models at level 3 (see above). Root length densities of most crops decrease with depth. Graphs of the logarithm of the root length density against depth normally show a linear trend, except for soils with specific layers restricting or stimulating root development. A two-parameter descriptive model based on an exponential decay can thus be used to describe $L_{rv}(h)$, the root length density as a function of depth h:

$$L_{rv}(h) = bL_{ra}e^{-bh} \qquad (9.7)$$

where L_{ra} = root length per unit of cropped area (cm cm^{-2}) and b is the slope of the regression line of $\log(L_{rv})$ on h. Exceptions from this exponential pattern can be found in relatively deep-rooted trees such as *Dactyl-*

adenia (*Acioa*) on acid soils (Ruhigwa *et al.* 1992) or *Eucalyptus camaldu-lensis* (Jonsson *et al.*, 1988).

Root density normally also decreases with increasing distance to the plant in the horizontal direction. The combined effect in a two-dimensional plane radial to a soil cylinder with the plant in its centre can be described by elliptical models of the general form:

$$L_{rv} = ae^{-b\sqrt{h^2+cr^2}} \tag{9.8}$$

where *r* is radial distance to the plant and *a*, *b* and *c* are parameters. The parameter *c* indicates whether root length density decreases faster with radial than vertical distance ($c > 1$) or vice versa ($c < 1$).

Branching models

A number of parameters are used as indicators of different root functions (van Noordwijk and De Willigen, 1991):

1. length of the longest (deepest) root, roughly indicating the *exploration* of soil zones;
2. total length or surface area of live roots, governing the *exploitation* of most nutrients and water from the soil zones explored;
3. number of root tips and associated young unsuberized root sections, governing cytokinin production and Ca uptake; and
4. root dry weight, indicating the amount of carbon in the root system and giving an initial estimate of the C costs of making and maintaining roots.

Relationships between these parameters, such as 'specific root length' or length per unit dry weight (van Noordwijk and Brouwer, 1991), indicate the constraints that plants face in combining these functions. The relationships can be studied in the actual shapes of the root systems, but can also be derived once the underlying morphogenetic branching rules are known. A combination of an easily observable indicator of root system size and knowledge of the morphogenetic rules will be of value for practical root studies.

Fitter and co-workers (Fitter, 1986; Fitter *et al.*, 1988; Fitter and Stickland, 1991) have described topological and fractal aspects of branched root systems. Fitter (1991) specified five types of information which are needed to reconstruct a three-dimensional model of a root: (i) the number of *internal* and *external* links (without and with top meristem, respectively); (ii) the lengths of the links; (iii) the distribution of branches within the root, i.e. the *topology*; (iv) the branching angles; and (v) the diameter per link. If one is interested in total size, rather than three-dimensional distribution, the branching angles are not relevant. For the total length, rather than volume or weight, the diameters can be left out and only the first three types of information are needed.

Leonardo da Vinci (Mandelbrot, 1983) claimed that the cross sectional area (csa) of the main stem is equal to the sum of the cross sectional areas of tree branches. The same rule might apply to rivers (at least in a landscape with constant slope), and may be based on the approximately constant volume of water passing through the river system from the sum of all sources to the final sink. A constant sum of squared diameters in trees might indicate a constant resistance to longitudinal water flow, if individual xylem cells have a constant diameter (the maximum of which is determined by the risk for cavitation in large cells, Milburn, 1979) and functional xylem forms a constant proportion of total stem diameter. For tree stems, stability and strength requirements may be as relevant as water transport capacities in determining stem diameters, but the 'constant squared diameter rule' or 'pipe-stem model' (Shindzaki *et al.*, 1965) at least forms a valuable point of reference in studying trees. Empirically a close relationship between cross sectional area of sapwood and total leaf area has been established (Waring and Schlesinger, 1985). A similar rule might apply to tree root systems and this assumption forms the basis of fractal branching models (Spek and van Noordwijk, 1994; van Noordwijk *et al.*, 1994b). According to these models, a relationship can be expected between the diameter of roots at the stem base ('proximal roots') and the total length of that root, given a few parameters of the branching pattern which can be obtained from small samples at some distance from the tree (Fig. 9.5). Tests of the assumptions underlying these models should be made under field conditions.

Measuring the 'proximal' diameter of roots, i.e. the diameter of the root segment connected to the stem base, is relatively simple (Fig. 9.5b), and can be done after careful excavation, e.g. of a half sphere of 0.3 m radius, without damaging the tree (van Noordwijk *et al.*, 1991a). Relationships between proximal root diameters and the total length of all root links obviously depend on the root branching pattern (Fig. 9.5a), but we may hope to identify this 'branching pattern' from a few relatively small samples, if we can find a suitable system for quantification.

Santantonio *et al.* (1977) reported a highly significant correlation of root end diameter and total root fresh weight in Douglas fir, which could be used to estimate the biomass of roots broken off in windthrows, but which also indicates that the proximal root diameter may be a good indicator of root size. Some attempts have been made to relate root biomass to total stem diameter at breast height as well (Santantonio *et al.*, 1977; Kuiper *et al.*, 1990). A close relationship between (the logarithm of) total root biomass and (the logarithm of) tree diameter at breast height was confirmed for tropical trees by Freezaillah and Sandrasegaran (1969) and Sanford (1989), but not for fine root biomass (Egunjobi, 1975).

Analysing the architectural rules underlying root development (Atger, 1991; Francon, 1991) opens perspectives for visualizing and predicting three-dimensional structures as they develop in time, but a major difficulty

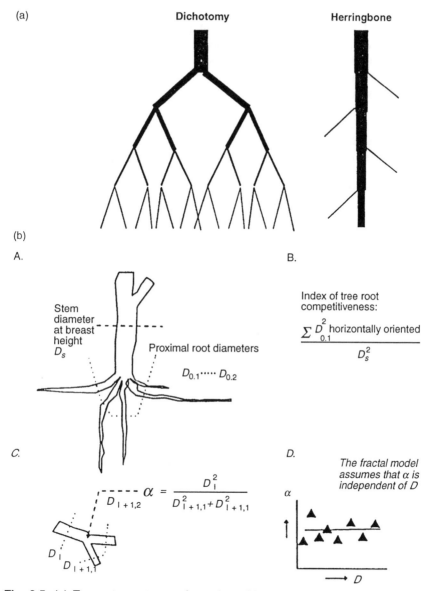

Fig. 9.5. (a) Two extreme types of root branching: dichotomous and herringbone. Under the pipestem model the ratio between initial diameter and number of links is the same for both patterns (and all intermediary ones) (van Noordwijk *et al.*, 1994b). (b) Measurement of proximal tree root diameters.

still is formed by difficulties of incorporating the large plasticity in response
to local soil conditions into the analytical framework. Still, to a considerable
extent secondary thickening of transport roots occurs in response to and in
coordination with fine branch root development, so the branching pattern
present at any time is likely to contain more regularity and predictability
than one would expect from the way it is formed.

Empirical relation between root pattern and tree growth rate
Coster (1932a) studied a large number of tree species as potential understorey trees for teak (*Tectona grandis*) plantations. Considerable variation
was found in root patterns of different species growing on the same (deep,
neutral) soil in Java. No simple relations between above- and below-ground
dimensions existed, contrary to widespread beliefs that crown diameter and
root spread are related. Hairiah and van Noordwijk (1986) re-analysed the
data and classified the trees in three groups (Fig. 9.6).

Trees with a deep tap root and few superficial, horizontally oriented
roots generally showed a slow initial growth of the shoot and had a
shoot : root ratio on a dry weight basis of 0.4 to 2.5. Trees with a deep tap
root as well as extensive horizontal root development in the topsoil showed
a faster shoot growth and had shoot : root ratios of 2 to 6. A group of trees
and shrubs with only shallow rooting had shoot : root ratios of 2 to 30
(Hairiah and van Noordwijk, 1986). Figure 9.7 shows that shoot dry weight
at 6 months is related to tap root length in trees without strong lateral root
development. In trees with more than two long lateral roots, however, there
is no relation between tap root length and shoot dry weight (if we consider
the point in the upper right corner of the graph (*Sesbania sesban*) as an
outlier). Average shoot weight is much higher for trees with at least two
horizontal branch roots of at least 1 m length than for trees without such
exploration of the topsoil. The often heard requirement of 'fast growing
trees with deep root development, causing little competition with shallow
rooted crops', i.e. squares in the upper right corner of the graph, seems thus
to ask for the impossible, at least based on the initial growth.

The data from Coster (1932a, b) showed that *Leucaena leucocephala* and
Acacia villosa gave the best chances to complement the relatively shallow
Tectona grandis root system: after a moderately rapid establishment phase
with some horizontal roots in the topsoil as well as a deep tap root, later root
development was largely confined to the subsoil.

Unfortunately, the statement of Howard (1924) has not lost any of its
actuality: 'it is remarkable... that no detailed information on the distribution of root activity during the year is available. Nevertheless it is clearly
essential in the ecological studies of the future.' Yet many efforts and some
real progress have been made. Srivastava *et al.* (1986) found maximum root
growth of teak in each of the rainy seasons in a bi-modal pattern with the
greatest seasonality in the finest roots.

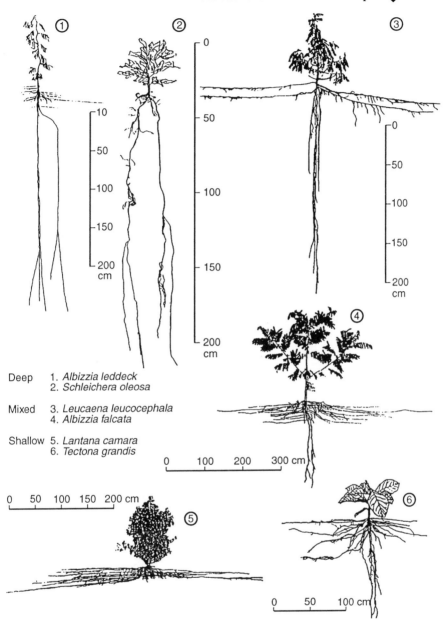

Deep   1. *Albizzia leddeck*
        2. *Schleichera oleosa*

Mixed   3. *Leucaena leucocephala*
        4. *Albizzia falcata*

Shallow 5. *Lantana camara*
        6. *Tectona grandis*

Fig. 9.6. Three root distribution types of young trees tested as accompanying species for teak (*Tectona grandis*) plantations (Coster, 1932a).

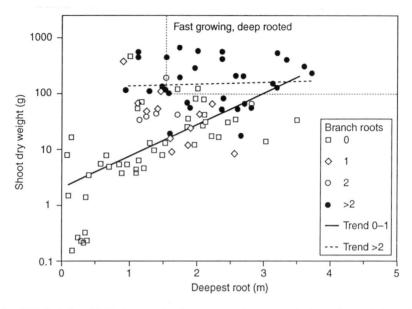

Fig. 9.7. Relationship between depth of the main root and shoot dry weight at 6 months, for a large number of trees and shrubs, classified according to the number of horizontal branch roots of more than 1 m length (data from Coster, 1932a).

Tree management effects on root distribution

Few trees in agroforestry systems are allowed to grow undisturbed. Branches are lopped, trees are pollarded or pruned to obtain fodder, fuelwood or green manure and/or to reduce shading of crops. All leaves of apple trees in East Java (Indonesia) are regularly stripped off to induce flowering in a tree which does not get the environmental trigger from its original area of distribution. Most of these tree management practices are on the above-ground parts, but on the basis of the functional shoot–root equilibrium concept we may expect below-ground effects as well. Reducing the leaf canopy reduces the transpirational demand and thus the 'need' for new root growth, but also the current carbohydrate supply to the root system which is required for root growth and maintenance. Under more severe pruning regimes, recovery of the tree depends on the remobilization of stored energy reserves in parts of the stem or storage roots not affected by pruning. Reduced carbohydrate supply to the roots after removing part of the tree foliage may be expected to cause a dieback of fine roots and nodules, but few hard data exist on such effects (Fownes and Anderson, 1991; Smucker *et al.*, 1994). Root death and subsequent decay will increase nutrient mineralization in the soil, so crops can benefit from pruning the tree by reduced shading as well as improved nutrition from both above- and below-ground sources. The latter may be especially relevant for well-nodulated trees where direct transfer of N to crop roots is possible after dieback of the tree roots. Rapid transfer of P from dying roots to living ones has been found in mycorrhizal roots, perhaps through direct hyphal links (Ritz and Newman, 1985). The decomposition rate of roots is likely to be slower than that for leaves, due to higher fibre and lignin content, lack of easy access via stomata and perhaps because of higher concentrations of toxic metals such as Al (Bloomfield *et al.*, 1993). Decomposition rates for roots were found to decrease with root diameter (Fahey *et al.*, 1988; compare Chapter 7).

The effect of partial pruning or lopping off branches is inadequately known and will partly depend on the stem anatomy. In trees with well-integrated transport tissue, the loss of a few branches will only moderately reduce total carbohydrate supply and no effects on the root system may be noticeable. In trees with a direct connection between individual branches and roots, cutting branches will directly affect the associated roots.

Pruning the trees may affect subsequent root distribution as well (Rao *et al.*, 1993). Van Noordwijk *et al.* (1991a) reported that the lower the height of pruning *Peltophorum dasyrachis* trees the greater the number and the smaller the diameter of proximal roots. The hypothesis was formulated that a reduced stem height after pruning reduces the chances of survival of, and maintenance of apical dominance by, meristems of main root axes. Regrowth of the root system during and after recovery of the shoot thus increasingly depends on new roots starting at the stem base. A further experiment (Hairiah *et al.*, 1992) confirmed the hypothesis for a number of tree species (*Calliandra calothyrsus*, *Senna siamea*, *Gliricidia sepium*, *Paraserianthes falcataria*, *Peltophorum dasyrrachis*), although *Gliricidia* forms thick fleshy storage roots at reduced stem pruning height. The larger number of proximal roots formed at reduced pruning height is, however, associated with a more superficial root distribution. Thus, while a lower tree pruning height may be desirable to reduce above-ground competition and/or to induce death of fine rootlets to increase nutrient transfer to crops, it thus also tends to increase subsequent competition between trees and crops in the topsoil.

Roots and their symbionts

Any account of root ecology, however brief, has to mention the major root symbionts. Mycorrhiza (fungus + root) formation rather than root development is the norm in most trees as well as crops, although there are notable exceptions in a number of plant families. Janse (1896) was among the first to describe the morphological structures that give evidence of this symbiosis, based on plants growing in the botanical garden in Bogor. In the century after that, enormous progress was made in understanding the function of these structures and the way mycorrhizas can be managed (Bowen, 1984; Sieverding, 1991). The literature still tends to emphasize 'infection percentages' rather than 'live hyphal length' as the main parameter, partly due to methodological problems in quantifying the latter. Thus uptake possibilities of mycorrhizal systems are more difficult to quantify as yet than it is for systems consisting of roots only.

The symbionts responsible for N_2 fixation in *Leguminosae* and a number of other plant families have also received due attention. Giller and Wilson (1991) reviewed nitrogen fixation in tropical cropping systems, and include references on tree crops and multipurpose trees.

Concepts for tree–crop interactions

Sequential versus simultaneous agroforestry systems

The relevant root parameters for predicting uptake efficiency depend not only on the resource studied, but also on the complexity of the agricultural system. In intensive horticulture with nearly complete technical control over nutrient and water supply, fairly small root systems may allow very high crop productions in a situation where resource use efficiency ranges from very low to very high, depending on the technical perfection of the often soilless production system (van Noordwijk, 1990). In field crops grown as a monoculture, the technical possibilities for ensuring a supply to the crop of water and nutrients where and when needed are far less; the soil has to act as a buffer, temporarily storing these resources. Adjustment of supply and demand in both time and space (synchrony and synlocation) become critical factors. In mixed cropping systems (including grasslands) the below-ground interactions between the various plant species add a level of complexity to the system; on one hand it opens possibilities of complementarity in using the space and thus the stored resources, hence im-proving overall resource use efficiency, on the other hand, it means that root length densities which would be sufficient for efficient resource use in a monoculture may not be sufficient in a competitive situation. Agroforestry systems are yet another more complex step, as the perennial and annual components have separate time frames in which to interact.

The supply of nutrients such as nitrogen from organic sources will never be completely synchronous with nutrient demand by crops. In so far as supply precedes demand, temporary storage of mineral nitrogen is required in the crop root zone. In climatic zones without rainfall surplus during the cropping season, such storage will be possible and there will be no compelling need for improving synchrony in order to achieve a high uptake efficiency. In climates such as the humid tropics, however, where rainfall exceeds evapotranspiration during the growing season, products of early mineralization will be washed into deeper layers of the soil (Fig. 9.8). If crop rooting is shallow, as is common on the acid soils typical of this climatic zone, nutrients will be leached beyond the crop roots. Deep rooted components of mixed cropping systems can then act as a 'safety net',

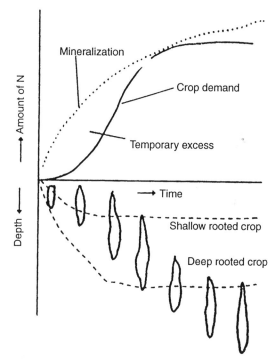

Fig. 9.8. Synchrony hypothesis: the time pattern of mineralization and crop demand (both shown in cumulative form) generally do not match; a temporary stock of mineral N in the soil will leach to deeper layers, depending on rainfall, and can be out of reach of shallow rooted crops by the time they need it.

intercepting N on its way to deeper layers (van Noordwijk and De Willigen, 1991; see below).

A distinction should be made between agroforestry systems where trees and crops use the same land simultaneously, and sequential systems such as improved fallows (E. Torquebiau, 1993, unpublished). Trees with abundant superficial roots may not be suitable for the first, but may be desirable for the second type of system. In sequential systems soil conditions at the time of transition of the tree to crop phase are the most important criterion. The tree may have left a considerable litter layer on the soil surface and a network of decaying tree roots in the soil. Effects on the subsequent crop may be based on the total soil organic matter and nutrient mineralization potential of the soil, but also on more specific facilitation of crop root development by using the old tree root channels. The latter is especially relevant on soils where soil compaction or Al^{3+} toxicity restrict crop root development. Old tree root channels provide easy pathways into a compact soil and a coating of organic matter which may help to detoxify

Al^{3+} (van Noordwijk *et al.*, 1991b). In simultaneous agroforestry systems, below-ground interactions are likely to be dominated by competition for water and nutrients. Complementarity in resource use is possible, however, especially under conditions of high leaching rates.

The soil water balance, as affected by climate, irrigation and drainage, has a major influence on root functions. In the temperate climatic zone of the northern hemisphere, the main crop growing season normally has a rainfall deficit: drying soil conditions hamper diffusive transport and hence increase the root length density required for uptake, but it also means that leaching is mainly confined to the autumn and winter period, after the growing season. A lack of synchrony between N mineralization and N demand which would lead to a build-up of mineral N in the topsoil is not a real problem under these conditions. In fact the main problem is that mineralization is too slow in spring. In the humid tropics, however, with a net rainfall surplus during most of the growing season, any accumulation of mineral N will be leached rapidly from the topsoil to deeper layers. Under such conditions synchrony of N mineralization and N demand is essential for obtaining high N use efficiencies.

'Nutrient pumps' and 'safety nets'

A letter to the *Tropical Agriculturalist* (Colombo, Ceylon) in 1887 stated that:

> Grevillea is valuable in the field, as its light shade if planted at, say, 30 to 36 feet apart, is rather beneficial to tea. But the great good it does is the bringing up of plant food from the subsoil, and distributing the same in the form of fallen leaves, ... which, too, are useful in preventing surface wash while decomposing on the ground.
>
> (Harwood and Getahun, 1990)

The idea that trees act as a 'nutrient pump' has thus been around for at least a century. Few hard data have accumulated, however, as it is no easy task to identify which part of the net nutrient uptake of a tree comes from deep or superficial soil layers (see Chapter 4). A large amount of circumstantial evidence is available, however. The nutrient pump hypothesis could be valid for both sequential and simultaneous agroforestry systems. A number of conditions need to be met, however, for trees to act as nutrient pumps:

1. the tree should have a considerable amount of fine roots and/or mycorrhiza in deep soil layers;
2. deep soil layers should contain considerable nutrient stocks in directly available form or as weatherable minerals in the soil or in a saprolite layer;
3. soil water content at depth should be sufficient to allow diffusive transport to the roots.

These conditions indicate that the possible role of deep-rooted trees as nutrient pumps is likely to be small on soils with limited weatherable minerals in the subsoil (most oxisols and ultisols fall in this category) or in climates with a limited annual depth of wetting. Uptake activity from deeper layers may be expected especially where nutrient stock and root development in deeper layers is larger than that in more superficial layers of the soil and total demand cannot be met from the topsoil.

If trees or shrubs develop a root system under the main crop root zone and with sufficient horizontal spread, these roots may act as a 'safety-net', intercepting mineral nutrients leaching from the crop root zone (Fig. 9.9). Through litterfall or pruning such nutrients may return to the topsoil later on and have a new chance of uptake by crops. In contrast to the 'nutrient pump' hypothesis, the 'safety-net' hypothesis is not restricted to specific soil types, but it depends on a rather specific root distribution pattern of the tree and crop component of an agroforestry component and on a water balance leading to leaching of nutrients beyond the crop root zone.

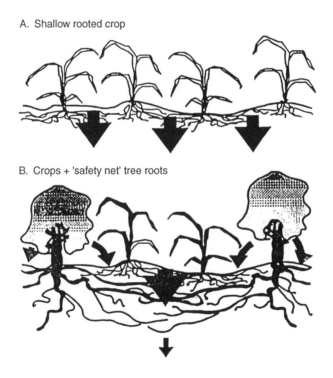

Fig. 9.9. Safety net hypothesis of tree roots intercepting nutrients leaching from a shallow crop root zone.

The safety-net role seems particularly valid for simultaneous agroforestry systems, but under certain conditions may apply to sequential systems as well. Van Noordwijk (1989) used a simple leaching model (related to time–depth curves) to analyse under what leaching rates (and consequently for which combinations of net precipitation surplus and apparent nutrient adsorption constants, K_a) a deep rooted component can intercept nutrients leached beyond the reach of a previous, shallow rooted component (Fig. 9.10). A limited window of opportunity exists for such interception, but only when the rooting depth of the fallow vegetation substantially exceeds that of the crop (Table 9.2). The chances for recovery of leached nutrients increase when K_a increases with depth, as may occur in soils with substantial nitrate adsorption capacity in deeper layers.

Hydraulic lift and heterogeneous water infiltration
Roots have a physiologically determined resistance to entry of water, but they can also leak water to their environment, if the water potential in the surrounding soil is more negative than in the plant. This condition can be expected when the topsoil dries out while roots in deeper layers are still able to take up water; it is most likely at night when plant water potential can

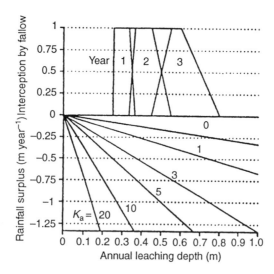

Fig. 9.10. Possibilities for deep rooted fallows to intercept nutrients leached from a shallow rooted crop. The lower part of the diagram gives a nomogram of nutrient leaching depth as a function of rainfall surplus (rainfall – runoff – evapotranspiration) and apparent adsorption constant K_a (ml cm^{-3}). The upper part of the graph shows the chances of recovery by a deep rooted fallow vegetation of nutrients lost from a shallow crop root zone, given this annual nutrient leaching depth (van Noordwijk, 1989).

Table 9.2. Range of values for the annual excess of rainfall over evapotranspiration $L_w(l)$, approximate annual rainfall zone and apparent adsorption constant K_a which allow a deep rooted fallow (crop rooting depth 0.3 m, fallow rooting depth 0.75, 1.5 and 2.5 m in years 1, 2 and 3, respectively) to intercept nutrients (van Noordwijk, 1989).

K_a (ml cm^{-3})	$L_w(l)$ (m)	Annual rainfall (m)
0	0.1–0.25	1.1–1.6
1	0.2–0.5	1.2–1.8
3	0.4–1.0	1.4–2.3
5	0.6–1.5	1.6–2.8
10	1.1–2.75	2.1–4.0
20	2.1–5.25	3.1–6.5

recover from the daytime stress. The total amounts of water leaking out of root systems by this 'hydraulic lift' phenomenon are generally small compared with the daily transpirational demand, but they can facilitate nutrient uptake from topsoil layers. The water leaking out of the roots of a deep rooted plant can be utilized by other plants as well. Caldwell and Richards (1989) used tritium (^{3}H) labelled water to show that such transfer occurred between a deep rooted shrub and a more shallow rooted grass. The phenomenon has been demonstrated in 'split-root' experiments as well (Baker and Van Bavel, 1988; Xu and Bland, 1993).

In semi-arid climates, trees may have a pronounced effect on the pattern of water infiltration (see Chapter 6). Their canopies intercept rainfall and, especially isolated trees with a 'funnel' shaped canopy, can have a high rate of stemflow, causing deep water infiltration under their stem (Knapp, 1973). Trees with umbrella-shaped canopies tend to have a high rate of water infiltration at the perimeter of the canopy.

Complementarity and competition
Without competition between plants environmental resources would probably not be used efficiently. Maximum light interception depends on a closed crop canopy, where all individual plants experience considerable competition, and reach a much smaller size than they would do in a more open stand. Competition between plant species is only a problem if its effects are more pronounced than those of intraspecific competition, and especially when this affects the plant component which is most highly valued (see Chapter 4). For light, plant canopy height is a simple index for the competitive strength of any plant. Below-ground resources cannot be treated in a similar one-dimensional way, however. Water and nutrients are

stored in the soil, so time of use should be considered, as well as at least two dimensions for describing horizontal and vertical stratification.

The general wisdom is that complementarity in root distribution is a key to the success of simultaneous agroforestry systems. Evidence for this hypothesis is widespread. *Paulownia* species are widely grown in China, intercropped with wheat, maize, groundnut etc. *Paulownia* has the majority of its fine roots in the layer 40–100 cm deep, below the crop root zone (Table 9.3). The apparent success of this intercropping system, similar to the *Grevillea* system in Kenya, coincides with a complementarity in fine root distribution, accompanied by a favourable above-ground tree morphology and phenology (Huxley *et al.*, 1994).

Nelliat *et al.* (1974) suggested that horizontally separated root systems could be the basis for complementarity in coconut–cocoa–pineapple multistorey agroforestry systems. On the basis of the older literature on coconut in Indonesia (reviewed by Wiersum, 1982), a shallow, but extensive, root system can be expected and the topsoil of any mixed stand would probably be nearly completely exploited by the coconut.

As a first approach to a process-based description of 'competitive strength', we may assume that the N_{res} term of Equation 9.3 indicates the amount left in the soil. If the combined demand A of all plants cannot be met, their relative 'competitive strength' may be based on their N_{res} value, and thus be related to local root length density.

As already discussed, uptake of water and nutrients is often directly related to above-ground demand, i.e. the size of the leaf canopy and the above-ground sink strength for nutrients. Also, the 'pipestem' model and similar approaches suggest that tree stem diameter can give a first indication of this, at least within a species. The fractal branching models suggest that the total amount of fine roots is related to proximal root diameters. Thus the ratio of superficial roots and stem diameter can be used as a simple indicator of the degree to which the tree can depend on topsoil resources, and thus for its competitive strength, when combined with shallow rooted crops (see below).

A number of hypotheses have been put forward for positive tree–crop root interactions. Stimulated nodulation of trees in the neighbourhood of N-depleting crops was postulated by van Noordwijk and Dommergues (1990), but the available evidence is not conclusive as yet. Complementar-

Table 9.3. Root distribution of 10-year-old *Paulownia elongata* stands (average for three sites) in China, according to Zhu Zsao-Hua *et al.* (1986).

Depth (cm)	0–20	20–40	40–60	60–80	80–100	100–150	150–200
% of fine root weight	1.9	12.9	38.9	18.1	19.1	6.7	2.3

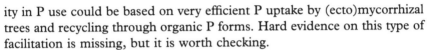

ity in P use could be based on very efficient P uptake by (ecto)mycorrhizal trees and recycling through organic P forms. Hard evidence on this type of facilitation is missing, but it is worth checking.

Table 9.4 summarizes the types of below-ground interactions which can occur in simultaneous as well as sequential agroforestry systems. The table indicates how to measure the various effects and which tree characteristics are desirable to optimize tree–crop species combinations.

Species selection is the major option to meet these desiderata, but tree management can have secondary effects. Above-ground tree management, including pruning, will have immediate effects on root function (demand for water and nutrients) and longer term effects on root distribution.

Methods for Root Studies in Agroforestry

Separating below- and above-ground interactions

A first question is how to separate above- from below-ground interactions experimentally. Figure 9.11 shows results of an experiment by Coster (1932b) on the interactions between mature *Tectona grandis* stands and neighbouring plots where trees and/or food crops were planted in a taungya system. The generally poor growth of tree seedlings and maize was dramatically improved when a trench was made around the old growth stand, excluding root competition. The shade effects only extended over a few metres from the old stand, while apparent root competition extended over approximately 20 m. The root trenches were refilled with soil and within 1 year *T. grandis* roots had re-invaded the neighbouring plots. Coster's approach to measure the below-ground competition effects by the positive yield effect of root separation is still used as a first approximation (see Chapter 4). A number of reports on root pruning effects have been published recently for hedgerow intercropping (Fernandes *et al.*, 1993; Nair *et al.*, 1994).

Some caution is needed in interpreting the results in the context of the tree–crop interaction equation (Chapter 1), because root pruning may reduce shoot growth and thus above-ground interaction as well. Positive crop responses to tree root pruning should, ideally, be compared with responses to equivalent above-ground tree biomass obtained by other means (e.g. shoot pruning, although this can have positive below-ground effects as such).

Putz (1992) tried to separate above- and below-ground competition effects between *Pinus* seedling growth and coppiced hardwoods and vines by a combination of root-trenching and guy-wiring back overtopping trees. In his situation below-ground competition was found more important than shading effects. In a similar analysis of the effects of the shrubs *Cornus*

Table 9.4. Types of below-ground tree–crop interactions and tree desiderata.

Interaction process	Measure of effect	Tree desiderata for agroforestry systems	
		Sequential	Simultaneous
Competition for water	Positive crop response on tree root pruning, especially in dry periods; measurement of water flow in horizontally oriented proximal roots	–	Deep rooted trees
Modified water infiltration	Water infiltration rates with and without trees and/or tree mulch	–	Slowly decomposing tree mulch for erosion prevention
Hydraulic lift (water transfer to topsoil)	Day–night cycles in soil water tension close to tree roots; water tracer movement	–	Deep rooted trees
Competition for N, P, K, etc.	Positive crop response on tree root pruning, esp. in dry periods	–	(Relatively) deep rooted trees
Vertical nutrient transfer to topsoil under the tree	Nutrient contents of prunings	Deep rooted trees	(Relatively) deep rooted trees
Horizontal nutrient transfer to topsoil under the tree	Nutrient contents of prunings	Efficiently scavenging trees	Rapid lateral spread; low root density, but large soil volume exploited
Arresting sediment flows ('erosion control')	Biological terrace formation by contour plantings	Creating effective terraces as high fertility zones	Non-competitive 'fertility traps'
Transfer of N etc. from root (nodule) turnover	Quantification of tree root nodule turnover		Rapid root decay (esp. after pruning)
Soil organic matter maintenance by root turnover, litterfall, etc.	Quantification of tree root turnover and litterfall; measurement of dead tree root decomposition rate	Abundant roots in topsoil, rapid root turnover, high content of lignin/polyphenolics	Rapid root turnover, high content of lignin/polyphenolics
Facilitation of crop root growth in old tree root channels (overcoming constraints of soil density or Al toxicity)	Visual check of crop root positions in the soil profile	Deep rooted trees, slow decomposing tree exodermis	–

(continued)

Table 9.4 (continued)

Interaction process	Measure of effect	Tree desiderata for agroforestry systems	
		Sequential	Simultaneous
Stimulation of root symbionts such as VAM fungi	Crop root VAM infection percentages with or without trees	Common VAM fungal partners	Common VAM fungal partners
Stimulation of root pathogens and pests	Crop root damage with or without trees	Lack of common pathogens and pests	Lack of common pathogens and pests
Stimulation of soil fauna (e.g. earthworms)	Faunal activity in crop root zone with or without trees	Year round food supply, by high lignin/polyphenolic content	

racemosa and *Rhus glabra* on the growth of tree seedlings (*Acer rubrum* and *Fraxinus americana*) below-ground competition dominated on a dry and nutrient poor soil, but above-ground interactions in a wet field with alluvial soil, and both were equally important on a more mesic field (Putz *et al.*, 1994).

The direct effects of root trenching when applied to existing stands can give evidence of direct competition effects, but not of the possible long term soil improvement by the trees which may compensate eventually for part of the competitive effect. To estimate such long term trends, trees have to be completely removed in some treatments and crop growth here should be compared with that on controls outside the reach of trees.

Below-ground interactions between trees and crops are fairly well understood for intensively managed hedgerow intercropping systems in humid areas. However, interactions are less pronounced in semi-arid parklands and natural rangelands. Circumstantial evidence exists on the added value of tree in these systems (Breman and Kessler, 1995) and the positive role of tree roots (Groot and Soumaré, 1995), but data to separate and to quantify above- and below-ground interactions are lacking. Nutrient accumulation by the tree can be easily quantified, but whether it is based on horizontal or vertical accumulation is less obvious (Kellman, 1979), and whether or not it is effectively recycled to other components is questionable.

Separation between above- and below-ground interactions requires a judicious and labour-intensive approach. An appropriate experimental design to elucidate these interactions can be achieved by a soil-transfer experiment, although there are many difficulties in undertaking this satisfactorily. At the onset of the growing (rainy) season, undisturbed soil columns, preferably PVC or stainless steel cylinders, are taken from

(a)   (b)

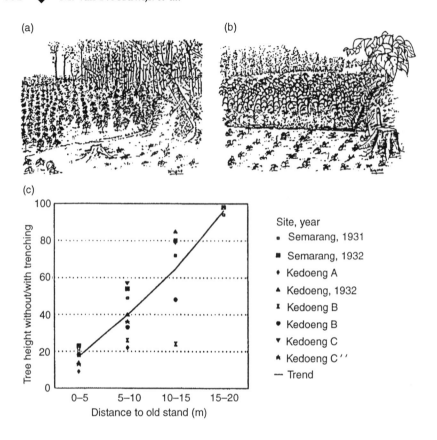

Fig. 9.11. Effects of root trenching (background = tree roots excluded, foreground = tree roots present) to prevent the roots of old *Tectona grandis* stands competing for water and nutrients with new *T. grandis* seedlings (a) and maize (b); (c) Growth of young *T. grandis* trees with and without root trenching as a function of distance to old *T. grandis* stands (Coster, 1933; (a) and (b) artist's impression (Mr Wihyono), based on old photographs).

beneath the tree canopy (Fig. 9.12, zone a) and outside the tree canopy but within the zone exploited by tree roots (Fig. 9.12, zone b). These columns can then be interchanged (positions 3 and 6). Next to the positions 3 and 6, cylinders are driven into the soil (positions 2 and 5), thus cutting the tree roots. Underneath and outside the tree, a dense, homogeneous annual crop or grass is sown.

This results in seven different crop production situations or treatments:

1. on 'native' soil underneath the tree in the presence of tree roots,
2. on 'native' soil underneath the tree without active tree roots,

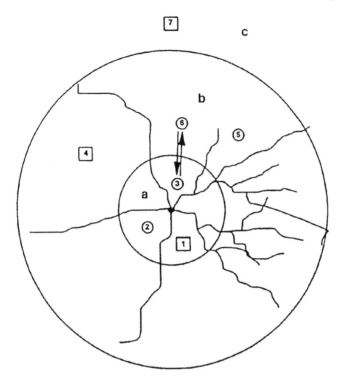

Fig. 9.12. Different cropping positions in so-called soil-transfer experiment.

3. underneath the tree on soil from outside the canopy without active tree roots,

4. on 'native' soil outside the canopy in the presence of tree roots,

5. on 'native' soil outside the canopy without tree roots,

6. on soil from underneath the tree placed outside the canopy without the presence of tree roots,

7. on native soil outside the influence of tree roots (zone c).

Comparison of crop production and nutrient uptake for the different production situations enables the effects of microclimate and soil fertility and effects induced by the presence of tree roots to be distinguished. Relevant differences in microclimate underneath and outside the canopy (air and soil temperature, radiation level, soil humidity) need to be measured. Although theoretically simple, practical and statistical aspects may complicate the experiment. The small surface of the cylinders and the small differences in crop growth between treatments necessitate many replicates, but these will be limited by the space available under each tree.

Different trees cannot always be considered as repetitions as canopy characteristics etc. may differ among trees, so that allowance must be made for orientation under the tree. To enable satisfactory data interpretation, the root distribution of the studied tree species needs to be known.

Quantifying root distribution

Basic methods for observing and quantifying tree and crop root biomass and length are given by Anderson and Ingram (1993). New developments are based on the fractal root branching models and the allometry between proximal root diameter and other, more functional root characteristics.

To study interactions between trees and the herbaceous layer in semi-arid tropical savanna rangelands in Mali, West Africa, observations were made on root systems of two Sahelian tree species, growing on deep sandy or sandy loam soils, and the herbaceous layer surrounding them (Groot and Soumaré, 1995). To characterize the structural root system, for each species the root systems of five trees were excavated, by digging a pit around each trunk with a radius of 2 m. Soil was removed using shovels, small picks, hoes and knives; soil around roots was removed using paint brushes and toilet brushes. Only roots with diameter > 2 mm were considered. For each tree, four primary roots were excavated and total root length was determined. Cross-sections of representative structural root systems of two species are given in Fig. 9.13a, b.

Acacia seyal (*Mimosaceae*) was characterized by a thin deep tap root, reaching depths down to 6 m. Lateral roots were mainly concentrated in the upper 40 cm of soil (Fig. 9.13a) and their diameter decreased gradually with length. Several secondary roots branched vertically downwards, sometimes reaching depths of over 5 m. Lateral roots reached lengths up to 25.7 m, being seven times the crown radius; on average, the land exploited was 25 times the area covered by the tree crown. *Sclerocarya birrea* (*Anacardiaceae*) (Fig. 9.13b) has a thick, relatively short tap root, reaching depths of 2.4 m. Lateral roots branched at the upper 100 cm of the tap root and gradually curved towards the soil surface with increasing distance from the trunk. Lateral roots were thick at the trunk base but their diameter decreased sharply with length within the first 100 cm. Sinker roots were absent. Maximum lateral root length was 50 m (10.2 times the crown radius) and exploited land area 34 times the tree crown area.

A relationship between diameter of main roots as measured near the trunk base and their length was established, allowing estimation of the surface occupied by tree roots (Fig. 9.14). Root diameter measured at different distances from the trunk as a function of this distance was best described by a negative exponential regression curve ($D_L = D_0 e^{-k^* L}$, where D_0 is base diameter and L represents the distance from the trunk).

To investigate fine-root distribution of both trees and herbs, so-called

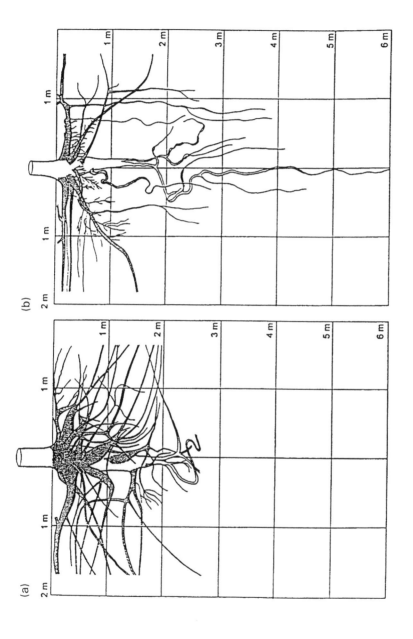
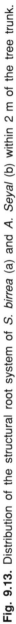

Fig. 9.13. Distribution of the structural root system of *S. birrea* (a) and *A. Seyal* (b) within 2 m of the tree trunk.

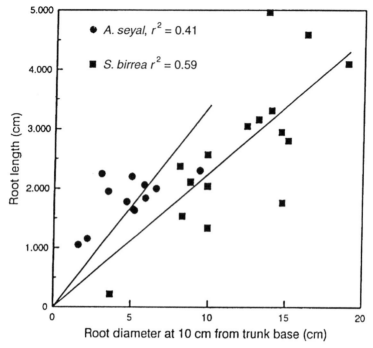

Fig. 9.14. Relationship between proximal root diameter (at 10 cm from the stem) and root length for *Sclerocaryea birrea* and *Acacia seyal* in Mali.

fakir-bed or pinboard monolith soil samples were taken perpendicular to the tree radius on two trees of comparable size for each species, at the crown limit (*r*), at 2*r* and at 3*r*. After washing the monolith, roots were cut in 10-cm compartments corresponding to soil layers and were separated into tree roots and herbaceous roots, and oven-dried. Only roots with a diameter < 2 mm were considered. The obtained root weights were averaged for the two trees. Results are presented in Fig. 9.15. For both species, biomass of fine roots under trees was considerably higher as compared with the open field, which explains partly the higher levels of organic matter and soil fertility often observed under and near trees. The majority of fine tree roots were located below those of herbaceous roots, showing that trees might recycle nutrients which, in the absence of trees, are lost from the system by leaching.

Need for simple, farmer-level criteria and observation methods

Current root research methods on trees are laborious and cannot be directly related to a farmer's criteria for selecting and judging the performance of trees. If root research stays in the domain of 'experts' it will not

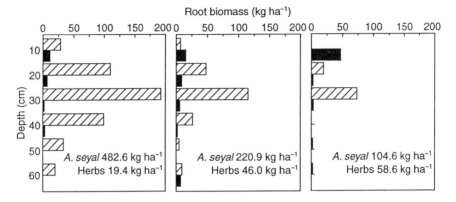

Fig. 9.15. Weights of fine roots (< 2 mm) of *A. seyal* and the herbaceous layer (from left to right) at the crown limit (*r*), and at two zones away from the crown (2*r* and 3*r*). ■, Herbs; ▨, *A. seyal*.

Table 9.5. A. Protocol for quantifying proximal tree root diameters and the index of tree root competitiveness. B. Protocol for test of fractal characteristic of root branching and measurement of parameters for predicting total root system size from proximal root diameters.

A. Proximal roots

1. Carefully excavate the first part of the proximal roots at the stem base (Fig. 9.5b). For a small tree a 0.3 m half sphere may be sufficient, for larger trees a 0.5–1 m half sphere will be needed. While excavating all major roots should be left intact; destruction of most of the fine roots cannot be avoided. Check for 'sinker' roots (vertically oriented roots starting from horizontal roots, often close to the tree stem).

2. Measure the root diameter of all proximal roots (i.e. roots originating from the stem base or as laterals from the top part of the tap root) and classify them by orientation (angle with a horizontal plane). Root diameter measurements should be made outside the range of obvious thickening close to the branching point or buttress roots (they normally taper off rapidly).

3. Measure stem diameter (either as 'root collar' diameter or as stem diameter at breast height, depending on the size of the tree).

4. Calculate the sum of root diameter squares for roots with a horizontal (angle with horizontal plane less than 45°), ΣD_{hor}^2, and vertical orientation, ΣD_{vert}^2.

5. A tentative *index of root shallowness* is then calculated as $\Sigma D_{hor}^2 / D_s^2$.

B. Test of fractal branching assumptions

1. Expose parts of the root system, by tracing roots from the stem base. For each branching point where both the previous and subsequent 'links' have been exposed, measure the diameter of each link (halfway from the link or at 5 cm from the previous branching point, outside the thickening often accompanying branching).

2. Analyse data by sorting the roots belonging to a common previous link and calculate the α parameter as $D_{before}^2 / \Sigma D_{after}^2$. Then analyse the regression of α and link length on root diameter. If neither of these regressions has a significant slope, the basic assumptions of fractal branching models are met. The average value of α and link length can now be used in the equations for total length, surface area and volume given by van Noordwijk *et al.* (1994b); if either of the regressions has a significant slope, modified equations will have to be developed (e.g. on the basis of the numeric model given by Spek and van Noordwijk (1994)).

---

contribute to the development of agroforestry systems in the real world. The validity of generalizations about deep- or shallow-rooted trees, competitive and beneficial ones, is likely to be vastly overestimated, unless we develop simple, non-destructive observation methods to check this. As a first step in that direction, the proximal root method and the 'index of tree root shallowness' is presented in Table 9.5. It definitely needs further testing, but it appears to bridge farmer level criteria ('trees where you can see roots on the surface are competitive') through the allometrics of root branching patterns to the functioning of roots in uptake, and thus to one of the major components of tree–soil–crop interactions.

References

Anderson, J.M. and Ingram, J.S.I. (eds) (1993) *Tropical Soil Biology and Fertility, a Handbook of Methods.* CAB International, Wallingford.

Anderson, L.S. and Sinclair, F. (1993) Ecological interactions in agroforestry systems. *Agroforestry Abstracts* 6, 58–91.

Atger, C. (1991) L'architecture racinaire est-elle influencée par le milieu? In: Edelin, C. (ed.) *L'Arbre, Biologie et Développement.* Naturalia Monspeliensia No. h.s., pp. 71–84.

Baker, J.M. and Van Bavel, C.H.M. (1988) Water transfer through cotton plants connecting soil regions of different water potential. *Agronomy Journal* 80, 993–997.

Barber, S.A. (1984) *Soil Nutrient Bio-availability, a Mechanistic Approach.* John Wiley, New York, 398 pp.

Berish, C.W. (1982) Root biomass and surface area in three successional tropical forests. *Canadian Journal of Forest Research* 12, 699–704.

Bloomfield, J., Vogt, K.A. and Vogt, D.J. (1993) Decay rate and substrate quality of fine roots and foliage of two tropical tree species in the Luquillo experimental forest, Puerto Rico. *Plant and Soil* 150, 233–245.

Bowen, G.D. (1984) Tree roots and the use of soil nutrients. In: Bowen, G.D. and Nambiar, E.K.S. (eds) *Nutrition of Plantation Forests.* Academic Press, London, pp. 147–179.

Breman, H. and Kessler, J.J. (1995) *Woody Plants in Agro-Ecosystems of Semiarid Regions.* Springer-Verlag, Heidelberg/Berlin.

Brouwer, R. (1963) Some aspects of the equilibrium between overground and underground plant parts. *Jaarboek IBS Wageningen* 1963, 31–39.

Brouwer, R. (1983) Functional equilibrium: sense or non-sense? *Netherlands Journal of Agricultural Science* 31, 335–348.

Caldwell, M.M. and Richards, J.H. (1989) Hydraulic lift: water efflux from upper roots improves effectiveness of water uptake by deep roots. *Oecologia* 79, 1–5.

Caldwell, M.M., Dudley, L.M. and Lilieholm, B. (1992) Soil solution phosphate, root uptake kinetics and nutrient acquisition: implications for a patchy soil. *Oecologia* 89, 305–309.

Coe, R. (1994) Through the looking glass: ten common problems in alley-cropping research. *Agroforestry Today* 6 (1), 9–11.

Coster, Ch. (1932a) Wortelstudiën in de tropen (Root studies in the tropics). I. De jeugdontwikkeling van het wortelstelsel van een zeventigtal boomen en groenbemesters (Early development of the root system of seventy trees and green manure species). *Korte Mededelingen van het Boschbouw Proefstation* No. 29.

Coster, Ch. (1932b) Wortelstudiën in de tropen (Root studies in the tropics). II. Het wortelstelsel op ouderen leeftijd (Root systems of older trees) and III. De zuurstofbehoefte van het wortelstelsel (Oxygen demands of root systems). *Korte Mededelingen van het Boschbouw Proefstation* No. 31.

Coster, C. (1933) Wortelstudiën in de tropen (Root studies in the tropics). IV. Wortelconcurrentie (Root competition). *Tectona* 26, 450–497.

Daubenmire, R.F. (1959) *Plants and the Environment*. Wiley, New York.

Deans, J.D. (1984) Deep beneath the trees in Senegal. In: *Institute of Terrestrial Ecology Annual Report*. Natural Environment Research Council, Swindon, pp. 12–14.

De Jager, A. (1985) Response of plants to a localized nutrient supply. PhD thesis, Rijks Universiteit Utrecht, 137 pp.

De Willigen, P. and van Noordwijk, M. (1987) Roots for plant production and nutrient use efficiency. Doctoral thesis, Agricultural University Wageningen, 282 pp.

De Willigen, P. and van Noordwijk, M. (1989) Rooting depth, synchronization, synlocalization and N-use efficiency under humid tropical conditions. In: J. van der Heide (ed.) *Nutrient Management for Food Crop Production in Tropical Farming Systems*. Institute for Soil Fertility, Haren, pp. 145–156.

De Willigen, P. and van Noordwijk, M. (1991) Modelling nutrient uptake: from single roots to complete root systems. In: Penning de Vries, F.W.T., van Laar, H.H. and Kropff, M.J. (eds) *Simulation and Systems Analysis for Rice Production (SARP)*. Simulation Monographs, PUDOC, Wageningen, pp. 277–295.

De Willigen, P. and van Noordwijk, M. (1994) Diffusion and mass flow to a root with constant nutrient demand or behaving as a zero-sink. *Soil Science* 157, 162–175.

Dhyani, S.K., Narain, P. and Singh, R.K. (1990) Studies on root distribution of five multipurpose tree species in Doon Valley, India. *Agroforestry Systems* 12, 149–161.

Diggle, A.J. (1988) ROOTMAP – a model in three-dimensional coordinates of the growth and structure of fibrous root systems. *Plant and Soil* 105, 169–178.

Egunjobi, J.K. (1975) Dry matter production of immature stands of *Pinus caribea* in Nigeria. *Oikos* 26, 80–85.

Fahey, T.J., Hughes, J.W., Pu, M. and Arthur, M.A. (1988) Root decomposition and nutrient flux following whole-tree harvest of northern hardwood forest. *Forest Science* 34, 744–768.

Fernandes, E.C.M., Garrity, D.P., Szott, L.T. and Palm, C.A. (1993) Use and potential of domesticated trees for soil improvement. In: Leakey, R.R.B. and Newton, A.C. (eds) *Tropical Trees: the Potential for Domestication and the Rebuilding of Forest Resources*, HMSO, London, pp. 137–147.

Fitter, A.H. (1986) The topology and geometry of plant root systems: influence of watering rate on root system topology in *Trifolium pratense*. *Annals of Botany* 58, 91–101.

Fitter, A.H. (1991) Characteristics and functions of root systems. In: Waisel, Y., Eshel, A. and Kafkafi, U. (eds) *Plant Roots, the Hidden Half.* Marcel Dekker, New York, pp. 3–25.

Fitter, A.H. and Stickland, T.R. (1991) Architectural analysis of plant root systems 2. Influence of nutrient supply on architecturally contrasting plant species. *New Phytologist* 118, 383–389.

Fitter, A.H., Nicholas, R. and Harvey, M.L. (1988) Root system architecture in relation to life history and nutrient supply. *Functional Ecology* 2, 345–351.

Fownes, J.H. and Anderson, D.G. (1991) Change in nodule and root biomass of *Sesbania sesban* and *Leucaena leucocephala* following coppicing. *Plant and Soil* 138, 9–16.

Francon, J. (1991) Sur la modélisation informatique de l'architecture et du développement des végétaux. In: Edelin, C. (ed.) *L'Arbre, Biologie et Développement.* Naturalia Monspeliensia No. h.s., pp. 231–249.

Freezaillah, C.Y. and Sandrasegaran, K. (1969) Preliminary observations on the rooting characteristics of *Pinus caribea* grown in Malaya. *Research Pamphlet* 58, Forest Research Institute of Malaya, Kepong.

Gijsman, A.J., Floris, J., van Noordwijk, M. and Brouwer, G. (1991) An inflatable minirhizotron system for root observations with improved soil/tube contact. *Plant Soil* 134, 261–269.

Giller, K.E. and Wilson, K.J. (1991) *Nitrogen Fixation in Tropical Cropping Systems.* CAB International, Wallingford.

Gillespie, A.R. (1989) Modelling nutrient flux and interspecies competition in agroforestry interplantings. *Agroforestry Systems* 8, 257–265.

Gower, T. (1987) Relations between mineral nutrient availability and fine root biomass in two Costa Rican tropical wet forests: a hypothesis. *Biotropica* 19, 171–175.

Grime, J.P. (1979) *Plant Strategies and Vegetation Processes.* John Wiley, Chichester.

Groot, J.J.R. and Soumaré, A. (1995) Root distribution of *Acacia seyal* and *Sclerocarya birrea* in sahelian rangelands. *Agroforestry Today* 7(1), 9–11.

Hairiah, K. and van Noordwijk, M. (1986) *Root Studies on a Tropical Ultisol in Relation to Nitrogen Management.* Instituut voor Bodemvruchtbaarheid, Haren. Rapport 7–86.

Hairiah, K., van Noordwijk, M. and Setijono, S. (1991) Tolerance to acid soil conditions of the velvet beans *Mucuna pruriens* var. *utilis* and *M. deeringiana.* I. Root development. *Plant Soil* 134, 95–105.

Hairiah, K., van Noordwijk, M., Santoso, B. and Syekhfani, M.S. (1992) Biomass production and root distribution of eight trees and their potential for hedgerow intercropping on an ultisol in Lampung. *AGRIVITA* 15, 54–68.

Hairiah, K., van Noordwijk, M., Stulen, I., Neijboom, F.W. and Kuiper, P.J.C. (1993) P nutrition effects on aluminium avoidance of *Mucuna pruriens* var. *utilis. Experimental and Environmental Botany* 33, 75–83.

Hairiah, K., van Noordwijk, M. and Setijono, S. (1995) Aluminium tolerance and avoidance of *Mucuna pruriens* at different P supply. *Plant and Soil* 171, 77–81.

Harwood, C.E. and Getahun, A. (1990) *Grevillea robusta*: Australian tree finds success in Africa. *Agroforestry Today* 2(1), 8–10.

Hauser, S. (1993) Root distribution of *Dactyladenia (Acioa) barteri* and *Senna (Cassia) siamea* in alley cropping on Ultisol. I. Implication for field experimentation. *Agroforestry Systems* 24, 111–121.

Howard, A. (1924) The effect of grass on trees. *Proceedings of the Royal Society of London* 97B, 284–321.

Huxley, P.A., Pinney, A., Akunda, E. and Muraya, P. (1994) A tree/crop interface orientation experiment with a *Grevillea robusta* hedgerow and maize. *Agroforestry Systems* 26, 23–45.

Janse, J.M. (1896) Les endophytes radicaux de quelques plantes javanaises. *Annales Jardin Botanique Buitenzorg* 14, 53–212.

Jonsson, I., Fidjeland, L., Maghembe, J.A. and Högberg, P. (1988) The vertical distribution of fine roots of five tree species and maize in Morogoro, Tanzania. *Agroforestry Systems* 6, 63–69.

Kellman, M. (1979) Soil enrichment by neotropical savannah trees. *Journal of Geology* 67, 565–577.

Kerfoot, O. (1963) The root system of tropical forest trees. *Empire Forestry Review* 42, 19–26.

Knapp, R. (1973) *Die Vegetation von Africa*. Gustav Fischer Verlag, Stuttgart.

Kooistra, M.J. and van Noordwijk, M. (1995) Soil architecture and distribution of organic carbon. In: Carter, M.R. and Stewart, B.A. (eds) *Structure and Organic Carbon Storage in Agricultural Soils. Advances in Soil Science*, pp. 15–57.

Kuiper, L.C., Bakker, A.J.J. and van Dijk, G.J.E. (1990) Stem and crown parameters related to structural root systems of Douglas fir. *Wageningen Agricultural University Papers* 90-6, 57–67.

Kutschera, L. and Lichtenegger, E. (1982) *Wurzelatalas Mitteleuropäische Grünlandpflanzen. Band 1: Monocotyledonae*. Gustav Fischer, Stuttgart, 516 pp.

Lambers, H. (1983) 'The functional equilibrium', nibbling on the edges of a paradigm. *Netherlands Journal of Agricultural Science* 31, 305–311.

Mandelbrot, B.B. (1983) *The Fractal Geometry of Nature*. Freeman, New York.

Milburn, J.A. (1979) *Water Flow in Plants*. Longman, London.

Nair, P.K.R., Rao, M.R. and Fernandez, E.C.M. (1994) Tree–crop interactions in sustainable agroforestry systems. *Proceedings 15th World Congress of Soil Science*, Acapulco, Mexico. Vol. 7a, pp. 110–137.

Nelliat, E.V., Bavappa, K.V. and Nair, P.K.R. (1974) Multistoreyed, a new dimension in multiple cropping for coconut plantations. *World Crops* 26, 262–266.

Nye, P.H. and Tinker, P.B. (1977) *Solute Movement in the Soil–Root System*. Blackwell, Oxford, 342 pp.

Pagès, L., Jordan, M.O. and Picard, D. (1989) A simulation model of the three-dimensional architecture of the maize root system. *Plant and Soil* 119, 147–154.

Putz, F.E. (1992) Reduction of root competition increases growth of slash pine seedlings on a cutover site in Florida. *Southern Journal of Applied Forestry* 16, 193–197.

Putz, F.E. and Canham, C.D. (1992) Mechanisms of arrested succession in shrublands: root and shoot competition between shrubs and tree seedlings. *Forest Ecology and Management* 49, 267–275.

Putz, F.E., Canham, C.D. and Ollinger, S.V. (1994) Root foraging efficiencies of trees and shrubs. *Functional Ecology* (in press).

Rao, M.K., Muraya, P. and Huxley, P.A. (1993) Observations of some tree root systems in agroforestry intercrop situations and their graphical representation. *Experimental Agriculture* 29, 183–194.

Ritz, K. and Newman, E.I. (1985) Evidence for rapid cycling of phosphorus from dying roots to living plants. *Oikos* 45, 174–180.

Ruhigwa, B.A., Gichuru, M.P., Manbani, B. and Tariah, N.M. (1992) Root distribution of *Acioa barteri*, *Alchornea cordifolia*, *Cassia siamea* and *Gmelina arborea* in acid Ultisol. *Agroforestry Systems* 19, 67–78.

Sanford, R.L. (1985) Root ecology of mature and successional Amazonian forests. PhD thesis, University of California, Berkeley.

Sanford, R.L. (1989) Root systems of three adjacent old growth Amazonian forests, and associated transition zones. *Journal of Tropical Forest Science* 1, 268–279.

Santantonio, D., Hermann, R.K. and Overton, W.S. (1977) Root biomass studies in forest ecosystems. *Pedobiologia* 17, 1–31.

Schuurman, J.J. (1983) Effect of soil conditions on morphology and physiology of roots and shoots of annual plants, a generalized vision. In: Kutschera, L. (ed.) *Wurzelökologie und ihre Nutzanwendung*. Bundesanstalt Gumpenstein, Irdning, pp. 343–354.

Shindzaki *et al.* (1965) A quantitative analysis of plant form – the pipe model theory. *Japanese Journal of Ecology* 14, 97–105.

Sieverding, E. (1991) *Vesicular-Arbuscular Mycorrhiza Management in Tropical Agrosystems*. GTZ, Eschborn, 371 p.

Smucker, A.J.M., Ellis, B.G. and Kang, B.T. (1995) Alley cropping on an Alfisol in the forest savanna transition zone: root, nutrient and water dynamics. In: Kang, B.T., Osiname, O.A. and Larbi, A. (eds) *Alley Farming Research and Development*. IITA, Ibadan, Nigeria, pp 103–121.

So, H.B. and Nye, P.H. (1989) The effect of bulk density, water content and soil type on the diffusion of chloride in soil. *Journal of Soil Science* 40, 743–749.

Spek, L.Y. and van Noordwijk, M. (1994) Proximal root diameters as predictors of total root system size for fractal branching models. II. Numerical model. *Plant and Soil* 164, 119–128.

Srivastava, S.K., Singh, K.P. and Upadhayay, R.S. (1986) Fine root growth dynamics in teak (*Tectona grandis*). *Canadian Journal of Forest Science* 16, 1360–1364.

Toky, O.P. and Bisht, R.P. (1992) Observations on the rooting pattern of some agroforestry trees in an arid region of North–Western India. *Agroforestry Systems* 18, 245–263.

Vandenbeldt, R.J. (1991) Rooting systems of western and southern African *Faidherbia albida* (Del.) A. Chev. (syn. *Acacia albida* Del.) – a comparative analysis with biogeographic implications. *Agroforestry Systems* 14, 233–244.

van Noordwijk, M. (1983) Functional interpretation of root densities in the field for nutrient and water uptake. In: Kutschera, L. (ed.) *Wurzelökologie und ihre Nutzanwendung*. Bundesanstalt Gumpenstein, Irdning, pp. 207–226.

van Noordwijk, M. (1989) Rooting depth in cropping systems in the humid tropics in relation to nutrient use efficiency. In: van der Heide, J. (ed.) *Nutrient Management for Food Crop Production in Tropical Farming Systems*. Institute for Soil Fertility, Haren, pp. 129–144.

van Noordwijk, M. (1990) Synchronization of supply and demand is necessary to increase efficiency of nutrient use in soilless horticulture. In: van Beusichem, M.L. (ed.) *Plant Nutrition – Physiology and Applications*. Kluwer Academic, Dordrecht, pp. 525–531.

van Noordwijk, M. (1992) Three levels of complexity in root ecological studies. In: Kutschera, L., Hübl, E., Lichtenegger, E., Persson, H. and Sobotik, M. (eds) *Root Ecology and its Practical Application*, pp. 159–161.

van Noordwijk, M. and Brouwer, G. (1991) Review of quantitative root length data in agriculture. In: Persson, H. and McMichael, B.L. (eds) *Plant Roots and their Environment*. Elsevier, Amsterdam, pp. 515–525.

van Noordwijk, M. and Brouwer, G. (1993) Gas-filled root porosity in response to temporary low oxygen supply in different growth stages. *Plant Soil* 152, 175–185.

van Noordwijk, M. and Brouwer, G. (1995) Roots as sinks and sources of carbon and nutrients in agricultural systems. *Advances in Agroecology* (in press).

van Noordwijk, M. and De Willigen, P. (1987) Agricultural concepts of roots: from morphogenetic to functional equilibrium. *Netherlands Journal of Agricultural Science* 35, 487–496.

van Noordwijk, M. and De Willigen, P. (1991) Root functions in agricultural systems. In: Persson, H. and McMichael, B.L. (eds) *Plant Roots and their Environment*. Elsevier, Amsterdam, pp. 381–395.

van Noordwijk, M. and Dommergues, Y.R. (1990) Root nodulation: the twelfth hypothesis. *Agroforestry Today* 2 (2), 9–10.

van Noordwijk, M. and Wadman, W. (1992) Effects of spatial variability of nitrogen supply on environmentally acceptable nitrogen fertilizer application rates to arable crops. *Netherlands Journal of Agricultural Science* 40, 51–72.

van Noordwijk, M., de Willigen, P., Ehlert, P.A.I. and Chardon, W.J. (1990) A simple model of P uptake by crops as a possible basis for P fertilizer recommendations. *Netherlands Journal of Agricultural Science* 38, 317–332.

van Noordwijk, M., Hairiah, K, Syekhfani, M. and Flach, B. (1991a) *Peltophorum pterocarpa* – a tree with a root distribution suitable for alley cropping. In: Persson, H. and McMichael, B.L. (eds) *Plant Roots and their Environment*. Elsevier, Amsterdam, pp. 526–532.

van Noordwijk, M., Widianto, Heinen, M. and Hairiah, K. (1991b) Old tree root channels in acid soils in the humid tropics: important for crop root penetration, water infiltration and nitrogen management. *Plant Soil* 134, 37–44.

van Noordwijk, M., Widianto, Sitompul, S.M., Hairiah, K. and Guritno, B. (1992) Nitrogen management under high rainfall conditions for shallow rooted crops: principles and hypotheses. *AGRIVITA* 15, 10–18.

van Noordwijk, M., Brouwer, G. and Harmanny, K. (1993a) Concepts and methods for studying interactions of roots and soil structure. *Geoderma* 56, 351–375.

van Noordwijk, M., Brouwer, G., Zandt, P., Meijboom, F.W. and Burgers, S. (1993b) Root patterns in space and time: procedures and programs for quantification. *IB-Nota* 268, IB-DLO, Haren.

van Noordwijk, M., Brouwer, G., Koning, M., Meijboom, F.W. and Grzebisz, W. (1994a) Production and decay of structural root material of winter wheat and sugarbeet in conventional and integrated arable cropping systems. *Agriculture Ecosystems Environment* 51, 99–113.

van Noordwijk, M., Spek, L.Y. and De Willigen, P. (1994b) Proximal root diameters as predictors of total root system size for fractal branching models. I. Theory. *Plant and Soil* 164, 107–118.

Vitousek, P.M. and Sanford, R.L. (1986) Nutrient cycling in moist tropical forests. *Annual Review of Ecology and Systematics* 17, 137–167.

Waring, R.H. and Schlesinger, W.H. (1985) The carbon balance of trees. In: *Forest Ecosystems, Concepts and Management*. Academic Press, Orlando, 340 pp.

Wiersum, L.K. (1982) Coconut-palm root research in Indonesia. *FAO Consultancy Report* INS/72/007, 33 pp.

Wiersum, L.K. and Harmanny, K. (1983) Changes in the water-permeability of roots of some trees during drought stress and recovery, as related to problems of growth in urban environment. *Plant and Soil* 75, 443–448.

Xu, X. and Bland, W.L. (1993) Reverse water flow in sorghum roots. *Agronomy Journal* 85, 384–388.

Zhu Zhaohua (1991) Evaluation and model optimization of *Paulownia* intercropping system – a project summary report. In: Zhu Zhaohua, Cai Mantang, Wang Shiji and Jiang Youxu (eds) *Agroforest Systems in China*. Chinese Academy of Forestry and IDRC, pp. 30–43.

Zhu Zhaohua, Chao Ching-Ju, Lu Xin-Yu and Xiong Yao Gao (1986) *Paulownia in China: Cultivation and Utilization*. Asian Network for Biological Sciences and IDRC, Singapore and Ottawa, 65 pp.

10

---◆---

Woody–Non-woody Plant Mixtures: Some Afterthoughts

P. HUXLEY

Flat 4, 9 Linton Road, Oxford OX2 6UH, UK

Agroforestry Expectations: Understanding Systems Design and Plant Management

As a gross generality agroforestry differs from most agricultural practices, including many forms of intercropping, because of the wider choice of plant components and their management that often has to be made, particularly with respect to the multipurpose trees. Some of the biological issues that may need to be taken into account in making such choices have been raised, particularly in Chapters 7 and 9. When one considers that the design and management of individual agroforestry systems also depend on the precise nature of the outputs and services required as well as the specific characteristics of each site, we can appreciate that farmers, even within a small area, may well be separately practising different forms of agroforestry. Even if they are all undertaking the same practice (e.g. alley cropping), there can be many different variations according to farmers' precise needs. Under such circumstances it becomes mandatory to elucidate some guidelines about how such systems work. Some of the issues needed to establish these have emerged from the preceding chapters but, inevitably, we have had to point out not only what is known but, also, what is not.

If they are not already actively involved in agroforestry, farmers may adopt it for two basic reasons: (i) if they have sufficient land and labour for their needs they may wish to increase production and, by having a diversity of outputs, reduce risk; or (ii) they may be concerned that the sustainability of their production system is under threat at present output levels and they are, therefore, seeking a low-cost, long term solution to this problem, even at the expense of diminishing crop production somewhat. Where farmers have little land they *must* grow their main food or cash crops without loss but they may believe that adding multipurpose trees for short or long term benefits is, nevertheless, still a possibility. The authors of the various chapters have all drawn on well-established existing knowledge, and have emphasized the development and extension of known concepts to the different dimensions in space and time that we find in agroforestry. Such a rational approach is essential, and the enticing suggestion that merely adding a woody component to crop systems can in some way *automatically* produce benefits has, hopefully, been dispelled.

Preceding chapters (Chapters 2, 3, 4, 6, 9) have emphasized that, as in general agricultural practice, to design and manage agroforestry systems effectively we need to start with an understanding of the potentials of the assembled plants to capture the maximum amounts of available environmental resources. In agroforestry the added dimension, stressed throughout the book, is then to see how our mixtures of very different kinds of plant components can be made to interact optimally so as to fulfil their individual yield potentials. Much of what has been discussed illustrates how our understanding of the extent and degree of these interactions at the 'tree–crop interface' (TCI) has progressed over the last 10 years or so, starting with more observational assessments (Huxley *et al.*, 1989) and now involving targeted experimentation. Alley cropping may seem to be a good starting point because it represents a relatively 'simple' practice involving a regular plant arrangement. However, there are still many problems in sampling and in fully understanding the interactions involved even in such systems. Moreover, the concepts and issues arising need to be used to evaluate more complex practices such as scattered trees in crops or pastures, improved fallows, etc., and to help address real world problems in systems following these practices. Alley cropping itself has frequently been promoted as a solution to landuse problems in unsuitable environments or in inappropriate forms (Sanchez, 1995) and this misconception emphasizes the need for more fundamental research to ensure that extension efforts are not misplaced.

For low input agroforestry in the tropics the maintenance of a successful system will usually mean *maximizing biomass*, largely because of the various benefits that plant residues can bring and the need to recycle plant nutrients and not to lose them. As explained here, this may be done by attending to canopy features and by maximizing the use of available soil water for transpiration in the system as a whole throughout the year. However, any system will become exploitive and non-sustainable if too much biomass is exported. This balance of productivity versus sustainability is an underlying issue for *all* landuse systems and agroforestry is no exception. If agroforestry 'works' it is only because it offers a potential to combine a greater sum of beneficial processes than is likely to occur in other types of managed landuse. Agroforestry gives us the opportunity, *should we understand what these processes are and how we may manipulate them*, to combine and utilize them collectively. Many of these benefits are to be derived from the woody component itself and/or from the interactions between woody and non-woody components (as Chapter 7 emphasizes), and from interactions between the woody component and the soil (Young, 1989). The extent to which these potential benefits are achieved depends on how cleverly we arrange and manage the plant associations (and the animals, in some systems, e.g. Baumer, 1991), and thus on the extent of our knowledge of the biophysical characteristics of the system, and our perception of the trade-offs that need to be made between productivity and sustainability. The penalty to be paid lies in the time it will take to investigate and unravel the interactions in these highly complex systems, but this field offers the most stimulating area for research to have presented itself for many decades.

To provide suitable advice about biological possibilities under a multiplicity of site-specific conditions we need to know what levels of environmental resources are available, and how best they can be captured by arranging and managing the plant components that are to be assembled. The choice of these components may well be made at field level, but the scientific basis underlying the options, especially with regard to the multipurpose trees and their possible interactions at the tree–crop interface, has to be available. In order to assess the output from different components (and to understand the effects of the trees on the soil), we need to know not only how these resources may be shared but whether or not the resource use efficiency of some or all of the plant components can be enhanced. Limits can then be set on production if sustainability is threatened. These are not easy points to answer as the various chapters in this book have indicated. Nevertheless, there clearly *is* a caucus of theory and validating data that can be drawn upon to further our understanding of what is likely to happen at the tree–crop interface under any particular set of circumstances. Because woody plants offer many more opportunities for different forms of management than crop species, this is a lively challenge which, again, demands perceptive insights of how species, environment and management all interact.

Many of the ideas presented in this volume will, hopefully, encourage further development of this analytical approach.

Some Limitations

The initiatives taken in this book to compile and elucidate some of the more important plant issues underlying the functional nature of agroforestry practices are not of course complete; they are part of an ongoing process of consolidation. In particular, tree–grass combinations have not been dealt with so that the particular problems relating to grass establishment and the maintenance of green leaf area in swards in the presence of trees have not been addressed. Nor have pasture problems such as the maintenance of grass–legume and tree mixtures (e.g. Lefroy *et al.*, 1992), forage quality (Leng, 1995) and, overall, the specific responses of plants to herbivores in such systems. Mycorrhizal and nitrogen fixing associations are mentioned in Chapter 9, but they could well occupy a volume of their own. The interactions of plants and pests and pathogens in agroforestry systems are an area where there is a crucial need for a summation of current knowledge (Huxley and Greenland, 1989), but which is clearly outside the scope of this volume. In its early days, agroforestry was promoted as most suitable for 'wasted lands' (King and Chandler, 1978). Indeed, many examples of agroforestry in the tropics are to be found on poor or shallow soils where nutrients may well be a major limiting factor, although the main emphasis in this book is on light and water. We certainly still need to explore further the links between root activity, water and nutrients. The model presented in Chapter 3, which explores the interactions of nitrogen (from mulch) and shade in alley cropping in some detail, represents a major clarification of these issues. Further elaboration of the ways in which light, nutrients and water interact awaits attention.

The various chapters provide a conceptual background with supportive examples from a variety of sources where these are relevant. These do not, as yet, represent a set of definitive conclusions or a comprehensive collection of guidelines. The concepts require further elaboration and their application to the whole range of agroforestry situations needs to be explored. For example, the tree–crop interface equation in Chapter 1 has intentionally been kept as a simple, additive form although, in reality, it may well not be. Extending the application of Production Possibility Frontiers (Chapter 2) to interacting plant components in agroforestry systems intriguingly merges and extends what were originally economic and ecological concepts (Total Factor Productivity and Carrying Capacity) to analogous situations, but it has to be remembered that it applies only to iso-resource situations and that maximum yield–density relationships for different crop species at the same site will, of course, often not be the same, or even consistently rank in the same order season by season. We also face the problem that, although plant biomass often reflects an asymptotic yield–density relationship, crop yields may not and, furthermore, Spitters's original equation applied only over a limited range of disparity in plant height (Wallace, 1995). Again, in Chapter 6 there is clearly a need to go further and to establish *coupled* models of water and light (e.g. see Wallace, 1995), as various groups are now working to do (e.g. at the Institute of Terrestrial Ecology, Scotland, with ODA support; Cannell, 1994). The need to establish meaningful 'functional types' for multipurpose trees

(Chapter 7) has been given an additional emphasis by the topics we have covered here, but it still remains an elusive objective. Tree management and genotypic selection will only become truly effective and clearly defined when both form *and* function are related to the ways in which we intend to use these plants.

In many cases the chapters emphasize how much we still lack appropriate data. In Chapter 1 assessments of M, P, and L (respectively the effects of changes in above-ground microclimate, soil properties and any reduction or avoidance of losses of water and/or nutrients, especially on sloping land) are still poorly or inadequately documented for agroforestry situations, especially for M and L. Chapter 5 has dealt with the principles underlying microclimatic changes and their effects on plant growth in general, and the example of windbreaks has been elaborated because our knowledge of the physical and physiological factors involved in providing plant structures of this kind in crop land or pastures is well established. Current problems are more of a practical nature, i.e. concerning the choice, establishment and subsequent management and final replacement of species (Bird *et al.*, 1992) and the hydrological implications of establishing potentially deep rooting trees, as well as the effects of this on soil salinity (Bell, 1988; Greenwood, 1988). There are many instances in the agricultural and horticultural literature of positive effects of 'shelter', e.g. from windstrips and physically small barriers of different kinds, and mostly expressed as circumstantial evidence without accompanying microclimatic data. It is often very difficult, therefore, to ascribe the outcome to particular microclimatic causes. In agroforestry, however, we are likely to meet very many situations where there is a potential for the woody component to afford some degree of shelter to lower canopy plants, or to function as physically small barriers to windflow (Fig. 10.1). There are, clearly, some occasions where this might be of some importance (e.g. Huxley *et al.*, 1994) but we need to ascertain where and under what circumstances.

Some Issues Arising

Very little land is actually flat; usually there are at least some changes in microrelief at varying scales which encourage water and/or soil movement over relatively small distances. Conventional methods of preventing the loss of soil and water on slopes can usefully be supplemented by tree planting (Young, 1989). Chapter 8 explores alley cropping (hedgerow inter-cropping) as a solely agroforestry solution. The detailed investigation of 'scouring' effects across small plots is a particularly important issue, especially as small departures of alignment from the contour, as might occur on farmers' fields, can cause unexpected results (Fig. 10.2).

370 ◆ P. Huxley

Fig. 10.1. Shelter effects on the growth of maize at Machakos, Kenya, by a small hedge of *Senna siamea* – prevailing winds come from the south-east (right) (Huxley *et al.*, 1994).

Below-ground interactions have emerged as one of the most important issues that need to be investigated if we are to fully understand and manipulate agroforestry plant associations. The comprehensive account given in Chapter 9, supplemented by information in Chapters 4 and 7, indicates the complexities of the problems. Some of these relate to the control of plant dry matter distribution, including fine root turnover and the environmental factors affecting this, and some to the form and nature of the root system as a whole and the opportunities that these offer for sharing or not sharing space, and hence below-ground resources (Fig. 10.3). By analogy above ground, the early prediction of crop losses by weed competition indicates that the competitive strength of a species can be determined by its share of leaf area at the time the canopy closes and (above-ground) competition starts (Kropff and Spitters, 1991). In agroforestry, however below ground, the woody component already has an established rooting volume at the start of each cropping season and so, unless the roots of the crop components fill unoccupied volumes (i.e. in microspaces), or there are phenological differences in the times of root growth and activity, a high level of below-ground competition is inevitable. Clearly, until we know much more about the rooting habits of multipurpose trees, explanations of the causation of competition/complementarity and our understanding of the ways in which water and nutrients are utilized in these systems will remain incomplete (Ong and Black, 1994).

Fig. 10.2. Hedgerow intercropping (alley cropping) with *Grevillea robusta* and maize on a slope (top of slope to right). Because these hedgerows were not accurately aligned on the contour, surface water flowing across the fertilized plot in rainstorms and running off at the bottom of the 'alley' carried off soil and nutrients from the bottom half of the plot. This resulted in better growth at the top than at the bottom of the alley in contrast to what might be expected (see Chapter 8). This plot was part of a parallel row systematic spacing demonstration (Huxley *et al.*, 1989), and this effect started in the widest 'alleys' and progressed, season-by-season, to the narrower ones up the slope. This example emphasizes a potential problem when using such systematic spacing designs.

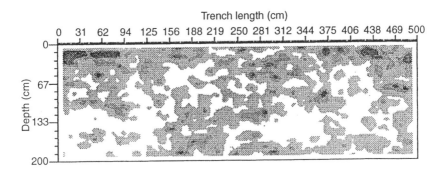

Fig. 10.3. 'Patchy' rooting pattern of 4-year-old *Citrus sinensis* cv, Washington navel at ICRAF's Experiment Station, Machakos, Kenya. Darkest shading represents most fine roots (≥ 90 per 25 cm^2). Such a root system leaves plenty of empty spaces.

The interactions between roots and shoots remain a key issue because of the ways in which environmental factors, including management considerations, affect carbon allocation, nutrient storage and phenological behaviour. Growth regulation and flowering are controlled by hormone 'signals' and by changes in the receptivity of different plant tissues to these but, once initiated, the transfer of carbon components and nutrients can be accounted for by mechanistic models (Cannell and Dewar, 1994). Even being able to describe the behaviour of above- and below-ground parts in simple phenological terms can help put growth and development into a framework suggesting functional opportunities or disadvantages. But here we have to be cautious because competition takes place in both resource-poor and resource-rich environments, and we need to think of spatial complementarity as either the outcome of all positive features *or* of a net benefit where both positive and negative features exist (and vice versa for competition). This makes the identification of the plant traits (or sets of traits involved) more difficult. However, if we are to have a rational framework for the design and management of agroforestry systems, and especially for selecting the multipurpose trees, the physiological basis must ultimately be translated into identifiable plant characteristics which can be readily classified. Most chapters in this book contribute to this aim.

The number of possible spatial arrays, the diversity of plant arrangements within these and the high level of interaction between different plant components make taking measurements within agroforestry systems a much more complex task than in conventional agricultural systems. Sampling procedures need especial consideration and, where some average measure of systems behaviour is needed, the problem can be extremely onerous, as is emphasized in several chapters, either explicitly or by example. The experiment in Chapter 4 in which fractional radiation interception was measured is a case in point since reliable estimates of the light conversion coefficient depend on accurate measurements of both biomass/yield and of light interception by the various components of the system. As the authors point out, radiation *reflected* by the canopies should be measured and discounted, because it is not utilized. With well-established tree species the fraction of the radiation intercepted by non-photosynthetic organs (e.g. woody stems and branches) may amount to as much as 40% or so of total interception, and should similarly be accounted for. Making the whole range of measurements necessary within the spatial complexities of even a hedgerow intercropping system over extended periods requires substantial resources. Even more complex are the sampling problems that obtain when estimating water balances and responses to water stress in these plant mixtures (Chapters 5 and 6). This is especially the case if we wish to observe how soil water status and/or plant stress change over short distances (Fig. 10.4).

Horizontal distance (m)

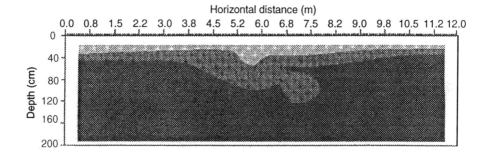

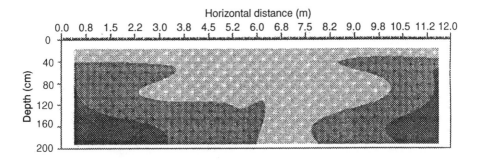

Fig. 10.4. Change in soil water status under a three-row *Grevillea robusta* hedge (centre) grown contiguously with maize on either side from end July at maize harvest time (top) and end August (bottom). Darkest shading near field capacity, lightest near wilting point. Because this was a wet year at Machakos (1990) much of the profile was still filled at the end of the cropping season in July and subsequent drying was by the hedgerow alone. Sampling can be a problem in such circumstances.

Our ability to observe, measure and understand what is happening at the tree–crop interface is obviously crucial. However, the operation and intensity of competition between different competing plant species is sometimes predictable, sometimes not, depending on whether the environment is orderly or disorderly. This also depends on the number of factors on which a competitive outcome depends, and the size of the stochastic component of the interactions. The greater these are, the more inconsistent will be the outcome (Fowler, 1990). A problem in identifying and measuring the effects of competition in agroforestry systems is that, as we have seen from previous chapters, agroforestry environments are invariably patchy, a condition, both above and below ground, which the very components themselves help to bring about and enhance. This condition needs to be understood (Caldwell and Pearcy, 1994) in order to be exploited effectively, as explained in Chapters 3, 4, 6, 7 and 9, especially.

In addition, depending on the scale on which it occurs, patchiness represents disorder and it may weaken the processes of competitive exclusion. Phenotypic plasticity may also do this, again depending on scale, the intensity of the environmental changes that occur and the ability of specific genotypes to respond to them.

Agroforestry systems usually exhibit a high degree of vertical integration and this is, again, a characteristic that needs to be explored and exploited in the field once the basic biophysical factors involved have been established. A continuing theme throughout this book addresses the influences and consequences of 'verticality'. Above ground it is likely to be important in determining radiation distribution, and, hence, water use (Chapters 4 and 6), as well as affecting environmental gradients and providing shelter (Chapter 5). Below ground the ways in which roots are integrated vertically can also be a key issue (Chapters 7 and 9).

Foresters are well aware of a problem we also face in agroforestry, and one we have not addressed; our systems are dynamic. More particularly, they will change structurally and functionally as the woody components grow older. The consequences are accentuated as lower storey crop species are raised in sequential seasons against a background of changing stature, form and environmental exploitation of the trees, shrubs, hedges, etc. Above ground these changes can be moderated by management, especially pruning, but this is not usually the case below ground. Even where the mixtures are trees and grass swards, these two parts of the system will age differently. Very little attention has been given, as yet, to considering when to replace old agroforestry systems, or how best to do so.

Conclusions

We hope that this book contributes to a more precise approach to the consideration of the biological feasibility of agroforestry practices, and helps take our understanding of some of the factors underlying plant management in specific systems a step further. Particular agroforestry systems will always be highly site-specific and, in such situations, there are two ways forward: empiricism and analysis. Both have a part to play as, in the broader context, the history of agricultural development has shown.

Considerable strides are now being made towards understanding the complex plant–soil–environment interactions on which the success or failure of agroforestry systems depends. Encouragingly, in the not too distant future, it is likely that the fruits of these efforts will add significantly to our ability to evaluate, compare and ultimately predict the outcome of specific agroforestry enterprises, and thus form a sound scientific foundation to support the considerable amount of exploratory and adaptive research that has been undertaken to date.

In this book we have explored what is known about the outcome of intercropping woody and non-woody plants and attempted also to expose what is not. The ways in which we can begin to achieve further progress and remedy our requirements for more understanding and information have also, hopefully, been clarified. Further progress in establishing the appropriate conceptual background in this and other research fields can only lead to considerable advantages for the practical implementation of sound agroforestry initiatives in the field.

References

Baumer, M. (1991) Animal production, agroforestry and similar techniques. *Agroforestry Abstracts* 4, 179–198.

Bell, A. (1988) Trees, water and salt – a fine balance. *Ecos* 58, 2–9.

Bird, P.R., Bicknell, D., Bulman, P.A., Burke, S.J.A., Leys, J.F., Parker, J.N., Van der Sommen, F.J. and Voller, P. (1992) The role of shelter in Australia for protecting soils, plants and livestock. *Agroforestry Systems* 20, 59–86.

Caldwell, M.M. and Pearcy, R.W. (1994) *Exploitation of Environmental Heterogeneity by Plants: Ecophysiological Processes Above- and Below-ground.* Academic Press, San Diego.

Cannell, M.G.R. (1994) Note on agroforestry modelling and research co-ordination. *Agroforestry Forum* 5, 32.

Cannell, M.G.R. and Dewar, R. (1994) Carbon allocation in trees: a review of concepts for modelling. *Advances in Ecological Research* 75, 59–104.

Fowler, N.L. (1990) Disorderliness in plant communities: comparisons, causes and consequences. In: Grace, J.B. and Tilman, D. (eds) *Perspectives in Plant Competition.* Academic Press, San Diego, pp. 291–306.

Greenwood, A.N. (1988) The hydrologic role of vegetation in the development of dryland salinity. In: Allen, E.B. (ed.) *The Reconstruction of Disturbed Arid Lands: An Ecological Approach.* Westview Press, Boulder, pp. 205–233.

Huxley, P.A. and Greenland, D.J. (eds) (1989) Pest management in agroforestry systems. A record of discussions held at CAB International, Wallingford, UK, 28–29 July, 1988. *Agroforestry Abstracts* 2, 37–46.

Huxley, P.A., Darnhofer, T., Pinney, A., Akunda, E. and Gatama, D. (1989) The tree/crop interface: a project designed to generate experimental methodology. *Agroforestry Abstracts* 2, 127–145.

Huxley, P.A., Pinney, A., Akunda, E. and Muraya, P. (1994) A tree/crop orientation experiment with a *Grevillea robusta* hedgerow and maize. *Agroforestry Systems* 26, 23–45.

King, K.F.S. and Chandler, T. (1978) *The Wasted Lands.* ICRAF, Nairobi.

Kropff, M.J. and Spitters, C.J.T. (1991) A simple model of crop loss by weed competition from early observations on relative leaf area of the weeds. *Weed Research* 31, 97–105.

Lefroy, E.C., Dann, P.R., Wildin, J.H., Wesley-Smith, R.N. and McGowan, A.A. (1992) Trees and shrubs as sources of fodder in Australia. *Agroforestry Systems* 20, 117–139.

Leng, R.A. (1996) *Trees – Their Role in Animal Nutrition in Developing Countries in the Humid Tropics.* Animal Production and Health Paper. FAO, Rome (in press).

Ong, C.K. and Black, C.R. (1994) Complementarity of resource use in intercropping and agroforestry systems. In: Monteith, J.L., Scott, R.K. and Unsworth, M.H. (eds) *Resource Capture by Crops.* Nottingham University Press, Loughborough, pp. 255–278.

Sanchez, P.A. (1995) Science in agroforestry. *Agroforestry System* 30, 5–55.

Wallace, J.S. (1995) Towards a coupled light partitioning and transportation model for use in intercrops and agroforestry. In: Sihoquet, H. and Cruz, P. (eds) *Ecophysiology of Tropical Cropping.* INRA, Paris, pp. 153–162.

Young, A. (1989) *Agroforestry for Soil Conservation.* ICRAF Science and Practice Series, No. 4. CAB International, Wallingford, 276 pp.

Index

Note: page numbers in italic refer to figures and/or tables